Experimental Methods in Biomechanics

John H. Challis

Experimental Methods in Biomechanics

 Springer

John H. Challis
Biomechanics Laboratory
The Pennsylvania State University
University Park, PA, USA

ISBN 978-3-030-52258-2   ISBN 978-3-030-52256-8   (eBook)
https://doi.org/10.1007/978-3-030-52256-8

This Springer imprint is published by the registered company Springer Nature Switzerland AG.
The registered company address is: Gewerbestrasse 11, 6330 Cham, Switzerland

Preface

In biomechanics, the experimentalist is faced with a plethora of tools for collecting data and another plethora of techniques for processing the data. The primary objectives of this textbook are to introduce the reader to:

- The theory behind the primary data collection methods used in biomechanics
- The theory behind the primary methods of data processing and analysis used in biomechanics

Understanding these various theories means that the scientist is better able to design experiments and similarly better able to develop new data collection, data processing, and analysis protocols. Biomechanics is a broad domain including the analysis of bacterium (e.g., Berg, 2004), plants (e.g., Niklas, 1992), and living tissues (e.g., Fung, 1981); here, the focus will be data collection methods and data processing techniques with reference to whole-body human biomechanics. That stated, many of the topics are applicable to other areas of biomechanics.

The book consists of 12 chapters, each covering material related to experimental biomechanics. In common with other areas of science biomechanics follows the scientific method, which is presented and discussed in Chap. 1. The following chapter provides an overview of the tools used in experimental biomechanics. Collected data typically comes in the form of a signal, often as a function of time, so Chap. 3 examines how signal contents can be determined. The processing of these signals is described in Chap. 4. The collection and processing of the electrical signals from muscles, electromyography, are presented in Chap. 5. Methods for determining the kinematics of human movement are explored in Chap. 6. The collection and analysis of forces, with a focus on force plates, are delivered in Chap. 7. In biomechanics anthropometry, the topic of Chap. 8 is focused on determining segmental and whole body inertial properties. Chapter 9 explains how the kinematics of human movement can be described and determined. The computation of inverse dynamics and body, segment, and joint energetics are presented and discussed in Chap. 10. Irrespective of the data collected, errors will remain; therefore, Chap. 11 describes how errors arise and can be analyzed. Finally, the influence of body size on biomechanical variables will be described in Chap. 12; such analysis can be important when undertaking the statistical analysis of data.

For most chapters, there is a basic level of math required to understand the contents. The start of each chapter outlines the required math and directs the reader to a series of appendices that summarize the relevant math. These appendices cover topics including matrix algebra, calculus, cross and dot products, and the singular value decomposition. Pythagoras stated that "*Mathematics is the way to understand the Universe.*" The reason for the great utility of math for the analysis of problems in science is much debated (e.g., Wigner, 1960), but irrespective of the reason for its utility to understand data collection and processing methods, math is an essential tool.

It is not possible to avoid presenting equations – even though physicist Stephen Hawking warned that the size of the readership of a book is inversely proportional to the number of equations in a book. In presenting equations, it is impossible to avoid replication of a symbol, that is, a symbol is used to represent two different variables in different equations, the aim is to avoid doing so within a chapter. Here the following conventions will be used:

- i, j, k are used as integers for counting
- m and n are used as integers to designates the number of elements
- uppercase letters are used for matrices

Each chapter is divided into sections to make dipping back into a chapter easier, where appropriate equations in one section or subsection are often repeated in a subsequent subsection so that cross-referencing is not required. In addition, for each chapter, pertinent references are provided in a number of ways. These references are presented in the following categories:

Cited References – those cited in the chapter

Specific References – references that provide examples that are specific to the content of the chapter, for example, there are references listed which deal with the definitions of joint angles for specific joints

Useful References – those references which provide further reading, for those that want to dive deeper into a topic.

The relevant references are presented at the end of each chapter.

There are examples of pseudo-code in some of the chapters. These give hints on how some of the methods and data processing techniques can be implemented. Indeed a thorough understanding of a method or data processing technique can be demonstrated by writing relevant computer code. It is appreciated that there are software packages available for the collection and analysis of biomechanical data. However, even if someone uses one of these packages, it is important that the principles underlying the software is understood by the user. This book will provide the background necessary to understand these principles. Another advantage of understanding the principles is that it becomes feasible to develop your own software for data collection and analysis, leading to novel analysis and, hopefully, new insights.

University Park, PA, USA   John H. Challis

References

Berg, H. C. (2004). *E. coli in motion*. New York: Springer.
Fung, Y. C. (1981). *Biomechanics: Mechanical properties of living tissue*. New York: Springer-Verlag.
Niklas, K. J. (1992). *Plant biomechanics: An engineering approach to plant form and function*. Chicago: The University of Chicago Press.
Wigner, E. P. (1960). The unreasonable effectiveness of mathematics in the natural sciences. *Communications on Pure and Applied Mathematics, 13*(1), 1–14.

Acknowledgments

I would like to acknowledge my sincere thanks to:

- My students, they have helped shape my thinking
- My mentors, they helped guide my thinking
- My colleagues, they helped challenge my thinking
- My family, they gave me the space to think

Contents

Overview

The word biomechanics is derived from the Greek words bios (βίος) and *mēchanikē (μηχανική)*. The "bio" component refers to living organisms and living tissues, while "mechanics" refers to the science of the action of forces on bodies. Given these definitions, the focus of work in biomechanics is the examination of biological phenomena from a mechanical perspective, with a view of identifying the mechanisms responsible for the phenomena. Here reference is made to "mechanisms," which means the focus is the arrangement and action by which a result is produced. This distinction is important because there is much more to biomechanics than simply measuring the mechanical variables associated with biological phenomena; it requires the search for mechanisms.

The scope of biomechanics can include both plants and animals; here the focus will be animal biomechanics with an emphasis on the biomechanics of humans. The process of performing studies in biomechanics can be modeling studies or experimental studies. Here the focus will be the methods used for experimental studies in biomechanics. Experimental studies require the ingenuity of the scientist and to some extent are guided by the following quote from the Italian physicist and astronomer Galileo Galilei (1564–1642):

> Measure what is measurable, and make measurable what is not so.

Making things measurable can be designing new data collection or analysis protocols, or designing insightful experiments. This type of ingenuity is one of the reasons why research in biomechanics can be so rewarding.

In the following two sections, the scientific method will be outlined, and then the confusion over the definitions of theories, laws, and facts discussed. The next section addresses the necessity, or otherwise, of hypotheses. The following section briefly outlines the predominant tools used in biomechanics, and the next section describes the potential ethical problems which arise from using these tools. Finally, overviews of the remaining chapters are presented.

1.1 Scientific Method

Karl Popper (1902–1994) was an Austrian-British philosopher who presented a philosophy on the method of the scientific process; this philosophy has proven popular with scientists (Medawar 1982). A Popperian view of the scientific method suggests that scientific knowledge advances by phrasing of hypotheses and theories which can be tested. He advanced the concept of falsification as the process via which science progresses (Popper 1959). Therefore, positive support for hypotheses and theories is unobtainable, accepting in the sense that they have failed to be falsified. The cyclical process of science is illustrated in Fig. 1.1 where an experiment is performed and support for a theory comes when the theory has failed to be falsified, in which case support has been provided for the theory.

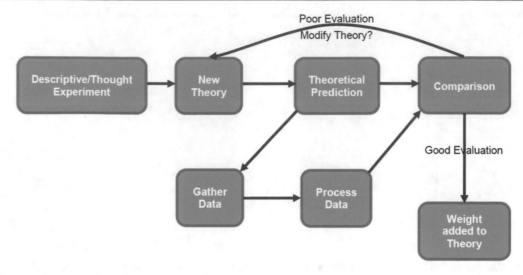

Fig. 1.1 The process of the scientific method from a Popperian perspective

A theory can be defined as a principle suggested to explain observed facts or phenomena, while a hypothesis is a provisional explanation of observed facts. A theory, or hypothesis, must have important characteristics to be considered scientific. To test if a hypothesis is scientific (similarly for a theory), you should:

1. Assume the statement is wrong
2. Imagine a result which would then prove it wrong

If both steps cannot be completed you must conclude that the statement is not scientific. Consider, for example, the following two hypotheses:

Hypothesis I: Gold is soluble in hydrochloric acid.
Scientific: Yes (though false).
Why is it scientific?: Because we can eliminate it if it is false.

Or, for example:

Hypothesis II: Some homeopathic medicine do work.
Scientific: No (though possibly true).
Why is it scientific?: It is unscientific because even if it were false we could not get rid of it by confronting it with an observation/report that contradicted it. The hypothesis as stated is too general for it to be testable.

In the Popperian view of science, proof is not provided by a theory, but support is gained as a consequence of failure to falsify. There are critiques to this approach. For example, Thomas Kuhn (1922–1991) argued that scientists work in a series of paradigms, where a paradigm is the current knowledge and understanding of a scientific problem (Kuhn 1962). The use of paradigms means the formulation of a simple hypothesis is not feasible, and therefore testing if it is falsifiable is problematic. It could be argued that a philosophical debate about the nature of science is not always relevant to the practicing scientist. The advantage of the Popperian approach is that it challenges the scientists to work hard to falsify their own hypotheses and theories. Such an approach helps provide faith in hypotheses that a scientist has advanced, and it is likely for this reason that the Popperian view of science is favored by scientists (Mulkay and Gilbert 1981).

Advances in science may occur because of scientists testing the hypotheses in Popperian sense, with their findings potentially causing a Kuhnian paradigm shift. While scientists deal with ideas, advances in science are often not just due to the wrestling with ideas, but they can be due to tool-driven changes in ideas. It is often the availability of new tools which produces profound new paradigms due to the new ability to collect and/or analyze data (Galison 1997). For example, in the twentieth century, the development of X-ray crystallography provided a powerful tool for many scientists to perform research, resulting in Nobel Prize recognitions (Galli 2014). The advent of the digital computer has impacted the majority of the areas of sciences, becoming for many an essential tool, as have many statistical analysis methods.

Of course the progress of science is not always as straightforward as Popper or Kuhn implies. For example, there may be multiple hypotheses which are consistent with a given set of experimental results. The existence of multiple hypotheses for the same sets of results provides a clear research agenda: the elimination and refinement of the existing hypotheses. If a theory is tested and fails to be falsified, there is no critical threshold of failed falsifications for a theory to become generally accepted. Indeed, different areas of science, because of the precision of their results, can have different thresholds. It should also be considered that a theory that is generally accepted in the scientific community may have a number of tests which did falsify its claims. This can occur due to unnoticed experimental errors and because when statistical tests are performed, statistical significance levels leave the chance that there is an erroneous conclusion, for example, the test fails to reject a false null hypothesis (Type II error, a false negative).

1.2   Spectrum of Theories

The French mathematician, theoretical physicist, and philosopher Jules Henri Poincaré (1854–1912) wrote:

> Scientist must set in order. Science is built up with facts, as a house is with stones. But a collection of facts is no more a science than a heap of stones is a house.

The order is provided by the testing of hypotheses and the formulations of theories. The theories used in any domain in science can be considered to be on a spectrum from the tentative to the well established.

Tentative   Well Established

Tentative theories have limited experimental evidence which lend support to the theory, while well-established theories might have sufficient support that they are considered laws. Even in the case of laws, these may also be considered as provisional as a result may eventually arise which requires their modification (see next section for an example). Both tentative theories and well-established theories can be replaced if a superior explanation of the experimental evidence is presented. Our knowledge can be put into four categories:

I.   Well-established theories
II. Supported theories
III. Tentative theories
IV. What we would like to know
     Can address
     Can probably address
     Can probably not address

The last category is perhaps the most intriguing as the role of the scientist includes making important questions addressable by designing new equipment, new analysis techniques, and elegant experiments. The most impactful papers are often those which start to address a question we would like to know the answer to, when previously the prevailing thought was that the methods were not available to address the question.

1.3   Hypotheses, Theories, and Laws

There is often confusion about how science works, in particular what is the difference between a hypothesis, theory, and law, and how does science discriminate? The confusion comes, in part, from the liberal use of the terms theory and law in the common vernacular but also in science. For example, Robert Hooke (1635–1703) described the mechanical behavior of a spring "ut tensio, sic vis" ("as the extension, so the force") which can be expressed in equation form as:

$$F = k\Delta L \tag{1.1}$$

where F is the force applied to the spring, k is the spring constant (stiffness), and ΔL is the change in length of the spring. Hooke described this as a law, but elastic materials do not behave strictly following this relationship, so at best it might be considered a poor law. In contrast, the Pythagorean theorem, to borrow from math, has been proved yet is labeled a theory. Even Newton's second law ($F = m\,a$) was accepted as correct for all cases until the work of Albert Einstein who showed that at very high speeds, the object being accelerated appears to get more massive and therefore harder to accelerate. Therefore, (even) Newton's second law required modification, due to developing knowledge, resulting in changing from being a general law to a more specific law.

It is possible to define the key terms in science which in turn provide a framework on how science works; with the caveat, there are exceptions to how the terms are used in science. Therefore:

Theory can be defined as a system of ideas or statements which explain a group of facts or phenomena.

Hypothesis can be defined as a provisional theory, awaiting additional support so it can be considered a theory.

If support for a hypothesis is not available, the hypothesis can be rejected. Additional support for a hypothesis comes from observations and experiments that fail to falsify the hypothesis. Once sufficient evidence is available, a hypothesis becomes a theory. After more testing, a theory may be considered a fact, but many are still labeled with the term theory.

Considerable debate can rage over certain hypotheses and theories, for example, the theories of evolution and climate change. Of course for the general public, the debate has the added confusion that the term theory in everyday vernacular is normally used for something which is a vague suggestion, not an idea that has been vigorously tested via the scientific method. Indeed, some theories are considered to be facts in science, and if they were not, we would be unable to make progress. The other source of confusion is that theories are not proven but become generally accepted due to the weight of scientific evidence, but there often exists a few papers which argue against a theory, while the preponderance of papers provides support for a theory. Understanding how science works, and its inability to typically provide proof, makes it easier to understand why a minor fractions of papers in an area may be at variance with the majority.

Using the view of Karl Popper, science does not prove things, but just fails to falsify hypotheses. In mathematics, a different scheme exists where it is feasible to provide a proof, for example, of the Pythagorean theorem. Confusingly, some of the same key terms are used in a different sense in math compared with science. In math, they have conjectures, which can be considered a proposition which has not (yet) been proved. The conjecture will become a theorem once it has been proved, which typically occurs in math by a logical demonstration. One of the more famous mathematical theorems is Fermat's last theorem which states that there are no three positive integers a, b, and c which can satisfy the following equation:

$$a^n + b^n = c^n \tag{1.2}$$

where n is an integer greater than 2. Although Pierre de Fermat (1604–1665) claimed he had a proof, one was not published until 1995 (Wiles 1995). Another famous problem in math is Goldbach's conjecture. It states that any even integer can be expressed as the sum of two primes. To date no proof has been provided of this conjecture, but if somebody does, it will become Goldbach's theorem. But even in mathematics, things are not as straightforward as they seem, in particular because of the work of Kurt Gödel (1906–1978), his incompleteness theorem (Gödel 1931). His theorem states that any formal axiomatic system containing arithmetic will have undecidable propositions; thus, mathematics is a system of logic which is not complete.

1.4   Are Hypotheses Necessary?

It is often assumed that all research is hypothesis driven, but this is not the case. For example, if a study examines the peak forces applied to a prosthesis during gait, no hypothesis is required. This information could be useful in a number of ways, and may form important background information for hypothesis formulation for a subsequent study. Winter (1987) has argued hypotheses were not necessary for his research but did advocate for inductive reasoning. Inductive reasoning involves generalizing from specific cases to produce general rules (deduction goes in the reverse direction). As Arthur Conan Doyle's famous hero Sherlock Holmes observed:

It is a capital mistake to theorize before one has data. Insensibly one begins to twist facts to suit theories, instead of theories to suit facts. (from A Scandal in Bohemia, published in 1891)

In new areas of scientific endeavor, it would be erroneous to hypothesize before there was some information to use as the basis for some inductive reasoning.

1.5 Tools Used in Biomechanics

Scottish mathematician and physicist James Clerk Maxwell (1831–1879) stated:

The most important step in the process of every science is the measurement of quantities... The whole system of civilized life may be fitly symbolized by a foot rule, a set of weights, and a clock.

So reference is made here to measures of position or distance (foot rule), mass (set of weights), and time (clock); of course from these three quantities, additional quantities can be derived (e.g., momentum). Although in some areas of science other quantities might be used, these three quantities do encompass many of the measures used in biomechanics. In the following chapters, with Newtonian mechanics as the basis, methods for measuring motion and force will be described, and the procedures for the analysis and processing of data obtained from these measurements will be presented. In addition, the measurement of the electrical signals produced by the muscles will be described (electromyography).

Biomechanics, as do many other domains in science, borrows from other areas of investigation to collect and process data. For example, many of the signal processing techniques used in biomechanics have been appropriated from methods used in electrical engineering.

1.6 Ethical Problems

Studies in human biomechanics typically use three vehicles for analysis: human subjects, cadaver material, and computers. In particular, in the use of human subjects, there are ethical issues to consider. On occasions, other animals are used in biomechanics studies, and these are subject to some of the same ethical concerns as the use of human subjects. Any research involving human subjects can present complex ethical, legal, and social issues. The research objective can sometimes be in conflict with the investigators' obligations to the subject. The National Commission for the Protection of Human Subjects of Biomedical and Behavioral Research identified three guiding principles for the ethical conduct of human subject research (Belmont Report 1979). These three principles are respect for persons, beneficence, and justice.

Respect for persons requires the protection of the autonomy of individuals and their treatment with courtesy and respect. For people who are incapable of making their own choices, they must be protected. Arising from this principle is that the researcher must be truthful and conduct no deception. In addition, there arises the requirement to obtain from study participants informed consent and to maintain subject confidentiality.

The principle of beneficence requires that participation in research be associated with a favorable balance of potential benefits from the research and the potential of harm to the subjects. A broad edict would be "Do no harm."

The principle of justice entails an equitable distribution of the burdens and benefits of the research. There are three broad categories related to justice: distributive justice, rights-based justice, and legal justice. Distributive justice is the fair distribution of scarce resources. Rights-based justice is respect for people's rights. Legal justice is respect for morally acceptable laws. For example, it should not occur that one group in society bears the burden of the research, but another group benefits from the research.

For research, there are other conditions which should also be satisfied:

- The study must have a scientifically valid design; poorly constructed studies are not ethical.
- The study must address a question of sufficient value to justify subjecting the participants to the study. For a simple survey, the bar of sufficient value would not be too high, but for a study where, for example, the subject has to undergo invasive measures, the bar would be much higher.
- The study must be conducted in an honest manner. The study has to follow approved protocols.

- The study outcomes must be reported both accurately and promptly. Accuracy includes reporting of methods, so others can replicate the study. The results should be reported without obfuscation. Failure to report findings can be considered to have wasted the contributions of the study participants.

There have been many studies in biomechanics which have challenging experimental designs due to their invasive nature but which have provided important information; there follows three examples. Milner-Brown et al. (1973) placed indwelling electromyography electrodes into the first dorsal interosseous of the hand, and they were able to describe the relative importance of recruitment and rate coding to the production of muscle force. The motion of the tibia and femur of humans walking was tracked from markers placed on the ends of intracortical pins screwed into those bones (Lafortune et al. 1992). Komi (1990), to explore the function of the human triceps surae, under local anesthetic placed a force transducer on the Achilles tendon of a subject and then made measurements of tendon forces during walking, running, and jumping. This study revealed the magnitudes of the forces in the Achilles tendon and aspects of the coordination of this muscle group during three common activities.

1.7   Overview of Chapters

Chapter 2 introduces some of the experimental tools used in biomechanics, including the dimensions and units used for biomechanical parameters and variables, and surveys the process of making the required computations.

Chapter 3 outlines the methods which can be used for determining signal content, including how the signal is sampled, Fourier analysis, and cross-correlation and autocorrelation analyses.

Signal processing is the focus of *Chap. 4*. This chapter presents methods for signal filtering and for data differentiation.

Chapter 5 introduces how data from electromyography can be collected and analyzed.

Three-dimensional image-based motion analysis is the focus of *Chap. 6*, with other methods for obtaining motion data outlined.

The methods available for collecting and analyzing forces are presented in *Chap. 7*. In particular, the focus will be on force plate data, including the analysis and processing of center of pressure data.

Chapter 8 presents methods for determining both whole body and segmental inertial properties. It also outlines the influence on the inertial properties of the operator, making the measures required to determine these properties.

The methods for the definition of the position and orientation of rigid bodies and series of rigid bodies are presented in *Chap. 9*. These definitions will be described for both two- and three-dimensional analyses.

The process of computing resultant joint moments is presented in *Chap. 10*, once again for both two- and three-dimensional analyses.

The errors arising in biomechanical data collection and their analysis are presented in *Chap. 11*.

Chapter 12 presents how data arising in biomechanical analyses can be scaled. This scaling can be important prior to performing statistical analyzes.

There is also a collection of appendices which predominantly review the fundamental math required for the materials covered in the chapters.

1.8   Review Questions

1. What is/are the essential properties of a scientific theory?
2. Describe the cyclical process of the scientific method.
3. Are hypotheses required for scientific research?
4. Define a hypothesis and a theory.
5. What are the three principles of the Belmont Report? Outline what each one means for conducting biomechanics research using human subjects.
6. Give an example of a study in biomechanics which uses an experimental protocol which presents challenges from an ethical perspective.

References

Cited References

Department of Health, Education, and Welfare. (1979). *The Belmont report: Ethical principles and guidelines for the protection of human subjects of research* (Vol. 2). Washington, D.C.: U.S. Government Printing Office.

Galison, P. L. (1997). *Image and logic: A material culture of microphysics*. Chicago: University of Chicago Press.

Galli, S. (2014). X-ray crystallography: One century of Nobel prizes. *Journal of Chemical Education, 91*(12), 2009–2012.

Gödel, K. (1931). Über formal unentscheidbare Sätze der Principia Mathematica und verwandter Systeme I. *Monatshefte für Mathematik, 149*(1), 1–29. [*Translated from the German "On formally undecidable propositions of Principia Mathematica and related systems I", there was no part II.*].

Komi, P. V. (1990). Relevance on in vivo force measurements to human biomechanics. *Journal of Biomechanics, 23*(Suppl. 1), 23–34.

Kuhn, T. S. (1962). *The structure of scientific revolutions*. Chicago: University of Chicago Press.

Lafortune, M. A., Cavanagh, P. R., Sommer, H. J., III, & Kalenak, A. (1992). Three-dimensional kinematics of the human knee during walking. *Journal of Biomechanics, 25*(4), 347–357.

Medawar, P. B. (1982). *Pluto's republic*. Oxford: Oxford University Press.

Milner-Brown, H. S., Stein, R. B., & Yemm, R. (1973). Changes in firing rate of human motor units during linearly changing voluntary contractions. *Journal of Physiology, 230*(2), 371–390.

Mulkay, M., & Gilbert, G. N. (1981). Putting philosophy to work: Karl Popper's influence on scientific practice. *Philosophy of the Social Sciences, 11*, 389–407.

Popper, K. R. (1959). *The logic of scientific discovery*. New York: Basic Books.

Wiles, A. (1995). Modular elliptic curves and Fermat's last theorem. *Annals of Mathematics, 141*(3), 443–551.

Winter, D. A. (1987). Are hypotheses really necessary in motor control research? *Journal of Motor Behavior, 19*(2), 276–279.

Useful References

Popper, K. R. (1999). *All life is problem solving*. London: Routledge.

Stewart, C. N. (2011). *Research ethics for scientists: A companion for students*. West Sussex/Hoboken: Wiley-Blackwell.

Experimental Tools

Overview

The analysis of any tasks can be qualitative or quantitative. Qualitative analysis deals with the nature of an activity, but does not involve measurement. This type of analysis is the one often utilized by scientists when planning experiments, and is the approach frequently used by clinicians and sports coaches. Quantitative analysis involves measurement, but of course the correct selection of the appropriate variables to measure is a crucial part of experimental design. Qualitative analysis is important because it is performed so often, and such analysis can only be strengthened by the insights provided by quantitative analysis.

Nobel Prize winner physicist Lord Rutherford (1871–1937) claimed:

Qualitative is poor quantitative.

While not advocating such a dogmatic perspective, in experimental biomechanics the focus is to measure various kinematic and kinetic quantities: quantitative analysis. In this chapter, the focus will be on an overview of the tools used in biomechanics, with a focus on the units of measurement, dimensional analysis, an overview of the primary measurement equipment used in biomechanics, and how computations required to determine the kinematics and kinetics quantities can be made.

Our ability to make measurements with accuracy and precision is often taken for granted, but even with something as simple as time our ability to standardize it and measure it accurately is a relatively new construct. The Inca Empire identified the time of day by looking at the position of the sun in the sky, which is convenient but not always available or accurate. As a measure of duration, the Incas referred to the time it took to boil a potato (McEwan 2006). The first watches were available in fifteenth-century Europe, but these were not very accurate. In the eighteenth century, the need for accurate naval navigation was a stimulus for the development of accurate timepieces, with the watchmaker John Harrison (1693–1776) producing famous timepieces which lost less than 3 seconds per day (Sobel 1995). But even in the nineteenth century, timekeeping was still imprecise with, for example, the American railways recognizing 75 different local times. Now we have the Coordinated Universal Time, which is a common reference for the world to set its clocks, and the second is now precisely and accurately defined by a characteristic frequency of a cesium atomic clock.

In performing biomechanical analyses, we exploit the Le Système International de Unités to define the relevant mechanical quantities; this system is outlined in the following section. This is followed by a description of dimensional analysis. As a precursor to the following chapters, the primary measurement tools used in biomechanics will be outlined. Finally, the process of making computations will be described and discussed.

The content of this chapter requires a working knowledge of matrix algebra (see Appendix A) and binary arithmetic (see Appendix C).

2.1 SI Units

The standard units and associated abbreviations for measurements are derived from Le Système International de Unités (SI). The convenience of the SI system arises from the consistent use of standard prefixes which can be used to designate different sizes of units of measurement. One of the issues with the US customary units is that it can be confusing because of the use of different bases. For example, inches are divided into 8ths or 16ths, and then there are 12 inches to the foot, 3 feet to the yard,

J. H. Challis, *Experimental Methods in Biomechanics*, https://doi.org/10.1007/978-3-030-52256-8_2

Table 2.1 The seven base units of the SI system

Base quantity	Name	Symbol
Length	Meter	m
Mass	Kilogram	kg
Time	Second	s
Electric current	Ampere	A
Thermodynamic temperature	Kelvin	K
Amount of substance	Mole	mol
Luminous intensity	Candela	cd

and 1760 yards to the mile (or 5280 feet). The variations in the bases in different measurement systems often reflect the evolution of these systems. The number of feet in the mile can help illustrate this, so why are there 5280 feet in the mile?

Historically, a rod has been used for measuring lengths and areas for the purpose of surveying. For example, in Revelations (11:1):

I was given a reed like a measuring rod and was told, 'Go and measure the temple of God and the altar, and count the worshipers there.'

By the Middle Ages in the western world, the rod had been standardized to be 16 1/2 feet long. Other standard measurements were based on the rod; the idea is that numbers such as 4, 8, 16, 32, and 64 are easy to manipulate (doubling or taking half).

Furlong	40 rods
Mile	8 furlongs
Mile	320 rods
Mile	5280 feet

The SI units can be broadly categorized into three sets: base units, derived units, and derived units with special names. The seven base units are all described in Table 2.1.

Each of these seven base units is as accurately defined as current technology permits. Here are the three examples:

Length – measured in meters, with a meter as the length traveled by light in a vacuum in 1/299 792 458 of a second.
Mass – measured in kilograms, which from 1889 until 2019 was defined as the mass of a platinum-iridium cylinder stored in Sèvres, France. Now it is defined relative to the Planck constant, an invariant constant from quantum mechanics.
Time – measured in seconds, with a second defined by the characteristic frequency of a cesium clock.

Given these seven base units, all other units can be derived. For example, if you are measuring the cross-sectional area of a circular section of muscle, the area of a circle is:

$$\text{area} = \pi . r^2 \tag{2.1}$$

The radius (r) would be measured in meters (m), and the constant π is dimensionless, so the unit for area is meters squared (m^2). The velocity of blood flow could be measured by determining the motion of blood over a certain distance for a measured period of time. The velocity is computed from:

$$\text{velocity} = \frac{\Delta d}{\Delta t} \equiv \frac{m}{s} \tag{2.2}$$

where Δd is the change in position or displacement measured for a time period Δt. These compound units can be formed by the multiplication or division of two or more units; in these cases, the compound unit should be written with either a space between the symbols or a period (.). Therefore, the units for velocity are:

$$\text{m.s}^{-1} \text{ or } \text{m s}^{-1}$$

Table 2.2 Some commonly derived SI units

Quantity	SI unit	Definition
Displacement	m	The change in location of a point
Area	m^2	Measure of a portion of a surface
Volume	m^3	Space encompassed
Density	$kg.m^{-3}$	Concentration of matter in an object
Velocity, speed	$m.s^{-1}$	The time rate of change of displacement
Acceleration	$m.s^{-2}$	The time rate of change of velocity

Table 2.3 Derived SI units with special names

Quantity	Name	SI symbol	SI units
Frequency	Hertz	Hz	s^{-1}
Force	Newton	N	$kg.m.s^{-2}$
Pressure	Pascal	Pa	$N.m^{-2} \equiv kg.m^{-1}.s^{-2}$
Energy	Joule	J	$N.m \equiv kg.m^2.s^{-2}$
Power	Watt	W	$J.s^{-1} \equiv kg.m^2.s^{-3}$
Charge	Coulomb	C	$A.s$
Electric potential	Volt	V	$m^2.kg.s^{-3}.A^{-1}$
Capacitance	Farad	F	$m^{-2}.kg^{-1}.s^4.A^2$
Electric resistance	Ohm	Ω	$m^2.kg.s^{-3}.A^{-2}$

but not ms^{-1}. When one unit is divided by another, there is another acceptable format, for example, the units for velocity can be expressed using:

$$m/s$$

Table 2.2 lists some commonly derived SI units; a more comprehensive list is provided in Appendix B.

Some SI-derived units have special names and symbols. These units are most often named after a scientist who had made some important contribution in science related to the derived unit. Having a unit named after you is a good way to immortalize your name, and you could be listed, for example, alongside Heinrich Hertz (1857–1894), Isaac Newton (1642–1727), Blaise Pascal (1623–1662), James Joule (1818–1889), James Watt (1736–1819), Charles-Augustin de Coulomb (1736–1806), Alessandro Volta (1745–1827), Michael Faraday (1791–1867), and André-Marie Ampère (1775–1836) (see Table 2.3).

For example, the units of force are named after Isaac Newton (1743–1827). Force is the product of mass and acceleration, so the derived SI units are:

$$\text{force} = m.a \equiv kg.m.s^{-2} \tag{2.3}$$

Therefore, one unit of force is $1\,kg.m.s^{-2}$ but is preferentially written as 1 N. By convention when using the abbreviation for the unit, we use uppercase (N) because it is derived from a proper noun, but if the units are written in full, then we use lower case (newton) to avoid confusion with the scientist associated with the unit.

There are a number of standard prefixes which can be used to designate units of measurement which are either larger or smaller than the base SI unit. The derivations of these prefixes come from a strange variety of sources (see Table 2.4). So, for example, the base unit for measurements of length is the meter; therefore, a centimeter is:

$$1\,m \times 10^{-2} = 0.01\,m = 1\,cm$$

Or, for example, a kilometer is:

$$1\,m \times 10^3 = 1000\,m = 1\,km$$

Table 2.4 Table of common SI prefixes

Prefix	Multiplier	Symbol	Source
Tera	10^{12}	T	"Teras" meaning monster in Greek
Giga	10^{9}	G	"Gigas" meaning giant in Greek
Mega	10^{6}	M	"Megas" meaning great in Greek
Kilo	10^{3}	k	"Chilioi" meaning 1000 in Greek
Hecto	10^{2}	h	"Hecaton" meaning 100 in Greek
Deca	10^{1}	da	"Deka" meaning 10 in Greek
	10^{0}	–	
Deci	10^{-1}	d	"Decimus" meaning 10 in Latin
Centi	10^{-2}	c	"Centum" meaning 100 in Latin
Milli	10^{-3}	m	"Mille" meaning 1000 in Latin
Micro	10^{-6}	μ	"Micros" meaning small in Greek
Nano	10^{-9}	n	"Nanos" meaning dwarf in Greek
Pico	10^{-12}	p	"Piccolo" meaning small in Italian
Femto	10^{-15}	f	"Femten" meaning 15 in Danish
Atto	10^{-18}	a	"Atten" meaning 18 in Danish

2.2 Dimensional Analysis

Dimensional analysis uses the base quantities to analyze the relationships between different physical quantities. The majority of quantities defined in biomechanics can be derived from the three base quantities:

$$\text{LENGTH}\,(L) \quad \text{MASS}\,(M) \quad \text{TIME}\,(T)$$

For example, force is the product of mass and acceleration (Newton's second law); therefore, in dimensions force is:

$$\text{force} = m.a \equiv (M)\left(L\,T^{-2}\right) \equiv M\,L\,T^{-2} \tag{2.4}$$

Prudent analysis of the dimensions influencing a mechanical process can provide insight into the important mechanical variables driving an observed phenomenon. Consider a simple pendulum consisting of a mass attached to a rotation point by a massless string (Fig. 2.1). The question is what does the period of swing of the pendulum depend on? An equation can be written relating the thing we want to know, which is the period of the swing (T), to the variables related to the system (see Table 2.5).

The generic equation for the period of the pendulum is:

$$t = f(k, l, m, \theta, g) \tag{2.5}$$

where k is a constant. We now use dimensional analysis to work out the relationship between the variables. Note that angles are dimensionless, as angles are defined as the ratio of two lengths: the length of a circular arc to its radius. Gravity is an acceleration, so its dimensions are the same as those of acceleration:

$$g \equiv L\,T^{-2} \tag{2.6}$$

An equation can be written describing the relationship between the variable we want to know (period of swing) and the dimensions of the mechanical variables:

$$T \equiv L^{\alpha}\,M^{\beta}\left(L\,T^{-2}\right)^{\gamma} \tag{2.7}$$

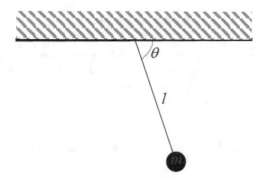

Fig. 2.1 Simple pendulum

Table 2.5 Key variables and their dimensions for the oscillation of a simple pendulum

Variable	Time	Length	Mass	Angle	Gravity
Abbreviation	t	l	m	θ	g
Dimensions	T	L	M	$(-)$	$L\,T^{-2}$

where the exponents (α, β, γ) have to be determined, and this can be achieved using some simple logic. As the units on the left-hand side do not include mass $\beta = 0$, therefore:

$$T \equiv L^{\alpha}\left(L\,T^{-2}\right)^{\gamma} \tag{2.8}$$

To eliminate L from the units on the left-hand side of the equation, then the following equation must hold: $0 = \alpha + \gamma$. To have time in the same dimensions on both sides of the equations, then $1 = -2\,\gamma$. Therefore, $\gamma = -\frac{1}{2}$ and so $\alpha = \frac{1}{2}$, which gives a dimensional equation as:

$$T \equiv L^{\frac{1}{2}}\left(L\,T^{-2}\right)^{-\frac{1}{2}} \equiv \sqrt{L\left(L\,T^{-2}\right)^{-1}} = \sqrt{\frac{L}{L\,T^{-2}}} \tag{2.9}$$

Therefore, we can now describe how the period of swing is a function of the identified variables:

$$t \propto \sqrt{\frac{l}{g}} \quad \rightarrow \quad t = k\sqrt{\frac{l}{g}} \tag{2.10}$$

This correctly describes the mechanical relationship between the period of a simple pendulum, its length, and the acceleration due to gravity, where $k = 2\,\pi$.

In a more formal sense, such dimensional analysis exploits Buckingham's Π-theorem (Buckingham 1914), which was developed and expanded by Bridgman (1922). For a system which has n physical quantities (a_i), the relationship between variables can be expressed in equation form:

$$f(a_1, a_2, \ldots, a_n) = 0 \tag{2.11}$$

The physical quantities can be described by k physical dimensions (for mechanical systems, $k = 3$ for length, mass, and time). Therefore, the equation can be expressed in terms of p physical dimensions, where $p = n - k$:

$$F\left(\Pi_1, \Pi_2, \ldots, \Pi_p\right) = 0 \tag{2.12}$$

where Π_i are groups of dimensionless parameters.

Taking the previous example of a simple pendulum, a table can be formed describing the relationship between the dimensions and the physical quantities (Table 2.6).

Table 2.6 The mechanical variables used to describe the simple pendulum and the dimensions of each variable

Dimensions	Mechanical variables			
	Time t	Mass m	Length l	Gravity g
Time T	1	0	0	−2
Mass M	0	1	0	0
Length L	0	0	1	1

There are four variables expressed in terms of three physical dimensions; therefore:

$$p = n - k = 4 - 3 = 1 \tag{2.13}$$

As $p = 1$ then,

$$\Pi_1 = t^\alpha \, m^\beta \, l^\gamma \, g^\delta = 0 \tag{2.14}$$

The theorem can be expressed in equation form:

$$\begin{bmatrix} 1 & 0 & 0 & -2 \\ 0 & 1 & 0 & 0 \\ 0 & 0 & 1 & 1 \end{bmatrix} \begin{bmatrix} \alpha \\ \beta \\ \gamma \\ \delta \end{bmatrix} = \begin{bmatrix} 0 \\ 0 \\ 0 \end{bmatrix} \tag{2.15}$$

which can be solved for the vector $a = [\alpha \, \beta \, \gamma \, \delta]^T$, and has a solution of:

$$a = \begin{bmatrix} 2 \\ 0 \\ -1 \\ 1 \end{bmatrix} \tag{2.16}$$

$$\Pi_1 = t^2 \, m^0 \, l^{-1} \, g^1 = t^2 \, l^{-1} \, g^1 = 0 \tag{2.17}$$

which lead us to how the period of swing is a function of the identified variables:

$$t^2 \propto \frac{l}{g} \quad \rightarrow \quad t \propto \sqrt{\frac{l}{g}} \quad \rightarrow \quad t = k\sqrt{\frac{l}{g}} \tag{2.18}$$

Note that the solution of the equation can often be often determined by visual inspection; if this is not feasible, the matrix manipulation techniques can be used (see Appendix A – Matrices).

Another example would be the time it would take for a sprinter to cover a given distance if running at a particular velocity. The example has a straightforward result but provides another illustration (Table 2.7).

There are three variables (time, displacement, velocity) expressed in terms of two dimensions (time, distance); therefore:

$$p = n - k = 3 - 2 = 1 \tag{2.19}$$

As $p = 1$, then:

| | Mechanical variables | | |
Dimensions	Time t	Displacement l	Velocity v
Time T	1	0	−1
Length L	0	1	1

Table 2.7 The mechanical variables used to describe the time it takes a sprinter to cover a given distance

$$\Pi_1 = t^\alpha \, l^\beta \, v^\gamma = 0 \qquad (2.20)$$

The theorem can be expressed in equation form:

$$\begin{bmatrix} 1 & 0 & -1 \\ 0 & 1 & 1 \end{bmatrix} \begin{bmatrix} \alpha \\ \beta \\ \gamma \end{bmatrix} = \begin{bmatrix} 0 \\ 0 \end{bmatrix} \qquad (2.21)$$

which can be solved for the vector $a = [\alpha \; \beta \; \gamma]^T$, and has a solution of:

$$a = \begin{bmatrix} 1 \\ -1 \\ 1 \end{bmatrix} \qquad (2.22)$$

$$\Pi_1 = t^1 \, l^{-1} \, v^1 = 0 \qquad (2.23)$$

which lead us to how the time to sprint a distance is a function of the displacement and the runner's velocity:

$$t = \frac{l}{v} \qquad (2.24)$$

2.3 Primary Measurement Tools

Measurements in biomechanics can be very simple, for example, aspects of gait have been quantified by measuring with a rule footprints obtained as subjects walked along a cardboard walkway with their feet covered in colored talcum powder (Wilkinson and Menz 1997). More frequently, data are collected using sophisticated electronics and computers. These measurement tools rely on transducers, devices which convert energy from one form to another. For example, a transducer can produce an electrical signal varying with the colors of light interacting with it, or a change in resistance proportional to the strain in a strip of metal foil which can then be related to the applied force.

Examining factors such as walking velocity, step length and width, hand accelerations are all studies of kinematic quantities. In classical mechanics, kinematics is that area of mechanics that describes the motion of points, bodies, and systems of bodies but with no concern for the forces or inertial properties which dictate that motion. In biomechanics, for example, electronic goniometers can be used to measure joint angles, and image-based motion analysis systems to measure the positions and orientations of body segments.

Examining factors such as peak ground reaction forces during running, the resultant joint moment profiles for the ankle, knee, and hip joints during walking are all studies of kinetic quantities. In classical mechanics, kinetics is that area of mechanics which describes the relationship between motion and its causes, that is, the forces and moments applied to points, bodies, or systems of bodies. In biomechanics, for example, force plates are used to measure ground reaction forces and inverse dynamics to compute the resultant moments at the joints.

Irrespective of the measurement tools being used, they should have sufficient accuracy, precision, and resolution for the analysis being performed (see Chap. 11 for details of error analysis).

2.4   Computations

To collect, process, and analyze biomechanical data requires sequences of computations. Some of these computations are performed in commercial software, sometimes in custom-written software and sometimes by hand. Irrespective of how the computations are performed, it is important that these computations are accurate; otherwise, interpretations of data may turn out to be erroneous.

The results from any computations we perform we want to be accurate, but to be able to judge that we need an estimate of what the result of the calculation should be. There are several approaches to confirm the computations are correct:

1. Have test data which has a known result.
2. Compare the results with values in the literature.
3. Make an estimate of the anticipated results.

Test Data – for certain applications, it is feasible to have test data with which to assess that the computations are correct; this may be as simple as, for example, applying known loads to a force plate and assessing if the computational process provides appropriate estimates of those known loads. Indeed, it may be possible to generate data with which to test computational processes.

Comparison with Literature – if others have performed similar studies, a common assumption is that the output of your analysis should be similar to those presented in published studies.

Making an Estimate – it is good practice to always make an estimate of what you expect the results of a computation to produce. A deviation from the estimate should prompt an analysis of the computational process. For example, if measuring the location of whole body center of mass from data from a motion analysis of a subject walking, then with markers placed at the anterior superior iliac spine, it would be unusual if the center of mass of the whole body deviated too far away from these markers. Consider another example; if calculating the mass of a subject's thigh, it might be reasonable to assume that the thigh mass should be no more than 15% of the total body mass and no less than 5% of the total body mass. If the calculated thigh mass lies outside of this range, the computations should be checked.

The Nobel Prize winning Italian-American physicist Enrico Fermi (1901–1954) was considered to be the master of making good estimates of the results of calculations. Therefore, these estimates are sometime called a Fermi estimate in response to a Fermi question. A Fermi estimate is based on an approximate approach to the problem. Therefore, you do not do the detailed calculations required to determine the answer but make a series of assumptions which hopefully capture the nature of the problem at hand but also keep the math simple. When the first atomic bomb exploded in the Jornada del Muerto desert (New Mexico, July 16, 1945), the scientists wanted an estimate of the amount of TNT which would give a similar size explosion. Fermi's estimate was 10 kilotons of TNT based on the distance traveled by pieces of paper he dropped during the explosion. His estimate was within an order of magnitude of the result from more sophisticated and time-consuming calculations (20 kilotons).

One way of using this approach is to make two Fermi estimates based on different sets of principles, and these may either provide a confirmation of one another or provide a region within which the actual result should probably fall.

2.5   Computer Code

To collect and analyze biomechanical data requires software. Force plates and motion analysis systems are the cornerstone tools of most biomechanical analyses. Another important tool is the software which is used both for collecting and analyzing the data. This software should provide results which are both accurate and precise and provide the required resolution.

2.5.1   Custom-Written Code

For many aspects of data collection and processing of data, there is commercial computer code available. If a researcher relies on commercially available code only, this can limit the biomechanical studies which can be performed. There is freely obtainable code available for certain analyses, but these often require some coding to integrate into a planned analysis. Therefore, many scientists in their efforts to perform experiments and analyses write their own software, but importantly their own software is an important stepping-stone to novel analyses. In a survey of researchers, they reported that 35% of their research time was spent on programming and developing software (Prabhu et al. 2011).

Computer code implements an algorithm or series of algorithms to perform some analysis. The algorithm can be thought of as a recipe, and the computer code is the implementation of that recipe. In the process of programming, some preexisting code may be used, equivalent to a cook using some prepared ingredients, but this must be integrated into a suite of software.

2.5.2 Key Features

The computer code should have the following qualities: reliable, robust, portable, and be easy to use, modify, and maintain.

Reliable – the software should give the same results each time it is run for the same data set.

Robust – code would be considered robust if small changes of the input data do not produce large changes in the output. Here small means changes which are at the level of noise, for example, changing the mass of the human thigh by 0.0001 kg.

Portable – portability refers to the ability to move software from one computer to another and get the same results. If one computer is the same as another, then the same output can be expected on both machines, but this is not always the case. For two computers to be considered the same, they would need to have the same processor, operating system, version of programming environment, etc. But not all computers have the same processor, often with the bit resolution being a crucial factor. Computers can only handle a finite number of digits when representing numbers, which over a series of mathematical steps can influence the accuracy of the output.

Easy to Use, Modify, and Maintain – even if the software is written for use by the writer of the code only, ease of use is important. If coming back to use the software after several months, ease of use will allow the user to exploit the software more promptly. As new knowledge is developed, it may be necessary to modify the software, and this should be straightforward. Maintenance might be the substitution of a more accurate algorithm into the software. Modification and maintenance of software is facilitated by well-structured code with good documentation. The documentation does not have to be independent of the code but can be embedded in the code.

2.5.3 Computer Representation of Numbers

Computers do not store numbers to infinite precision, but have to pack a number into a certain number of bits (binary digits). For example, the number 1/3 and irrational numbers such as π cannot be represented precisely in a decimal system, but as computers represent numbers in binary form (see Appendix C), one tenth in binary form is also not finitely defined:

$$\left(\frac{1}{10}\right)_{10} = (0.1)_{10} = (0.0\ 0011\ 0011\ 0011\ldots)_2 \qquad (2.25)$$

Within a computer, a number are typically represented in the following form:

$$-1^s \times (1.f)_2 \times 2^{e-E} \qquad (2.26)$$

where s is used to designate the sign of the number, f represents the bits in the significand, e is an integer exponent, and E is the bias exponent (127 for 32 bits). Notice that 1 is automatically used, so the significand does not need to include the first 1 (see Fig. 2.2).

For example, consider the number 176.125. The sign bit is 0:

$$-1^s = -1^0 = 1$$

The number 176.125 can be converted to binary:

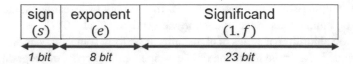

Fig. 2.2 Illustration of standard, IEEE 754 form, for presenting numbers with 32 bits, with 1 bit allocated to number sign, 8 bits to the exponent, and 23 bits to the mantissa

$$(176.125)_{10} = (10110000.001)_2$$

The binary version of the number can be expressed in scientific notation:

$$10110000.001 = 1.0110000001 \times 2^7$$

Remember the first 1 is given; therefore, the 23 bit significand is:

$$f = 0110\ 000\ 001\ 000\ 000\ 000\ 000\ 0$$

The required exponent is 7 but in the IEEE format for 32 bits, the bias is 127, so the computer representation of the exponent is 134 ($7 = 134 - 127$). The eight-bit exponent is:

$$(134)_{10} = (10000110)_2$$

The IEEE 754 form for the 32-bit representation of 176.125 is

Sign (1 bit)	0
Exponent (8 bits)	10000110
Significand (23 bits)	0110 000 001 000 000 000 000 0

So would occur as

$$01000011001100000010000000000000$$

Therefore, all numbers can be represented in this 32-bit format, which is efficient but imposes constraints on the precision of calculations. For each computer, there is a value, machine epsilon (ϵ), which describes the accuracy of the floating-point calculations. It is the smallest number which can be added to 1.0 and give a number other than 1. For numbers represented in 32 bits, there are 23 bits allocated to the significand, so:

$$\epsilon \approx 2^{-23} \approx 1.2 \times 10^{-7} \tag{2.27}$$

Giving ϵ about six decimal places. In 64 bits, we have 52 bits allocated to the significand, so $\epsilon \approx 2^{-52} \approx 2.2 \times 10^{-16}$, giving around 15 decimal places. Errors due to this imprecision are called round-off errors. These errors accumulate with the number of calculations, so if n arithmetic operations are performed, the total round-off error will be of the order $\sqrt{n}\,\varepsilon$, assuming the errors are randomly up or down. The errors follow a random walk pattern, and such patterns are also ubiquitous in biology (Berg 1983). Sometimes a sequence of calculations can occur in one direction in which case the total round-off error can grow in proportion to $n\,\varepsilon$.

For many calculations, machine precision and round-off errors are not influential, but it is important to be aware that these errors arise and can sometimes be problematic.

2.5.4 Programming Environments

There are a number of environments within which computations can be programmed, including MATLAB©, Python, C++/C, and FORTRAN. Some software packages also have their own scripting options embedded within them. Some analyses can be implemented in a spreadsheet environment such as Excel©. All of these environments have their own learning curve, but once one environment is mastered, others are more easily learned. MATLAB© is becoming the predominant programming environment, and its broad use brings with it the advantage that code for various tasks has been made available by researchers.

2.5.5 Coding Guidelines

Well-written code comes with a number of advantages including easier to follow, easier to modify if needed, easier to check if it is working appropriately, and more readily shared with colleagues. There are a number of guides for writing good computer code; one that is recommend is *The Elements of Programming Style* by Kernighan and Plauger (1978) which is the programmer's version of a guide to writing English (cf. *The Elements of Style*, by Strunk 1959). Their examples relate to FORTRAN, but the principles they emphasize generalize to any programming language and generally can be followed even if you do not use FORTRAN.

The following are useful guidelines:

General Writing Style
Make your code clear; do not make it complex or efficient just for the sake of it. Write the code to focus on the reader, so format a program to help the reader understand it; as long as you are accurate, the computer does not care.

Style
Whatever style you adopt, be consistent. Code can be easier to understand if parentheses are used to avoid ambiguity. Code can also be easier to read and follow if variables are given meaningful names which will not cause confusion. Another recommendation to consider is to use CamelCase in preference to under_score when naming variables; there is evidence that CamelCase positively influences the speed and accuracy of handling code (Binkley et al. 2009), but whichever you use, be consistent.

Comments
All programming languages allow the insertion of comments into the code; the computer ignores these comments but it helps the reader. These comments can make it clear what the following item of code is doing. These comments make it easy to come back to the code and modify it. In these comments, it is feasible, and good practice, to describe the format of the data that the routine will process.

Structure
Good code is written in modules, a main driving piece of code which systematically calls modules (functions, subroutines). The main code should read the data in and then systematically send it to subroutines for the analysis, and then it should output the results. All repetitive expressions should be replaced by calls to common subroutine/functions. If a subroutine/function seems to be getting too long, consider whether it should be divided into further subroutines/functions.

Writing and Testing Code
As the code is written, test each of the sub-modules individually. Work in small steps to develop the code; a complex analysis can be built of lots of tested smaller pieces. Plan for errors in the input data, so the data is checked, and if there are errors, they are flagged. Once the code is working, you can then optimize the code (if needed), being careful to keep notes of the versions of the code.

If modifying code, do so incrementally; this will help keep track of where an error might sneak in.

Use library functions whenever feasible; there is no need to reinvent the wheel unless it will help your understanding of the analysis.

Consider pair programming (two people, one computer, both working on the same code), it can be very efficient with the code written faster and with fewer errors (Padberg and Muller 2003).

Table 2.8 Basic convention used for pseudo-code

Notation	Meaning
$x \leftarrow y$	Assignment, x will now be equal to y
$x + y$	Addition
$x - y$	Subtraction
x/y	Division, will not work if $y = 0$
$x * y$	Multiplication
$x \wedge y$	Exponent, x to the power of y
x'	Transpose vector, or matrix
for $i = 1$ to n statement end for	For loop, which causes the statement to be performed n number of times. Counting does not need to start at 1
A * B	Element by element matrix or vector multiplication
sum(x)	Add all elements in x together

Other Considerations

Consider reserving certain letters for integers (e.g., i, j, k, l, m, and n). This means when counting within the code, it is easy to identify how the count is being tracked. Even if this recommendation is adopted, it does not mean you cannot use meaningful naming of variables (e.g., $n$subjects, $i$rows, $j$cols).

Depending on the programming environment, there are often features which can be useful. One is the editor, which, for example, might provide indenting of the code to reflect code structure; this can help with reading the code. Some environments have debuggers, which help you identify the source of an error in the code.

2.5.6 Pseudo-Code

Rather than present code in one programming language, pseudo-code will be presented to illustrate the implementation of an algorithm. Pseudo-code is a mixture of computer code and English. People might implement an algorithm in any of a number of computer programming environments, so pseudo-code looks like code but is easier to follow and then easy to implement in the programming environment of your choice. Therefore, pseudo-code should be a good description of the algorithm which can be easily translatable to computer code.

In the following chapters, the following conventions will be used for the pseudo-code (Table 2.8). In the pseudo-code, any more specific commands will be described within the code.

When presenting pseudo-code, the purpose of the code will be presented, and then the input(s) and output(s) of the code will be described. Then the pseudo-code is presented in one column with comments in an adjacent column. The following is a simple example, computing the mean of a sequence of numbers; it is designed to illustrate the general format of the presentation of pseudo-code.

Pseudo-Code 2.1: Computation of the Mean of a Series of Numbers

Purpose – given a series of numbers, compute the mean of those numbers.	
Inputs x – sequence (array) of numbers $(x_1, x_2, x_3, \ldots x_n)$ n – number of elements in array	
Outputs $x$mean – mean value of elements in array x	
$x$sum $\leftarrow$ sum(x)	Sum elements in x
$x$mean $\leftarrow$ $x$sum$/n$	Divide sum by n

For some analyses, there is a well-established code for performing computations. For example, there is long-standing code available for performing Fourier analyses of a time series; in such cases, pseudo-code will be not be presented as using tried and tested code is a better option than producing your own. Some programming is still, typically, required to integrate existing code into the planned analysis.

2.5.7 Use of Computer Code

Historically, the notebook has been an important way of scientists recording their progress through experiments. This is still an important step in the scientific process (Shankar 2004), but with extensive use of computer code to process data, there are other aspects of record keeping to consider. This record keeping is important irrespective of the source of the computer code. There follows three broad sets of recommendations, which are important to consider in support of the scientific process.

Data – make sure the raw and subsequently processed data are stored. These should be saved and retained. Even if the data is presented in graphical format, the data used to produce the image should be saved. These data are important, for example, should you chose to reanalyze the data or if a reviewer queries some aspect of your data. It is important to select a data format that is easy to follow and easy to understand and that is not too software environment specific.

Code – you should track what version of software was used. For commercial software, this may be simply a note. If using custom-written code, then store the used version of the software with the data. Make sure all code, driving routine and subroutines, have their versions clearly marked. By doing this, it is easy to compare across studies if used versions of software vary.

Access – although there may be reasons why this might not be feasible, but where possible make the computer code and data publically available. Most journals allow supplemental online material to compliment published papers. Such a philosophy can accelerate the process of scientific discovery.

2.6 Review Questions

1. What is the difference between quantitative and qualitative research?
2. What are study of kinematics and kinetics concerned with?
3. For the following, express in terms of their SI units and dimensions:

Linear displacement	Angular displacement
Linear velocity	Angular velocity
Linear acceleration	Angular acceleration
Mass	Moment of inertia
Force	Moment of force

4. Why is customized software for data analysis important?
5. How might you check your software is working appropriately?
6. What is meant when software is described as reliable, robust, and portable?
7. What influence might the finite precision of computer arithmetic have on calculations?

References

Cited References

Berg, H. C. (1983). *Random walks in biology*. Princeton: Princeton University Press.

Binkley, D., Davis, M., Lawrie, D., & Morrell, C. (2009). To CamelCase or Under_score. In *ICPC: 2009 IEEE 17th international conference on program comprehension* (pp. 158–167). IEEE.

Bridgman, P. W. (1922). *Dimensional analysis*. New Haven: Yale University Press.

Buckingham, E. (1914). On physically similar systems; illustrations of the use of dimensional equations. *Physical Review, 4*(4), 345–376.

Kernighan, B. W., & Plauger, P. J. (1978). *The elements of programming style*. New York: McGraw-Hill Book Company.

McEwan, G. F. (2006). *The Incas: New perspectives*. New York: W. W. Norton & Company.

Padberg, F., & Muller, M. M. (2003, 3–5 Sept). Analyzing the cost and benefit of pair programming. *Paper presented at the 9th International Software Metrics Symposium*.

Prabhu, P., Kim, H., Oh, T., Jablin, T, B., Johnson, N. P., Zoufaly, M., Raman, A., Liu, F., Walker, D., & Zhang, Y. (2011). *A survey of the practice of computational science*. Paper presented at the SC'11: Proceedings of 2011 International Conference for High Performance Computing, Networking, Storage and Analysis.

Shankar, K. (2004). Recordkeeping in the production of scientific knowledge: An ethnographic study. *Archival Science, 4*(3/4), 367–382.

Sobel, D. (1995). *Longitude: The true story of a lone genius who solved the greatest scientific problem of his time*. New York: Penguin.

Strunk, W. (1959). *The elements of style*. New York: Macmillan.

Wilkinson, M. J., & Menz, H. B. (1997). Measurement of gait parameters from footprints: A reliability study. *The Foot, 7*(1), 19–23.

Useful Reference

Weinstein, L., & Adam, J. A. (2008). *Guesstimation: Solving the world's problems on the back of a cocktail napkin*. Princeton: Princeton University Press.

Overview

Jean-Baptiste Joseph Fourier (1768–1830) was 30 years old when he accompanied Napoleon on his Egypt campaign, becoming the Governor of Lower Egypt. While considered a mathematician and physicist during his time in Egypt he was in charge of collating the scientific and literary discoveries made in Egypt, making important scholarly contributions on Egyptian antiquities. His later research was on how heat propagated through materials, which resulted in a famous reading of his work to the Paris Institute on December 21, 1807. Both Joseph-Louis Lagrange (1736–1813) and Pierre-Simon Laplace (1749–1827) initially objected to this work, but his mathematical description of heat transfer has become the cornerstone of the analysis and processing of signals collected in many scientific domains. It lead the Scottish physicist Lord Kelvin (1824–1907) to state:

> Fourier's Theorem is not only one of the most beautiful results of modern analysis, but it is said to furnish an indispensable instrument in the treatment of nearly every recondite question in modern physics.

Fourier analysis provides a tool for understanding the contents of a signal and for understanding the various data processing techniques used on collected signals. A signal is a function that conveys information about the state of a physical system. Signals mathematically consist of an independent variable (e.g., time) and a dependent variables, for example, in biomechanics, they could include:

- The muscle electrical activity, measured using techniques from electromyography
- The vertical ground reaction forces during running, measured using a force plate
- The displacement of the knee joint center, measured using a motion analysis system

While there are more advanced methods than Fourier analysis for analyzing signals, Fourier analysis is still used extensively and is a good basis for understanding other methods, for example, wavelet analysis. This chapter will focus on Fourier analysis while touching on other methods for determining the contents of a signal.

The content of this chapter requires a working knowledge of binary arithmetic (see Appendix C) and trigonometry (see Appendix D).

3.1 Analog and Digital Signals

Many studies require the sampling, processing, and analysis of signals. While these signals convey information about the state of a physical system, to be able to analyze the signal we need to sample it. The signals we seek to analyze are analog signals, where an analog signal is a signal that has continuous independent and dependent variables. For example, time might be our independent variable and the dependent variable the horizontal ground reaction force applied to the foot during a walking stride. Even the height of an adolescent boy or girl can be considered a continuous function of time. We typically analyze data using a computer that cannot handle a continuous signal, and it requires discrete packets of information; these signals are digital signals. A digital signal is represented by discrete independent and dependent variables. To return to our examples, we might collect force data 4800 times per second (e.g., Lieberman et al., 2010), while if examining growth curve of an

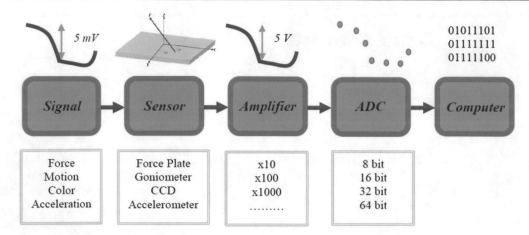

Fig. 3.1 A schematic representation of the data sampling process, where eventually the analog signal is digitized by the analog-to-digital converter (ADC) so the data can be stored and analyzed in digital form on a computer. The green boxes, with text, give examples of the corresponding component of the data sampling process

adolescent the height might be measured daily (e.g., Lampl et al., 1992). Most of our measurement equipment are electronic, so the sensed signal is an analog electrical signal which must be converted into digital format for storage and analysis on a computer. The data collection process is outlined in Fig. 3.1.

An analog-to-digital converter (ADC) is a device that takes an analog (continuous) signal and produces a digital signal. The outputs from a typical ADC are numbers in binary format (see Appendix C and Sect. 2.5.3). The ADC produced numbers are each represented by a single packet of binary information (word); for an 8-bit ADC, only eight digits comprise each word and therefore represent each digitized value, 16 digits for a 16-bit ADC, and so on. The number of bits limits the resolution of the measurements and their potential accuracy and influences the magnitude of the noise corrupting the sampled signal (see Sect. 11.6 for more details). The process of digitization introduces errors into the measurements, quantization error. These quantization errors are discussed in detail in Chap. 11, but a key property is that these errors have the properties of white noise.

Notice that in the data sampling process, the signal is amplified, and this occurs for two reasons. The first is if the signal is amplified any noise that is added to the signal in subsequent stages in the measurement process will be small relative to the size of the signal, and if the signal was to remain at its original size then the signal to noise ratio of the eventually recorded signal would be poorer. The second reason is to match the signal amplitude to amplitude range the ADC can tolerate. For example, if the ADC has a voltage range of 10 V but if the input signal has a range, after amplification, of 5 V then only 50% of the resolution of the system is being used. Therefore, it is a common practice to amplify the signal as much as it is feasible to exploit the maximum range of the ADC. Of course, setting the level of amplification must also be performed to ensure that the amplified signal does not exceed the range of the ADC. In some measurement processes it is not unusual to amplify the signal twice, for example when collecting electromyographic data the signal is often amplified close to the electrodes sensing the muscle electrical signal and once again just prior to analog-to-digital conversion.

The target is for the digital signal to be as accurate representation as possible of the sampled analog signal. Two factors have been identified which influence this: the appropriate amount of signal amplification, and the resolution of the ADC. Another factor is the number of samples taken of the analog signal, and this is guided by sampling theorem. To understand the selection of the appropriate sample rate, it is important to know the frequency content of the signal. The following section describes how Fourier analysis can be used to determine signal frequency contents.

3.2 Fourier Analysis

Data is often collected and analyzed in the time domain, for example, we collect force data and view it as a function of time. Frequency analysis allows us to view the frequencies composing a signal; the signal can be viewed and analyzed in the frequency domain. Being able to view collected data in both the time and frequency domains can help provide additional insight into the signal than would not be obtained if only one domain was used. One class of methods for analyzing signals in the frequency domain is Fourier analysis, the basics of Fourier analysis will be presented in the following sub-sections.

3.2.1   Sine and Cosine Waves

The basis of Fourier analysis is the modeling of signals using sine and cosine waves. Figure 3.2 gives the basic anatomy of a sine wave.

For a function, f, with values varying with time, t, a sine wave can be described by:

$$f(t) = b \, \sin(\omega_0 \, t) \tag{3.1}$$

where b is the maximum function amplitude and ω_0 is angular frequency (radians/second). The angular frequency is related to the period of the function (T) and the frequency of the sine wave (f_0):

$$f_0 = \frac{1}{T} = \frac{\omega_0}{2\pi} \tag{3.2}$$

The frequency of the signal is the number of cycles per second, measured in hertz (Hz). Figure 3.3 shows three different sine waves each with different frequencies and amplitudes.

The equations representing these three signals can be written as:

$$f_1(t) - 1.00 \, \sin(2 \, \pi \, t) \tag{3.3}$$

Fig. 3.2 Sine wave, with period and amplitude identified

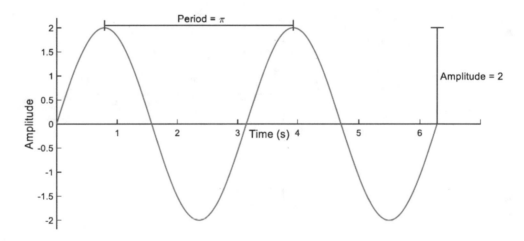

Fig. 3.3 Three different sine waves, (**a**) 1 Hz signal with amplitude of 1, (**b**) 2 Hz signal with amplitude of 0.5, and (**c**) 3 Hz signal with amplitude of 0.25

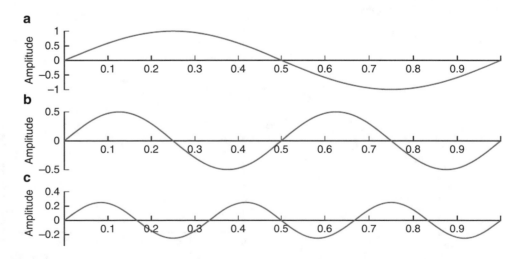

$$f_2(t) = 0.50 \ \sin{(4 \ \pi \ t)} \tag{3.4}$$

$$f_3(t) = 0.25 \ \sin{(6 \ \pi \ t)} \tag{3.5}$$

These three signals can be summed to generate a composite signal, shown in Fig. 3.4. This can also be represented in equation form:

$$f_4(t) = f_1(t) + f_2(t) + f_3(t) \tag{3.6}$$

$$f_4(t) = 1.00 \ \sin{(2\pi t)} + 0.50 \ \sin{(4\pi t)} + 0.25 \ \sin{(6\pi t)} \tag{3.7}$$

$$f_4(t) = \sum\nolimits_{k=1}^{3} b_k \ \sin{(k \ 2 \ \pi \ t)} \tag{3.8}$$

where b_k is the sequence of amplitudes (*1, 0.5,* and *0.25*) for the different frequency sine waves (*1, 2,* and *3*).

This composite signal can be considered to consist of three frequencies each with a different amplitude, which can be represented in a frequency spectrum plot (see Fig. 3.5).

Fig. 3.4 A signal comprised of three sine waves

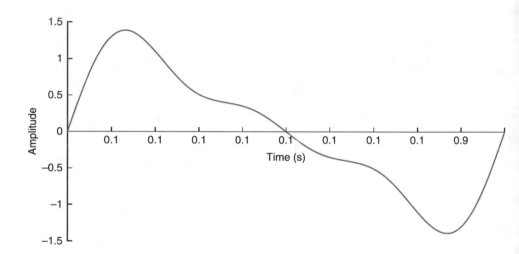

Fig. 3.5 The frequency spectrum of the signal comprised of three sine waves

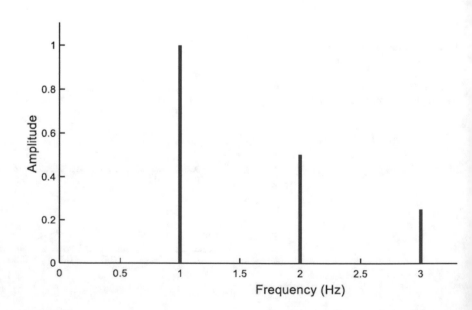

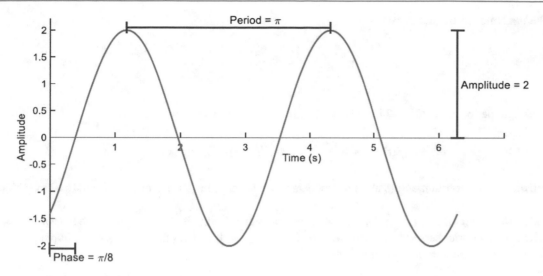

Fig. 3.6 Sine wave, with period, amplitude, and phase identified

The problem with the format we have used for representing a signal as a series of sine waves is constraint that when $t = 0$ the function must equal to zero ($f(t = 0) = 0$) which is clearly not the case for all signals. This problem can be accommodated with a sine wave by adjusting the phase (ϕ); therefore:

$$f(t) = b \, \sin(\omega_0 \, t - \phi) \tag{3.9}$$

This can be represented graphically in Fig. 3.6.

Using basic trigonometric relationships, the equation for a sine wave with a phase shift can be expressed as a combination of sine and cosine waves:

$$\begin{aligned}
f(t) &= b \, \sin(\omega_0 \, t - \phi) \\
&= b \, (\sin \omega_0 t \, \cos \phi - \cos \omega_0 t \, \sin \phi) \\
&= (b \, \cos \phi) \sin \omega_0 t - (b \, \sin \phi) \cos \omega_0 t \\
&- b' \sin \omega_0 t - c' \cos \omega_0 t
\end{aligned} \tag{3.10}$$

It is an equation of this format which is the basis of Fourier analysis. It is common to use Fourier series where data are collected as a function of time, in which case the frequency is in units of Hertz. But Fourier series can be fitted to other data, for example, if the collected data is a function of position, in which case, the dependent variable will be considered a function of the inverse of wavelength (cycles per meter). For example, Godfrey et al. (1987) examined the spatial frequency of the striped pattern of tigers and zebras using Fourier analysis.

3.2.2 Fourier Transform

The Fourier transform is a process that converts data from the time domain to the frequency domain. In their original format, the equations are infinite series, but when we collect data, it is for a finite time duration so the equations presented in the following are for the finite time series – the discrete Fourier transform (DFT). Somewhat confusingly, the equations describing the Fourier series are presented in a number of different forms, and all of these forms are equivalent, but the equations have been set up somewhat differently. Here is a finite Fourier series:

$$\begin{aligned}
f(t) = a_0 \\
+ a_1 \, \cos(\omega t) + b_1 \, \sin(\omega t) \\
+ a_2 \, \cos(2\omega t) + b_2 \, \sin(2\omega t) \\
+ a_3 \, \cos(3\omega t) + b_3 \, \sin(3\omega t) \\
+ \dots\dots\dots + \dots\dots\dots \\
+ a_N \, \cos(N\omega t) + b_N \, \sin(N\omega t)
\end{aligned} \tag{3.11}$$

Where the fundamental frequency is:

$$\omega = \frac{2\pi}{T} \tag{3.12}$$

and T is the sample period. Equation 3.11 can be written in a more compact form as:

$$f(t) = a_0 + \sum_{k=1}^{N} a_k \cos(k\,\omega\,t) + b_k \sin(k\,\omega\,t) \tag{3.13}$$

The first term, a_0, is a constant across all of the modeled time series. This term is often referred to as the DC term (direct current).

For a time series $(t_1, t_2, t_3, \ldots, t_{n-1}, T)$, there are n values. When fitting the Fourier series, we have n values that can be used to determine the Fourier series coefficients $(a_0, a_1, b_1, \ldots, a_N, b_N)$, which other than a_0 come in pairs.

Therefore, the maximum number of coefficients is:

$$N \le \frac{n-1}{2} \tag{3.14}$$

Each pair of coefficients represents a different frequency component of the signal. Therefore, for the highest-frequency component there must be at least two samples; this relationship has implications for rate at which data must be sampled (see latter Sect. 3.6).

If a requirement for the analysis is to examine the frequency spectrum, the amplitudes (c_i) can be computed from:

$$c_0 = a_0 \tag{3.15}$$

$$c_k = \sqrt{a_k^2 + b_k^2} \qquad k = 1 : N \tag{3.16}$$

The phase (θ_k) can be computed from:

$$\theta_k = \tan^{-1} \frac{b_k}{a_k} \tag{3.17}$$

The corresponding frequencies are:

$$f_k = k\,\frac{1}{T} \tag{3.18}$$

The general shape of sine and cosine waves is smooth, so does a Fourier series fit a less continuous function? Figure 3.7 shows a Fourier series fit to a saw tooth function.

The Fourier series can be written in a number of different forms; one exploits the following relationship:

$$e^{\pm i\theta} = \cos(\theta) \pm \sin(\theta) \tag{3.19}$$

where $i = \sqrt{-1}$. This equation is often referred to as Euler's after the Swiss mathematician and physicist Leonhard Euler (1707–1783). Euler was so prolific that a large number of equations are named in his honor. Exploiting Euler's formula, the Fourier series can be written as:

$$f(t) = \sum_{k=0}^{N} c_k\,e^{-i\,k\,\omega\,t} \tag{3.20}$$

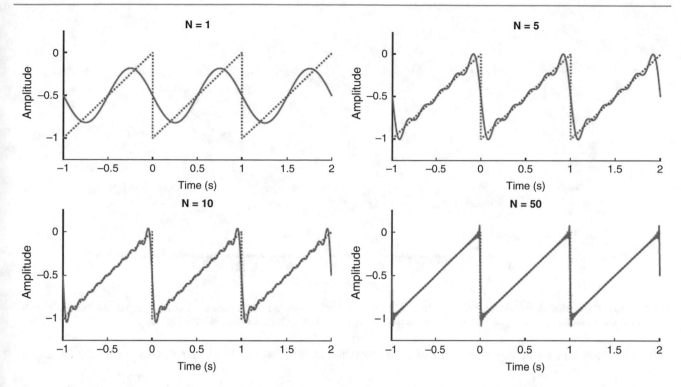

Fig. 3.7 Plots of the Fourier series fit to a saw tooth function. In each graph, the number of Fourier coefficient pairs (*N*) used in the reconstruction is noted

3.2.3 Determining Coefficients

The coefficients for the Fourier series are easy to compute but somewhat tedious. The pertinent equations for determining the discrete Fourier transform (DFT) are:

$$a_0 = \frac{1}{n} \sum_{i=1}^{n} f(t_i) \tag{3.21}$$

$$a_k = \frac{2}{n} \sum_{i=1}^{n} f(t_i) \cos\left((i-1)\omega t_i\right) \qquad k = 1 : N \tag{3.22}$$

$$b_k = \frac{2}{n} \sum_{i=1}^{n} f(t_i) \sin\left((i-1)\omega t_i\right) \qquad k = 1 : N \tag{3.23}$$

Computation of the coefficients associated with the DFT can be time-consuming; therefore, an alternative technique was introduced, the fast Fourier transform (FFT; Cooley and Tukey, 1965). This algorithm is so important that it was identified as one of the top ten algorithms of the last century (Dongarra and Sullivan, 2000; Rockmore, 2000); it still has important applications in the twenty-first century. Fast alternatives to the discrete Fourier transform had been around for a long time before the publications of Cooley and Tukey (Heideman et al., 1985), but it was their algorithm that unleashed the power of the Fourier transform for the analysis of problems in many areas of mathematics and engineering.

The FFT is much faster than the discrete Fourier transform, with if there are n data points to process a DFT requires $\sim n^2$ operations but the FFT only $\sim n \log_2 n$ (see Fig. 3.8). It turns out the FFT can be less susceptible to the rounding errors, making its estimates often more robust than the DFT. The speed of the algorithm is in part furnished by having the number of data points an integer power of 2 (e.g., 64, 128, 256, 512, 1024, 2048, 4096, etc.); this is fine if the collected data turns out to meet that constraint, but in reality, this is often not the case which means the data set has to be padded. Padding typically adds zeros to the data set until the total number of data points equals the next power of 2. There are issues with such padding (see Sect. 3.4). There are FFT algorithms for data sets which do not have to contain a sample comprising a power of two number of data points, but these enforce other constraints so the power of two algorithms dominate. For example, one algorithm requires the

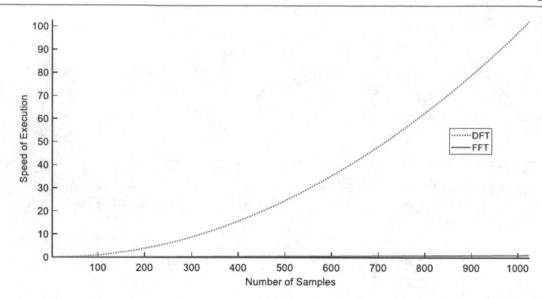

Fig. 3.8 Speed with which a discrete Fourier transform (DFT) and a fast Fourier transform (FFT) process data sets of different number of samples. The speed of execution expresses speed relative to the execution time for the FFT with maximum number of points

data length to be a prime number which generates similar issues to requiring the data length to be a power of 2 (Winograd, 1976).

3.3 Windowing of Data

Imagine a cosine wave with a frequency of 100 Hz and an amplitude of 1, sampled at 10000 Hz. If we take 128 samples, 256 samples, or 1024 samples, we would expect each to provide the same frequency spectrum. Figure 3.9 illustrates that the number of samples influences the accuracy of the estimated spectrum. With fewer samples, components of the signal not periodic over the sample interval leak out into adjacent portions of the spectrum. The DFT is the product of the signal and a rectangular window. The rectangular window reflects the act of sampling the signal over a given data window. The sharp corners of the rectangular window cause the leakage of the spectrum. With a greater number of samples the effect of the hard edges of the rectangular window can be greatly diminished.

Fourier analysis is based on the assumption of sampling the complete time series, but typically really only a window of the data is observed. This window is rectangular in shape potentially causing distortion at the edges of the time series, which therefore influences the analysis of the frequencies in the data. Prior to frequency analysis an additional windowing of the data can improve the estimation of signal frequencies. Therefore, to reduce the influence of the hard edges associated with the rectangular window we can apply a window to the data that has softer edges (Harris, 1978). Therefore, after sampling and before estimating the signal frequency spectrum a window function is used. There have been varieties of window functions proposed. Table 3.1 provides the equations for three different windows that can be used to preprocess data prior to computation of the signal spectrum. The nature of these windows is illustrated in the Fig. 3.10.

When sampling data we are taking a snapshot of the time series using a rectangular window; the influence of this rectangular window can be reduced by using a window function. Figure 3.11 illustrates the spectrum of rectangular windowed data and the same data after applying a Welch window. Notice from the figure that windowing of the sampled data can reduce leakage but the obtained spectrum has not precisely identified the frequency in the sampled signal.

Many software environments contain routines for performing the FFT. In many of the FFT routines, the data are automatically windowed with some default window, often the Welch window. The influence of this window on data analysis and interpretation should be considered rather than blindly assuming that the default window is the most appropriate for your data.

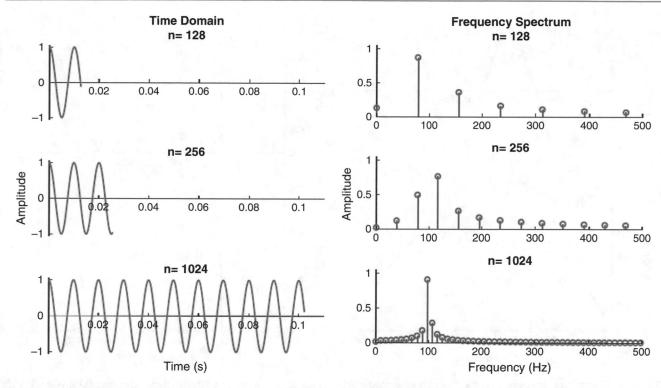

Fig. 3.9 The left-hand plots are the time domain plots of a 100 Hz sine wave, which has been sampled 128, 256, and 1024 times. The right-hand plots are the frequency spectrum of the sampled cosine waves. The frequency spectra illustrates leakage of frequency components

Table 3.1 Data window functions

Data window	Equation for sample interval $-T$ to T
Rectangular	$w(t) = \begin{cases} 1 & \|t\| \leq T \\ & \text{else } 0 \end{cases}$
Welch	$w(t) = 1 - \left(\frac{t - \frac{n-1}{2}}{\frac{n-1}{2}} \right)^2$
Bartlett	$w(t) = 1 - \left\| \frac{t - \frac{1}{2}n}{\frac{1}{2}n} \right\|$
Hanning	$w(t) = \begin{cases} 0.5 + 0.5 \cos\left(\frac{\pi t}{T} \right) & \|t\| \leq T \\ & \text{else } 0 \end{cases}$

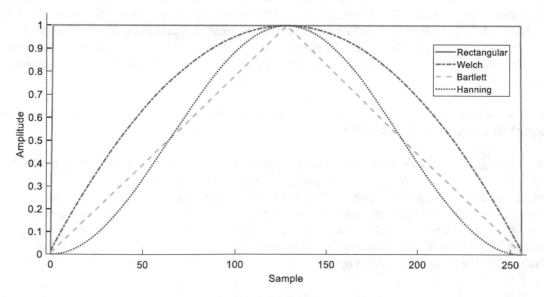

Fig. 3.10 Window functions used to soften edges before computing the discrete Fourier transform

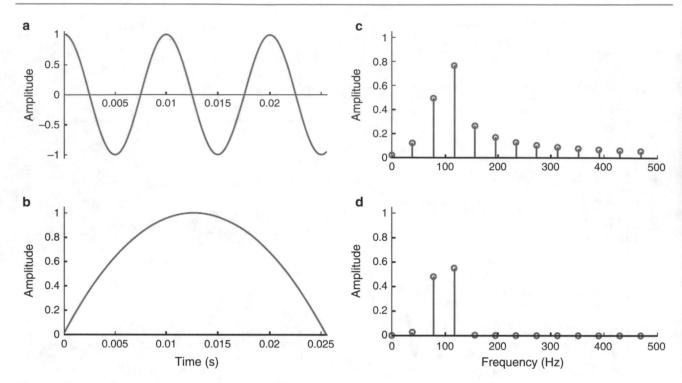

Fig. 3.11 Illustration of the reduction of leakage due to windowing data where (**a**) cosine plot of a 100 Hz cosine wave sampled at 1000 Hz with 256 data points, (**b**) the Welch window, (**c**) the frequency spectrum of data due to a rectangular window, and (**d**) the frequency spectrum of data after using a Welch window. These frequency spectra illustrate the reduction of leakage due to appropriate windowing of the data prior to spectrum computation

3.4   Gibbs Phenomenon

The Gibbs phenomenon refers to how a Fourier series fitted to a function behaves at a discontinuity in the time series. Figure 3.12 shows Fourier series of different numbers of Fourier coefficients fitted to a square wave. Notice that at the points where there is sudden transition from 0 to 1, the Fourier series has oscillations. The oscillations do not diminish with increasing number of Fourier coefficients, but it does tend to a finite limit. This phenomenon occurs because with the Fourier series, a finite series of continuous sine and cosine waves is fitted to a discontinuous function.

If a time series is sampled which because of the nature of the signal being analyzed has clear discontinuities, it would be incorrect to interpret high-frequency oscillations in a frequency spectrum as always arising from the signal, as the Gibbs phenomenon is a likely culprit. Most FFT algorithms require that the number of data points is an integer power of two, and if this is not the case, the data are padded typically with zeros so the signal has the appropriate number of data points. Depending on the nature of the signal being analyzed, this padding could introduce discontinuities in the time series leading to the Gibbs phenomenon.

3.5   Signal Spectrum and Signal Power

The spectrum of a signal is obtained from fitting a discrete (finite) Fourier series to a signal. The plot of signal amplitude against frequency is the spectrum. Sometimes a signal spectrum is referred to as harmonic analysis, where the harmonics are the integer multiples of the fundamental frequency. The first harmonic is the fundamental frequency, the second harmonic is twice the fundamental frequency, and so on. How these are plotted varies: if the sample rate is high, then the plot might be a line graph, but if this is not the case each harmonic might be represented as a vertical line. While the sample rate dictates the number of harmonics and the highest harmonic in the spectrum, it is the duration of the sampling that dictates the lowest-frequency component. For example, if a signal is sampled for 20 seconds the lowest feasible frequency that can be represented

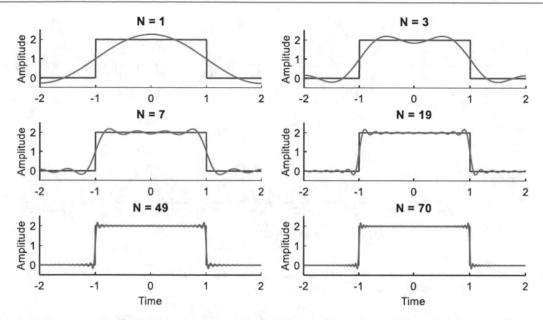

Fig. 3.12 Fourier series approximation to a square function using different numbers of Fourier coefficients. The fits demonstrate the Gibbs phenomenon (at times −1 and 1) for $N = 19$, 49, and 70 (where N is the number of Fourier coefficient pairs used to represent the signal)

in the signal is 0.05 Hz (i.e., 1 / 20), and this would be the fundamental frequency. The highest frequency present in a signal is crucial for determining the appropriate sample rate for a signal (see Sect. 3.6).

The power spectrum of a signal describes how the power of the signal is distributed over signal's frequencies. Power here is not defined in a mechanical sense (with unit watts) but refers to the squared value of the signal. The power values are simply computed as the modulus-squared of the discrete Fourier transform. It is worth noting that the total power in a signal is the same if the signal power is computed in the frequency or time domains (Parseval's theorem).

Examination of the frequency or power spectrum is most powerful if the signal being examined is stationary. For a stationary process, its signal properties are invariant irrespective of what portion of the signal is analyzed. These signal properties include the mean, variance, and frequency and power spectra. The DFT and FFT work well for stationary processes as they provide estimates of the signal power spectrum, but if the signal is not stationary and therefore the signal frequency components vary with time the DFT and FFT provide no information about the timing of signal frequency components. If you analyzed a signal that included high-frequency components at the start of a time series and only low-frequency components at the end, then this would not be stationary process, and the high-frequency components would appear in the computed spectrum but with no information about when they occurred. Many signals are analyzed assuming they are weakly stationary signals, and in which case, the spectrum can be informative. To analyze the phasing of the frequency components of a signal requires time-frequency analysis (see subsequent Sect. 3.8).

3.6 Sampling Theorem

In the frequency spectrum of a signal the lowest frequency is dictated by the sample duration but what is the highest frequency dictated by? Claude Elwood Shannon (1916–2001), an American mathematician and electrical engineer proposed a theorem that addresses this question (Shannon, 1949). Sampling theorem, or Shannon's sampling theorem, states that a signal should be sampled at a rate, which is at least as great as twice the highest-frequency component in the signal. This theorem sets the Nyquist frequency ($f_{Nyquist}$), which is related to the sample interval (Δt) and sampling rate (f_{Sample}):

$$f_{Nyquist} = \frac{1}{2\,\Delta t} = \frac{f_{sample}}{2}$$

(3.24)

There are two questions that emerge: What happens if a signal is not sampled at sufficiently a high rate? What happens to signal components above the Nyquist frequency if present in a signal before sampling? The following two paragraphs will address these questions in turn.

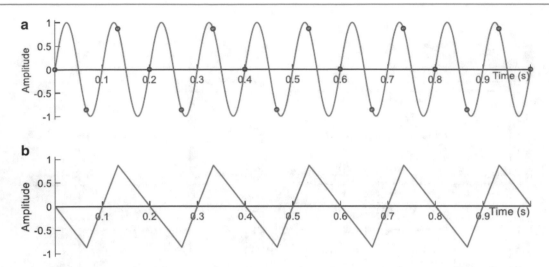

Fig. 3.13 Example of under-sampling of a signal, (**a**) a 10 Hz sine wave sampled at 1000 Hz solid line, and the red circle data points if sampled at 15 Hz, and (**b**) the 10 Hz signal sampled at 15 Hz

What happens if a signal is not sampled at a sufficiently high rate? Clearly some of the signal components are not (correctly) sampled, for example, in a golf swing if the sample rate is not sufficient there may only be data of the golf club at the top of the swing and after the ball is struck. Figure 3.13 illustrates a 10 Hz sine wave sampled at 1000 Hz and at 15 Hz. Clearly, with a sample rate of 1000 Hz, we see a sine wave with 10 distinct peaks in the 1-second sample period (top graph). Sampling theorem says a sample rate of at least 20 Hz is required for this signal, so when sampling at 15 Hz, the original signal is greatly corrupted (bottom graph). The 15 Hz sampled signal appears to have a frequency component at 5 Hz (5 peaks). In some circumstances it might be obvious the data has been sampled at too low a sample rate, but if for example, the movements to be analyzed are small the naked eye may not be able to identify that the movement seen does not match the signal collected. If the 10 Hz sine wave had been sampled at 10 Hz, so one sample every 0.1 seconds (0.0. 0.1, 0.2, etc.), the sampled signal would have been a straight line.

What happens to signal components above the Nyquist frequency if present in a signal before sampling? Any signal components above the Nyquist frequency assume a false identity, they are aliased. Aliasing occurs if the sampling theorem is not followed, which means that any frequency components higher than the half of the sampling rate will be folded back into the sampled frequency domain. These frequency components do not disappear rather they are reflected back into the sampled bandwidth corrupting lower-frequency components of the signal. Figure 3.14 illustrates this. In the left-hand column is a 1 Hz signal, which has a single frequency component at 1 Hz when the data is sampled at 1000 Hz or 8 Hz. In the right-hand column is an 8 Hz signal, which has a single frequency component at 8 Hz when the data is sampled at 1000 Hz, but this frequency component is reflected to 1 Hz when the signal is sampled at 8 Hz. This reflected or folding of signal frequency components greater than the Nyquist frequency is why this frequency is sometimes called the folding frequency. If the signal is sampled at 8 Hz the Nyquist frequency is 4 Hz, which means a 7 Hz signal frequency component will be reflected to appear as a 1 Hz signal component. This is evident in both the time and frequency domain versions of the 7 Hz sine wave sampled at 8 Hz. The example is for a signal with only one frequency component, in most sampled data the signal will be comprised of multiple frequency components. Therefore, if inappropriately sampled the aliased signal frequencies could interfere with actual frequency components in the range of frequencies adequately sampled.

An example of aliasing is often seen in film or video of wheels. Film is often recorded at 24 Hz and video at 60 Hz, so if the wheels rotate at a frequency greater than 12 Hz or 30 Hz, respectively, there will be aliasing in the film or video record. A wagon wheel on a moving wagon, for example, in movies, can often appear stationary or rotating backward due to aliasing (see Fig. 3.15 for an example).

But before you collect the data, how do you know what the highest frequency is? Theoretically, signals are not band-limited (Slepian, 1976), but in practice we have to assume that they are. Another common assumption is that the signal we are trying to measure is corrupted by white noise. One property of white noise is that this noise is across all of the frequency spectrum, of course only a portion of the spectrum, up to the Nyquist frequency can be collected, so some noise must be aliased. These problems aside sampling theorem still guides our collection of data. To avoid aliasing, there are two options:

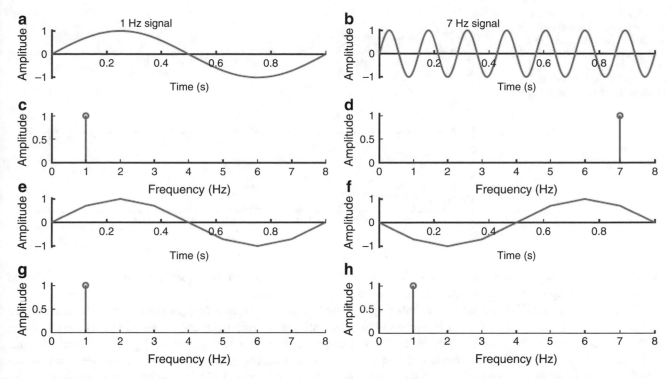

Fig. 3.14 Example of signal aliasing. In the left-hand column of figures is a 1 Hz signal sampled at 1000 Hz and 8 Hz, while the right-hand column is a 7 Hz signal sampled at 1000 Hz and 8 Hz. Figures (**a**) and (**b**) show the two time series sampled at 1000 Hz, (**c**) and (**d**) are the frequency spectra of these two signals, (**e**) and (**f**) are the 1 Hz signal and the 7 Hz signal both sampled at 8 Hz, and (**g**) and (**h**) are the frequency spectra of the two signals sampled at 8 Hz

Fig. 3.15 Sampled images of a wheel rotating clockwise at different frequencies but all sampled at 1 Hz. At a wheel frequency of 0.25 Hz, the images adequately represent the wheel motion, while for a wheel frequency of 0.5 Hz, some interpolation would be necessary to provide images that reflect the wheel motion, and for the other two wheel frequencies, signal aliasing has occurred

Wheel Frequency	0 seconds	1 second	2 seconds	3 seconds	4 seconds	Appearance
0.25 Hz	↑	→	↓	←	↑	Clockwise Rotation
0.50 Hz	↑	↓	↑	↓	↑	Arrow Flicks from Up to Down
0.75 Hz	↑	←	↓	→	↑	Counter-Clockwise Rotation
1.00 Hz	↑	↑	↑	↑	↑	Stationary

1. Know the bandwidth of the signal a priori, and select the sample rate accordingly.
2. Sample at a sufficiently high sample rate to avoid aliasing, by making assumptions about the maximum frequency content of the signal, and then use a sample rate at an order of magnitude (x10) of the size of the assumed maximum frequency.

Depending on the task it may be impossible to know signal bandwidth a priori, and it could be an error to assume you know it. A common recommendation is to sample the signal at a sample rate which is roughly 10x greater than the presumed highest

Fig. 3.16 Example of signal interpolation, (**a**) signal sampled at 100 Hz, (**b**) the same signal with points sampled at 20 Hz, and (**c**) solid blue line interpolated at 20 Hz signal, and dotted red line original signal sampled at 100 Hz

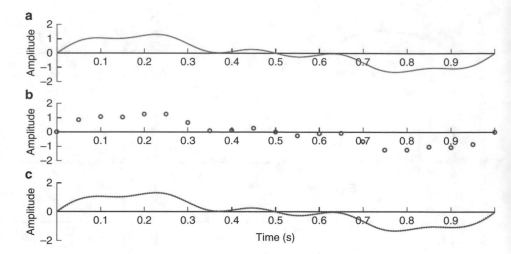

frequency in the signal, and such a strategy should ensure that anomalous behavior from a subject, resulting in unanticipated high-frequency components in the sampled signal, can be adequately sampled.

With some equipment it is feasible to band-limit a signal before sampling data from the equipment. For example, if the plan is to sample force plate data at 300 Hz it is prudent to filter the data before sampling using a filter that limits the maximum frequency components of the sampled signal to 150 Hz. This can be achieved by employing an analog filter, also called in this context an anti-aliasing filter, which filters out frequency signal components greater than the Nyquist frequency before sampling. Such an approach avoids aliasing but can mean that aspects of the signal of interest might not be sampled.

An often neglected part of Shannon's 1949 paper, where he presented sampling theorem, is how to increase the temporal resolution of data by signal interpolation. If a signal is sampled at the minimum sample rate required to accurately capture the signal in the time domain, the signal will not necessarily have sufficient temporal resolution to identify key time instants in the signal that are of interest. Shannon (1949) highlights how the sampled signal can be interpolated to increase temporal density. Such interpolation can be performed using a variety of interpolation techniques, here it will be demonstrated using an interpolating cubic spline. In Fig. 3.16, the top graph is a time series sampled at 100 Hz, and the middle graph is the same signal sampled at 20 Hz, which clearly has much lower temporal resolution. A cubic spline is used to interpolate the signal sampled at 20 Hz so that it has time intervals reflecting sampling at 100 Hz and the bottom graph shows that the original 100 Hz and interpolated signal are indistinguishable.

In the analysis of data, it is important to consider both the sample rate required to capture the signal using Shannon's sampling theorem, and once collected, a second consideration is the appropriate temporal resolution, which can be achieved by appropriate signal interpolation.

3.7 Cross-Correlation and Autocorrelation

Cross-correlation involves the correlation of one signal with a time shifted other signal. Autocorrelation is the correlation of a signal with a time-shifted version of itself (the same signal). These two methods can be useful for identifying signal properties. With cross-correlation analysis, the similarity of two signals can be identified, along with the extent to which the signals are phase-shifted versions of one another. An analysis using the autocorrelation function can help, for example, to identify the white noise or the periodicity of the signal. Somewhat confusingly, the way in which the cross-correlation and autocorrelation are both defined is somewhat different depending on which literature is being consulted. That stated the general principles are the same and the insights are the same irrespective of which formulations of the pertinent equations are used.

To compute the cross-correlation (r_{xy}) of two signals, for example, if the signals x and y consist of n equidistantly sampled data points, then the cross-correlation function is computed from:

$$r_{xy}(L) = \sum_{i=1}^{n-L} x_i\, y_{i+L} \tag{3.25}$$

where L is the lag number ($L = -(n-1), \ldots 2, -1, 0, 1, 2, \ldots n - 1$) and the maximum lag values are equal to $n - 1$ either side of zero. There are two common variants in the computation of the cross-correlation. One is to remove any constant trends in the two data sets by subtracting their means:

$$r_{xy}(L) = \sum_{i=1}^{n-L} (x_i - \bar{x}) \left(y_{i+L} - \bar{y}\right) \tag{3.26}$$

where $\bar{x}, \bar{y}$ are the means of the two signals. Another variant used is if the signals are of different amplitudes or measured in different units, then the cross-correlation is often normalized using:

$$\bar{r}_{xy}(L) = \frac{r_{xy}(L)}{\sqrt{r_{xx}(0)}\sqrt{r_{yy}(0)}} \tag{3.27}$$

Consider the two signals in Fig. 3.17: the signals look similar, and the cross-correlation can be computed to try and identify what relationship exists, if any. The cross-correlation at zero lag reveals the relationship between the two signals, $r_{xy}(0) = -0.93$, and that one signal is approximately the inverse of the other.

Examining a signal at multiple lag values can also reveal information about a signal, for example, if the cross-correlation has a high value at a particular lag value, it will indicate that the signals are similar in shape but time shifted relative to one another. In Fig. 3.18, there are two signals, and their cross-correlation function is plotted. The maximum value of the cross-

Fig. 3.17 Two signals with some apparent similarities. The cross-correlation at zero lag reveals the relationship between the two signals, $r_{xy}(0) = -0.93$, and that one signal is approximately the inverse of the other

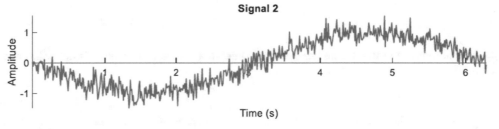

Fig. 3.18 Two signals and their cross-correlation function, where the peak cross-correlation value reflects the time lag between the two signals

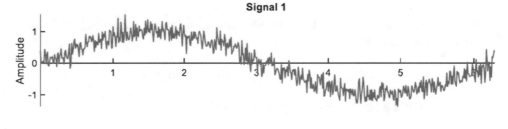

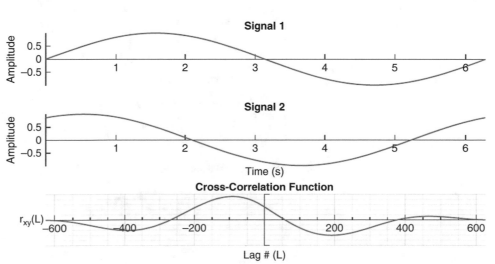

correlation indicates a time shift between the two "similar" signals, $r_{xy}(-92) = 0.91$. This lag value can then be used to identify the time shift between the two signals, in this example ~0.91 seconds.

There is an intrinsic relationship between the cross-correlation function and its Fourier transform. If the Fourier transform is applied to the cross-correlation function, the result is equal to the power spectrum of the signal, Wiener-Khinchin theorem. Similarly, the Fourier transformation of the power spectrum gives the cross-correlation of the signal.

If a signal is correlated (r_{xx}) with itself, then this is the autocorrelation function. The autocorrelation function does not provide any information about the phase of the signal, but can identify periodicity in a signal. The autocorrelation has several important properties:

1. The autocorrelation is an even function, therefore:

$$r_{xx}(L) = r_{xx}(-L) \tag{3.28}$$

2. All values of r_{xx} must be less than or equal to the value at the zero lag condition.
3. The autocorrelation function of a periodic signal is also periodic.
4. For two signals (x, y) that are uncorrelated, the autocorrelation of their sum ($x + y$) is equal to the sum of the autocorrelation of their sum:

$$r_{xy}(L) = r_{xx}(L) + r_{yy}(L) \tag{3.29}$$

5. The autocorrelation of a random signal has its maximum value at zero lag, with all other of r_{xx} values very small (see Fig. 3.19).

Figure 3.19 illustrates various functions and the corresponding autocorrelation functions.
For a cosine function:

$$x(n) = A \cos(n.\omega_0 + \varphi) \tag{3.30}$$

Its autocorrelation function can be determined analytically to be:

$$R_{xx}(L) = \frac{A}{2} \cos(L.\omega_0) \tag{3.31}$$

which demonstrates that the numerical approximation to the autocorrelation is not without some errors (see autocorrelation in the middle panel in Fig. 3.19).

Fig. 3.19 Time series and their corresponding autocorrelation functions

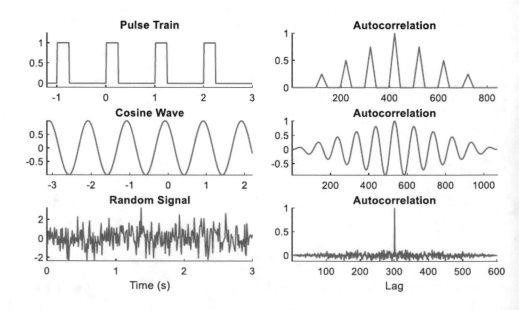

3.8   Time-Frequency Analyses

Fourier analysis identifies the frequency components of a signal and the amplitude of the frequency components. What it does not reveal is when these frequencies occur. A stationary signal is one for which its signal properties, such as the mean, variance, and frequency and power spectra, are invariant irrespective of what portion of the signal is analyzed. If a signal is stationary, then it is not important to identify when the frequencies occur, but many signals of interest in biomechanics are nonstationary. For example, the vertical ground reaction force in walking or running has high-frequency components at the start due to the impact but lacks these higher-frequency components in the latter part of the signal (e.g., Gruber et al., 2017). In Fig. 3.20, a signal is shown with two distinct frequencies that occur at different times, but the frequency analysis performed using Fourier analysis does not reveal the temporal sequencing of these frequencies. Time-frequency analyses are a group of techniques that endeavor to describe how the spectrum of a signal changes in time. There are two methods presented in this section for performing time-frequency analyses: short-time Fourier transform analyses and wavelet analyses.

A short-time Fourier transform performs its analysis on the sampled signal using a sliding window (Goodwin, 2008). The user selects the size of window and the amount of overlap of adjacent windows. For each data window a windowing function, for example, the Hanning window, is used to avoid the errors which arise from sampling using a rectangular window. The data within the window are then subjected to a Fourier transform, typically an FFT. Therefore, the frequency and their amplitudes are obtained as a function of time, with the time instants reflecting the selection of windows. If a signal spectral content changes rapidly the identification of the appropriate window is problematic, as ideally the signal should be stationary within the window. An obvious solution is to make the time window very short, but as the time window is reduced the frequency resolution is reduced. So using either a short-time Fourier analyses or wavelet analysis means there has to be a compromise between time and frequency resolution. Figure 3.21 shows the time-frequency analysis of the example signal obtained with the short-time Fourier transform.

The wavelet transform converts a time series into another function, which is defined by a series of wavelets (Strang and Nguyen, 1996). The wavelet is a small oscillating wave. A wavelet is a function with zero mean and defined time and frequency components. In Fourier analysis, the underlying fitting function is comprised of the sine and cosine waves, but in wavelet analysis, there are varieties of fitting functions that can be used. In contrast to a sine or cosine wave, a wavelet can be irregular in shape and typically lasts for only a limited time. The nature of the wavelet can be selected based on the nature of the signal to be analyzed. Figure 3.22 shows three wavelets.

In wavelet analysis, a time series is divided into sections, and a wavelet is fitted to each localized portion of the signal. This fit is achieved via a process of scaling of the wavelet: dilation (stretching or squeezing the wavelet) and translation (movement along time axis). For a wavelet function ψ, the dilation and scaling can be written as:

$$\psi_{s,\tau}(t) = \frac{1}{\sqrt{2}}\ \psi\left(\frac{t - \tau}{s}\right)$$

(3.32)

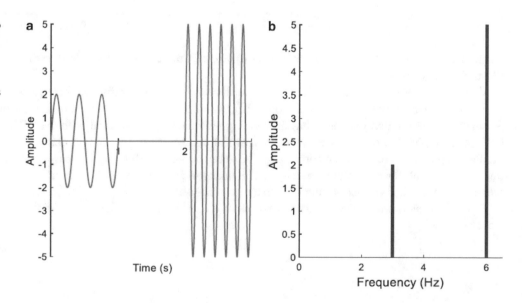

Fig. 3.20   A time series with two distinct frequencies, (**a**) the time domain version and (**b**) the frequency domain version of the signal obtained using Fourier analysis. Note the Fourier analysis does not reveal the temporal sequencing of these frequencies

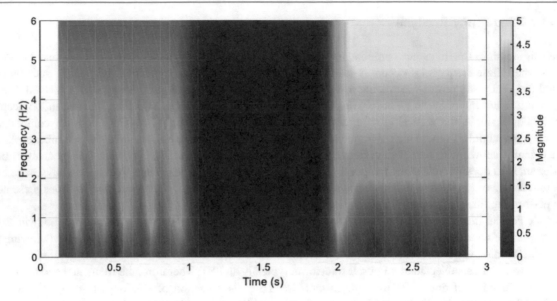

Fig. 3.21 A time-frequency analysis of a time series, from Figure 3.20a, performed using a short-time Fourier transform. The time series for the first second has a frequency of 3 Hz and an amplitude of 2; then for the middle second, the signal has a frequency of 0 Hz with 0 amplitude; and for the final second, it has a frequency of 6 Hz and an amplitude of 5. Note how the identified signal magnitudes are focused around the associated frequency

Fig. 3.22 Examples of three continuous wavelets

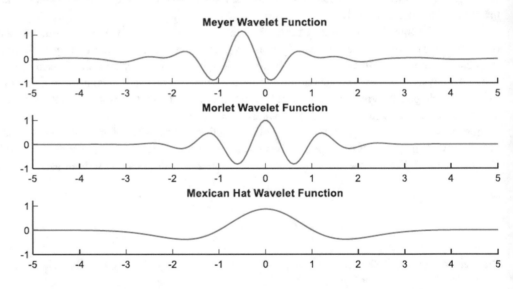

where s is the constant associated with dilation, and τ is the constant associated with translation (see Fig. 3.23 for an illustration of a scaled wavelet).

Wavelets can be continuous (continuous wavelet transform (CWT)) or discrete (discrete wavelet transform (DWT)). These wavelets differ in how they scale their wavelets (dilation), and with a CWT there is finer tuning of the wavelet scaling. Similarly, the translation parameter has finer tuning with the CWT compared with the DWT. The differences between the CWT and the DWT produce different advantages and disadvantages for the two classes of wavelet transforms. The DWT produces a parsimonious representation of signals, so can be used for storage producing "compressed" versions of signals, and this can be useful, for example, with images. The finer scales in the CWT allow high-resolution analysis of signals and are therefore better suited for performing time-frequency analyses of signals.

Figure 3.24 shows the time-frequency analysis of the example signal obtained with the wavelet analysis, using a Morlet wavelet.

Fig. 3.23 Example the scaling of a Morlet wavelet

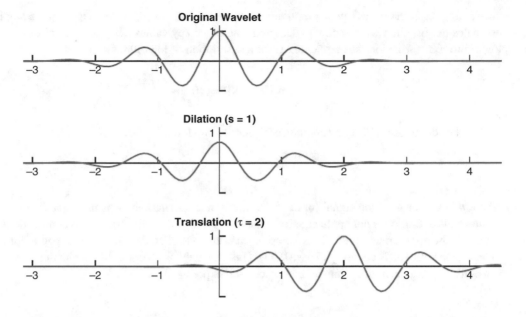

Fig. 3.24 A time-frequency analysis of a time series, from Figure 3.20a, performed using a wavelet transform. The time series for the first second has a frequency of 3 Hz and an amplitude of 2; then for the middle second, the signal has a frequency of 0 Hz with 0 amplitude; and for the final second, it has a frequency of 5 Hz and an amplitude of 5

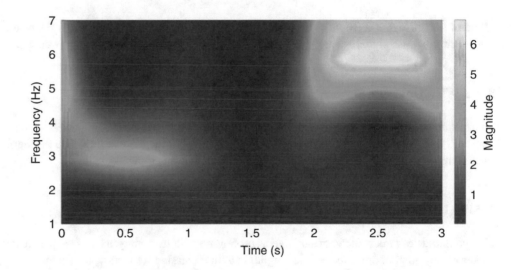

3.9   Signal Contents

A sampled (measured) signal can be considered to consist of two parts: the true signal and the noise that corrupts it:

$$\text{measured signal} = \text{true signal} + \text{noise} \tag{3.33}$$

Careful experimental protocols can help to minimize noise, but despite an experimenters' best efforts sampled signals will still contain noise. Our language comes from the early days of the voice communication, but here, the noise is not necessarily an unwanted sound but is the undesirable part of the sampled signal.

There is opportunity for noise to corrupt collected signals throughout the measurement chain, for example, all analog instruments (amplifiers and analog filters) will add noise, such as thermal noise. The target is to make the signal as large as possible and the noise as small as possible: a good signal to noise ratio. There are many definitions of the signal to noise ratio (SNR); one popular is the following:

$$\text{SNR} = \frac{MS_{\text{Signal}}}{MS_{\text{Noise}}} \tag{3.34}$$

where MS_{Signal} is the mean square amplitude of the signal and MS_{Noise} is the mean square amplitude of the noise. As with all estimates of noise, the issue is how to determine these two key values. Often this ratio is expressed on a decibel scale using a logarithm transformation (see Appendix E for a description of logarithms):

$$\text{SNR} = 10 \, \log_{10}\left(\frac{MS_{\text{Signal}}}{MS_{\text{Noise}}}\right) \tag{3.35}$$

Another definition of the SNR is the coefficient of variation:

$$\text{SNR} = \frac{\mu}{\sigma} \tag{3.36}$$

where μ is the mean of the signal (or expected mean of the signal) and σ is the standard deviation of the noise (or expected value). This definition is not applicable if the mean of the signal is zero. Irrespective of the definition of the SNR, the question is what is the appropriate SNR. The Rose criterion is an SNR of at least 5, which is required if features are to be distinguished in an image with 100% certainty (Rose, 1973). Unfortunately, for work in biomechanics no such criterion exists, and if it did, it would vary between measurement equipment and experiments.

3.9.1 White Noise

Much of the noise that corrupts measurements is considered white noise. White noise has the property that it is random in the sense that the noise signal is stationary and has a mean value of zero and a flat power spectrum across all of the spectrum.

3.9.2 Thermal Noise

Thermal noise is sometimes called Johnson or Nyquist noise. Thermal noise is caused by the random thermal motion of charge carriers, typically electrons, inside an electrical conductor. The factors that influence the thermal noise are the resistance in the system, its temperature, and the signal bandwidth. This noise is typically white.

3.9.3 Pink Noise

The magnitude of pink noise varies in inverse proportion with frequency, as a consequence it is often referred to as 1/f noise or flicker noise. As pink noise reduces in amplitude with increasing frequency, its influence is typically greater at the lower end of the frequency spectrum. It occurs in many electronic devices due to variations in the properties of the materials comprising the electronic device.

3.9.4 Shot Noise

The spectral density of shot noise decreases with increasing frequency. The magnitude of shot noise is typically less than that of thermal noise. It arises as the electric current is carried by electrons, which have discrete arrival times. There are variations in the number of particles arriving in each interval, it is this variation which causes the shot noise; the smaller the current the larger the fluctuations. Unlike thermal noise, shot noise is temperature- and frequency-independent.

3.9.5 Electromagnetic and Electrostatic Noise

The use of electronic equipment provides the opportunity for noise to be introduced to measured signals.

Electromagnetic noise, also known as electromagnetic interference, arises with all electronic equipment. Its influence can be minimized by simple measures in the experimental set-up. The three primary ways to reduce this noise is to use appropriately shielded cables, to appropriately route cables, and to have sensors that are grounded. Shielded cabling is more expensive than its non-shielded counterpart, but it does reduce noise influences. Cabling attached to sensors should be routed away from electrical cable, all power sources, and any electric motors. The looping and twisting of cables should be avoided as these can cause fluctuating magnetic fields. This noise is typically localized to the frequency of power supply of the equipment (e.g., 60 Hz or 120 Hz).

Electrostatic noise arises predominantly from hum from power lines and electric lights. This noise arises due to a parasitic capacitor, any two metallic areas separated by space. In a lab common sources of electrostatic noise are fluorescent lights, switching power supplies, and other electronics. The amount of coupling, and therefore noise, between the power source and the measurement equipment depends directly on the frequency of the power source and the voltage of the source and is in inverse proportion to the distance between them. Shielding of the source can reduce the noise transmitted. This noise is typically localized to the frequency of the source and so is normally contained in a limited bandwidth of frequencies (e.g., 60 Hz).

3.9.6 Quantization Error

Analog-to-digital converters (ADC) have a limit to the number of digits that are used to represent the sampled signal; the data values are quantized. They generally represent the sampled signal in binary form, so their resolution varies with the number of bits used to represent a number (e.g., 8 bits, 16 bits) The distance between adjacent quantization steps is the least significant bit (LSB), and this value reflects the resolution of the system. As outlined in Chap. 11, the quantization of the signal introduces white noise into the signal where the error has a magnitude which is ± 0.5 LSB.

3.9.7 Sampled Signal

The sampled signal is compromised by many noise sources, many of which have the properties of white noise, and others are focused around particular frequencies. But the other part of the signal is the signal that is of interest, the true signal underlying the sampled signal. Oftentimes, we have some knowledge of the properties of that signal. An important property is the maximum frequency content of that signal, and such information can guide the selection of sample rate. In human movement analysis, previous studies do give some indication of this maximum frequency component. In the analysis of human movement, the kinematics and kinetics tend to occupy the lower-frequency parts of the frequency spectrum. For example, in the analysis of ground reaction forces during human walking, Antonsson and Mann (1985) reported that the frequency components of the force signal were all contained below 50 Hz. While for the displacement of markers attached to the body during walking, the majority of their frequency content was below 5 Hz (Angeloni et al., 1994). In contrast, surface electromyography signals can have frequency components up to 300 Hz or greater (e.g., Komi and Tesch, 1979). The properties of the true signal and the noise that corrupt it can be exploited to improve the veracity of the sampled signal as a representation of the true signal; methods for achieving this will be outlined in the following chapter (Chap. 4).

3.10 Conclusions

The majority of data collected in biomechanics are time series. These times series contain the true (wanted) signal with noise superimposed. There are various tools available for analyzing the contents of a signal, including Fourier analysis and wavelet analysis. The processing of the sampled signal to reduce the influence of noise is described in the next chapter.

3.11 Review Questions

1. What are the properties of analog and digital signals?
2. What is frequency analysis? Explain an application.
3. State sampling theorem. What happens if a signal is inappropriately sampled?
4. What is the purpose of a window function?
5. What is noise?

6. What are the properties of white noise?
7. What are the potential sources of noise?
8. Compare what can be obtained from a Fourier analysis and a short-time Fourier analysis.
9. What is the nature of the trade-off when performing a time-frequency analysis of signal using wavelet analysis?

References

Cited References

Angeloni, C., Riley, P. O., & Krebs, D. E. (1994). Frequency content of whole body gait kinematic data. *IEEE Transactions on Rehabilitation Engineering, 2*(1), 40–46.

Antonsson, E. K., & Mann, R. W. (1985). The frequency content of gait. *Journal of Biomechanics, 18*(1), 39–47.

Cooley, J. W., & Tukey, J. W. (1965). An algorithm for the machine calculation of complex Fourier series. *Maths of Computation, 19*, 297–301.

Dongarra, J., & Sullivan, F. (2000). Guest editors introduction to the top 10 algorithms. *Computing in Science & Engineering, 2*(1), 22–23.

Godfrey, D., Lythgoe, J. N., & Rumball, D. A. (1987). Zebra stripes and tiger stripes: The spatial frequency distribution of the pattern compared to that of the background is significant in display and crypsis. *Biological Journal of the Linnean Society, 32*(4), 427–433.

Goodwin, M. M. (2008). The STFT, sinusoidal models, and speech modification. In J. Benesty, M. M. Sondhi, & Y. A. Huang (Eds.), *Springer handbook of speech processing* (pp. 229–258). Berlin, Heidelberg: Springer.

Gruber, A. H., Edwards, W. B., Hamill, J., Derrick, T. R., & Boyer, K. A. (2017). A comparison of the ground reaction force frequency content during rearfoot and non-rearfoot running patterns. *Gait & Posture, 56*(Supplement C), 54–59.

Harris, F. J. (1978). On the use of windows for harmonic analysis with the discrete Fourier transform. *Proceedings of the IEEE, 66*(1), 51–83.

Heideman, M. T., Johnson, D. H., & Burrus, C. S. (1985). Gauss and the history of the fast Fourier transform. *Archive for History of Exact Sciences, 34*(3), 265–277.

Komi, P. V., & Tesch, P. (1979). EMG frequency spectrum, muscle structure, and fatigue during dynamic contractions in man. *European Journal of Applied Physiology, 42*(1), 41–50.

Lampl, M., Veldhuis, J. D., & Johnson, M. L. (1992). Saltation and stasis: A model of human growth. *Science, 258*(5083), 801–803.

Lieberman, D. E., Venkadesan, M., Werbel, W. A., Daoud, A. I., D'Andrea, S., Davis, I. S., Mang'Eni, R. O., & Pitsiladis, Y. (2010). Foot strike patterns and collision forces in habitually barefoot versus shod runners. *Nature, 463*(7280), 531–535.

Rockmore, D. N. (2000). The FFT: An algorithm the whole family can use. *Computing in Science & Engineering, 2*(1), 60–64.

Rose, A. (1973). *Vision: Human and electronic*. New York: Plenum Press.

Shannon, C. E. (1949). Communication in the presence of noise. *Proceedings of the IRE, 37*(1), 10–21.

Slepian, D. (1976). On bandwidth. *Proceedings of the IEEE, 64*, 292–300.

Strang, G., & Nguyen, T. (1996). *Wavelets and Filter Banks*. Wellesley, MA: Wellesley-Cambridge Press.

Winograd, S. (1976). On computing the discrete Fourier transform. *Proceedings of the National Academy of Sciences, 73*(4), 1005–1006.

Useful References

Bloomfield, P. (2000). *Fourier analysis of time series*. New York: John Wiley & Sons.

Butz, T. (2006). *Fourier transformation for pedestrians*. New York: Springer.

Fulop, S. A., & Fitz, K. (2006). Algorithms for computing the time-corrected instantaneous frequency (reassigned) spectrogram, with applications. *The Journal of the Acoustical Society of America, 119*(1), 360–371.

Grinsted, A., Moore, J. C., & Jevrejeva, S. (2004). Application of the cross wavelet transform and wavelet coherence to geophysical time series. *Nonlinear Processes in Geophysics, 11*(5-6), 561–566.

Strang, G. (1994). *Wavelets. American Scientist, 82*, 250–255.

Torrence, C., & Compo, G. P. (1998). A practical guide to wavelet analysis. *Bulletin of the American Meteorological Society, 79*(1), 61–78.

Signal Processing

4

Overview

Giuseppe Piazzi (1746–1826) was an Italian Catholic priest who was active in astronomy. On January 1, 1801, while surveying the night sky, he identified what he thought was a comet moving across the sky. He was able to track its path for 40 days until its path could not be viewed due to the sun. The object was not a comet but Ceres which was eventually classified as a minor planet; it is the largest object in the asteroid belt between Mars and Jupiter. The question was where to look for Ceres when it re-emerged. Eventually the data of Piazzi caught the attention of German mathematician and physicist Johann Carl Friedrich Gauss (1777–1855). Gauss was able to model the path of Ceres and predict where it would re-emerge. What was significant about this work was that Gauss had to account for the errors in the measurements made by Piazzi. He used the method of least squares in modeling the trajectory of Ceres. There is a debate as to who first developed the method of least squares, but Gauss contributed significantly to this methodology (Stigler 1981). An important contribution here is the appreciation by Gauss that in processing the available data, he had to allow for errors corrupting the collected data. Errors are inevitable when data are collected; signal processing techniques can improve the veracity of the data and permit the computation of other signals from the collected data (e.g., velocity values from collected displacement data).

In the following sections, it is the intention to review various data processing methods which have applications for biomechanical analyses.

The content of this chapter requires a working knowledge of frequency analysis (see Chap. 3) and logarithms (see Appendix F).

4.1   What Is a Filter?

In signal processing, a filter transforms an input signal into another signal (the output). For example, the filter might be used to remove a 60 Hz mains hum which corrupts a sampled signal or remove the high-frequency vibrations of a force plate due to its natural frequency. Generally, the filter cannot perform its action perfectly so efforts to remove signal components either mean some of the undesired signal components remain, or that in removing the unwanted part of the signal other signal components are affected, or both. Often a filter is considered to help with the removal or reduction of sampled signal components (e.g., noise), but a filter might process a signal into another useful signal, for example, the computation of the first or second derivatives of a signal.

A filter could be implemented in hardware, for example, to process an analog signal; thus, it would be an analog filter. If the signal is sampled, then the filter might be implemented in software as could happen with a digital filter.

Many filters are designed in the frequency domain, in part because the signal components to be manipulated by a filter are more often easily visualized in the frequency domain.

A common application of a filter is to remove the high-frequency components of a signal, which typically makes the signal appear smoother. It is this smooth appearance of a signal which means that some filters are referred to as smoothers.

J. H. Challis, *Experimental Methods in Biomechanics*, https://doi.org/10.1007/978-3-030-52256-8_4

Fig. 4.1 A 3 Hz sine wave, which has been corrupted by noise (top graph), sampled ten times, and then the output of averaging these ten signals compared with the original (actual) signal

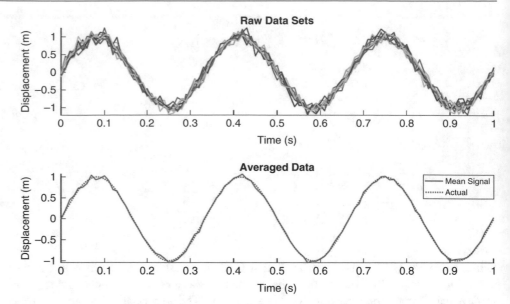

4.2   Signal Averaging

If the noise contaminating a signal is white, this means the noise is not correlated between samples and has a mean value of zero. Therefore, on summing repeat measures the true underlying signal is additive, and the noise tends to its mean value. As a consequence of taking the mean of a number (n) of repeat samples, the signal to noise ratio is improved; the signal amplitude to noise standard deviation ratio is increased by the square root of n. Such signal averaging has two applications: one as a method to estimate the signal noise and the other to produce a noise reduced estimate of a signal from repeat measures.

The influence of signal averaging is illustrated in Fig. 4.1, where a sine wave corrupted by noise is sampled ten times, and the signals then averaged.

Notice that the process of averaging produces a signal where much of the noise has been removed. While signal averaging is attractive as a way of reducing signal noise, it is often not typically feasible to sample a signal multiple times. With human movement, it is not feasible to simply have the subject repeat the task multiple times due to the inherent variability of human movement (e.g., Winter 1995). That stated in certain circumstances it makes sense to make multiple measures, for example, if measuring the circumference of a segment, as the mean of the measures will be less noisy.

4.3   Simple Signal Filtering

For a given time instant in a time series adjacent signal values can be used to compute an approximation to signal averaging and thus produce a signal containing less noise. One example of such a filter is the three-point moving average:

$$f'(t_i) = \frac{f(t_{i+1}) + f(t_i) + f(t_{i-1})}{3} \tag{4.1}$$

where $f'(t_i)$ is the filtered output value at time instant t_i and $f(t_i)$ is the unfiltered (raw) signal value. The influence of such a filter is illustrated in Fig. 4.2, where a sine wave is corrupted by noise.

For many filters using simple algebra, it is possible to work out the frequency magnitude response of the filter. The frequency magnitude response indicates the influence of the filter on signal magnitude at each frequency. To examine this for the three-point moving average rather than refer to the time instant of a sample, the sample number k will be used; therefore, Eq. 4.1 can be written as:

$$f'(k) = \frac{f(k+1) + f(k) + f(k-1)}{3} \tag{4.2}$$

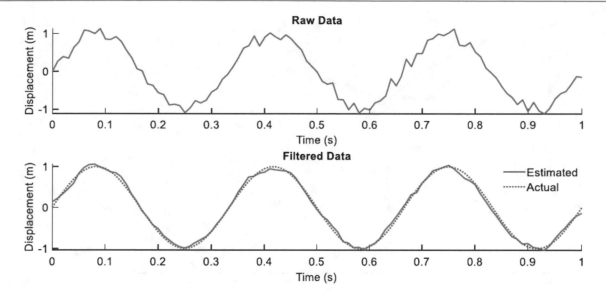

Fig. 4.2 A 3 Hz sine wave, which has been corrupted by noise (top graph), and then the output of a three-point moving average filter compared with the original (actual) signal

To determine the frequency response, assume a sine wave is filtered, where ω is the frequency of the sine wave:

$$f(k) = \sin(\omega\, k\, \Delta t) \tag{4.3}$$

where Δt is the sample interval. The filter output becomes:

$$f'(k) = \frac{\sin(\omega\, k\, \Delta t + \omega\, \Delta t) + \sin(\omega\, k\, \Delta t) + \sin(\omega\, k\, \Delta t - \omega\, \Delta t)}{3} \tag{4.4}$$

This can be simplified to:

$$f'(k) = \frac{(1 + 2\cos(\omega\, \Delta t))\, \sin(\omega\, k\, \Delta t)}{3} \tag{4.5}$$

The filter output is simply the product of the original signal and a frequency-dependent factor, so the frequency response can be written as:

$$H(\omega) = \frac{1 + 2\cos(\omega\, \Delta t)}{3} \tag{4.6}$$

The frequency magnitude response of the three-point moving average can be plotted (see Fig. 4.3).

The frequency magnitude of the filter shows that at low frequencies the signal amplitude is relatively unaffected but at higher frequencies the signal amplitude is attenuated, and as a consequence this would be called a low-pass filter. A low-pass filter passes the lower-frequency components of the signal relatively unaffected by the filter but attenuates the higher-frequency components of the signal. Assuming the sample rate is 100 Hz, after around 15 Hz there is a lot more signal attenuation until around 34 Hz. After 34 Hz, the amount of signal attenuation decreases up to the Nyquist frequency. If a signal is collected where the true signal is in the lower part of the frequency spectrum, this low-pass filter would reduce noise components in the higher-frequency range of the sampled signal. Ideally, the signal attenuation would be more aggressive than this filter produces, and the amount of attenuation would not increase for higher-frequency components (above 34% of the sample rate for this filter). Other filters exist which have these desirable frequency magnitude response characteristics.

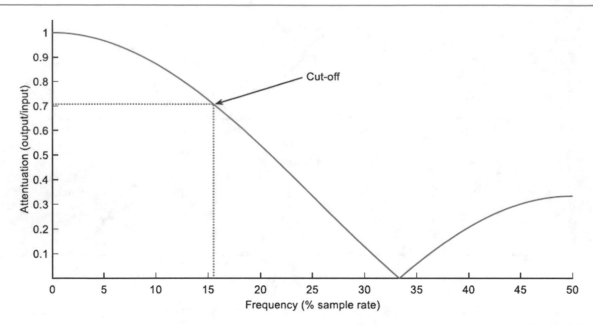

Fig. 4.3 The frequency magnitude response of the three-point moving average filter. Note the cut-off frequency has also been highlighted in the graph

By convention the filter cut-off is assumed to occur when signal attenuation is $\frac{1}{\sqrt{2}}$, so when this occurs can be computed for the three-point moving average:

$$H(2\,\pi\,\omega) = \frac{1 + 2\cos(2\,\pi\,f_0\,\Delta t)}{3} = \frac{1}{\sqrt{2}} \tag{4.7}$$

$$f_0 = \frac{1 + 2\cos(2\,\pi\,f_0\,\Delta t)}{3} = \frac{0.1553}{\Delta t} = 0.1553\,f_s \tag{4.8}$$

where f_s is the sample rate ($f_s = \frac{1}{\Delta t}$).

The relationship between the sample rate and cut-off frequency highlights a weakness of the three-point moving average which is that rather than being able to specify a desired cut-off, it is fixed as a function of a sample rate. This means if researchers wanted to increase the cut-off frequency, they would have to increase the sample rate and vice versa. There are classes of filters that allow specification of the cut-off frequency independent of the sample rate; the most common of these is the Butterworth filter which is described in a subsequent section.

The frequency magnitude response can be plotted either with a linear vertical axis or with a logarithmic scale. When using the log option, the units are decibels (dB), which are often used to the express the ratio between two numbers:

$$dB = 20\,\log_{10}\left(\frac{A_o}{A_I}\right) \tag{4.9}$$

where A_o is the amplitude of the signal after filtering (output) and A_I is the amplitude of the signal before filtering (input). Therefore, the cut-off frequency occurs at -3 dB:

$$-3\,dB = 20\,\log_{10}\left(\frac{1}{\sqrt{2}}\right) \tag{4.10}$$

The bel is named in honor of Alexander Graham Bell (1847–1922); it refers to a change which is a factor of 10. Table 4.1 shows the relationship between signal amplitudes and the corresponding value in decibels. For decibel values greater than zero there is signal gain, an increase in output signal amplitude relative to the input signal, while for decibel values less than zero there is signal attenuation, the output signal amplitude is reduced relative to the input signal.

Table 4.1 The relationship between output and input signals and their values in decibels

$\left(\frac{A_0}{A_I}\right)$	$20\log_{10}\left(\frac{A_0}{A_I}\right)$
0.00001	-100
0.001	-60
0.01	-40
0.1	-20
0.5	-6
0.707	-3
1	0
10	20
20	26

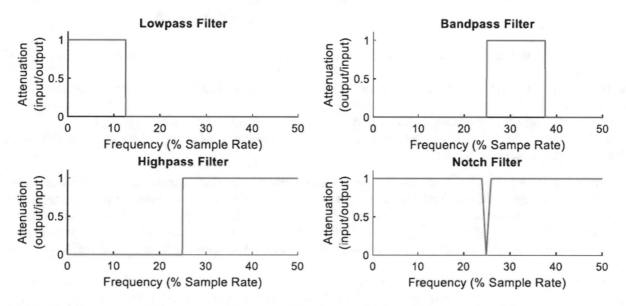

Fig. 4.4 The ideal frequency magnitude responses of four types of filter: low-pass, band-pass, high-pass, and notch filters

4.4   Types of Filter

The three-point moving average filter attenuates high-frequency components but only moderately affects low-frequency components of the signal, and it can be described as a low-pass filter. There are other types of filters classified based their influence on the frequency components of a signal (see Fig. 4.4 for schematic representations):

High-pass filter – a filter that passes high frequencies and attenuates low frequencies.
Band-pass filter – a filter that passes one frequency band and attenuates all frequencies above and below that band.
Band reject filter – a filter that rejects one frequency band and passes all frequencies above and below that band. This filter is often for a narrow band in which case is called a *notch filter*.

In reality, ideal filters cannot be formulated; there are compromises, for example, in the sharpness of the increase in signal attenuation after the cut-off frequency. Figure 4.5 shows a typical frequency response for a low-pass filter, although its general features also apply to band-pass, high-pass, and notch filters.

The different regions of a filter are defined in terms of where they occur on a decibel scale. The key regions are:

Passband – the band of frequencies for the input signal which the filter marginally influences.

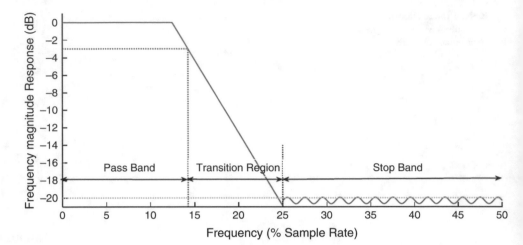

Fig. 4.5 The frequency magnitude responses of a low-pass filter

Cut-off frequency – this occurs at the −3 dB point of a filter's frequency magnitude response and marks an edge of the passband.

Transition region – the range of frequencies between the passband and the stopband. This is sometimes called the transition band.

Roll-off – is the steepness of the transition from the passband to the stopband.

Stopband frequency – this occurs at the −20 dB point of a filter magnitude response. It is the band of frequencies which the filter attenuates to maximum extent.

Different filters vary in their frequency magnitude response, for example, the Butterworth filter is either zero or negative in the passband, but a Type I Chebyshev filter has a ripple in its passband causing some low-frequency signal components to be magnified in value. Note the illustrated frequency magnitude response of a filter used decibels, but a linear scale can be useful, for example, for illustrating any passband ripple and roll-off, but the log scale (decibel) better illustrates stopband attenuation. The filter needs to be selected to match the characteristics of the signal to be processed; in biomechanics the Butterworth filter is typically the filter of choice.

Irrespective of the intent of the filter, e.g., low-pass, filters can be non-recursive or recursive. For a non-recursive filter, filter output values are a function of the weighted sum of input values. Therefore, a non-recursive filter could have the following format:

$$f'(t_i) = a_0 f(t_i) + a_1 f(t_{i-1}) + a_2 f(t_{i-2}) \qquad (4.11)$$

where $f'(t_i)$ is the filtered output value at time instant t_i, $f(t_i)$ is the unfiltered (raw) signal value, and a_0, a_1, a_2 are constants for the filter. With a recursive filter, output values are a function of the weighted sum of input values and previous output values. Therefore, a recursive filter could have the following format:

$$f'(t_i) = a_0 f(t_i) + a_1 f(t_{i-1}) + a_2 f(t_{i-2}) + b_1 f'(t_{i-1}) + b_2 f'(t_{i-2}) \qquad (4.12)$$

where a_0, a_1, a_2, b_1, b_2 are constants for the filter. Non-recursive filters require more coefficients to obtain the same frequency response as a recursive filter but have the advantage of not using the previous filter output values. Due to their relative compactness, in biomechanics a recursive filter is typically used (e.g., Butterworth filter).

Non-recursive filters are sometimes referred to as finite impulse response (FIR) filters, because their response to an impulse (brief burst of activity) is of finite duration; it settles to zero in finite time. Recursive filters are infinite impulse response (IIR) filters, as they can continue to respond indefinitely.

Non-recursive filters (FIR) are more stable than recursive filters and can be designed with a linear phase response. To their advantage, recursive filters require fewer terms than the non-recursive filters to produce the same frequency response; as a consequence, non-recursive filters require more data memory and take greater computation time.

Different applications require different filters. For example, a notch filter might be useful to remove mains hum from a sampled signal, while a high-pass filter can be used to remove the low frequencies associated with movement artifact in an electromyographic recording. The role of the filters can be easy to visualize; consider a low-pass filter compared with a high-pass filter. A low-pass filter is a weighted averaging process, so it smooths out the rapid fluctuations in a signal. These fluctuations are high-frequency components, which are what the low-pass filter tries to remove. In contrast a high-pass filter takes weighted differences in the signal. This process of taking differences highlights the rapid fluctuations but has little effect on the low-frequency parts of the signal where there are small changes.

4.5   Filtering in Biomechanics

In biomechanics, there are occasions where low-pass, band-pass, high-pass, and notch filters can each be used. The focus in the remainder of this chapter is on low-pass filtering, but for the other types of filter, most of the same principles apply. In the remaining chapters, examples of the use of band-pass, high-pass, and notch filters will be presented.

In the following three sections, three common ways of low-pass filtering data in biomechanics will be described: the Butterworth filter, spline smoothing, and truncated Fourier series. For each of these methods, the math behind the method is first outlined, then how the degree of filtering can be determined is presented, and finally how signal derivatives can be computed is described. As derivatives are required for many biomechanical analyses, a following section deals exclusively with methods of numerical data differentiation.

4.6   Butterworth Filter

The common digital filter used in biomechanics is the Butterworth filter. While originally implemented as an analog filter (Butterworth 1930), it is easily implemented for digital filtering. The Butterworth filter is a recursive filter; therefore, the output values are a function of the weighted sum of input values and previous output values. A second-order Butterworth filter has the following format:

$$f'(t_i) = a_0 f(t_i) + a_1 f(t_{i-1}) + a_2 f(t_{i-2}) + b_1 f'(t_{i-1}) + b_2 f'(t_{i-2}) \tag{4.13}$$

where $f'(t_i)$ is the filtered output value at time instant t_i, $f(t_i)$ is the unfiltered (raw) signal value, and a_0, a_1, a_2, b_1, b_2 are constants for the filter. The filter coefficients vary with the sample rate, the desired cut-off frequency, and the type of Butterworth filter (low-pass, high-pass, or band-pass). For example, for a low-pass Butterworth filter for data sampled at 100 Hz, and a desired cut-off of 5 Hz, the coefficients are:

$$a_0 = 0.020, a_1 = 0.040, a_2 = 0.020, b_1 = 1.561, b_2 = -0.641$$

Note that the sum of the coefficients is equal to 1.

For a low-pass Butterworth filter, the frequency magnitude response can be determined from:

$$H(f) = \left| \frac{V_{\text{out}}}{V_{\text{in}}} \right| = \frac{1}{\sqrt{1 + (f/f_0)^{2n}}} \tag{4.14}$$

where $H(f)$ is the frequency response of filter at frequency f, V_{out} is the signal output voltage, V_{in} is the signal input voltage, f is the signal frequency which ranges from 0 Hz to the signal Nyquist frequency, f_0 is the cut-off frequency, and n is the filter order. As the Butterworth filter is a recursive filter, the filter order specifies how many coefficients are there in the filter equation, $(2.n) + 1$ coefficients, but it also determines the sharpness of the transition band. The sharpness of the transition band increases with increasing order (see Fig. 4.6).

The temptation is to use a higher-order filter because of the sharper transition band, but because the Butterworth filter is recursive, it needs previous filter output values to filter a data point. This creates a problem at the start of a data set where there cannot be previous filter output values; the filter will therefore not operate as planned until the filter is processing data points at n data points into a sequence. Therefore, the problems with requiring previous filter output are larger the larger the filter order. In biomechanics, the second-order Butterworth filter is typically used.

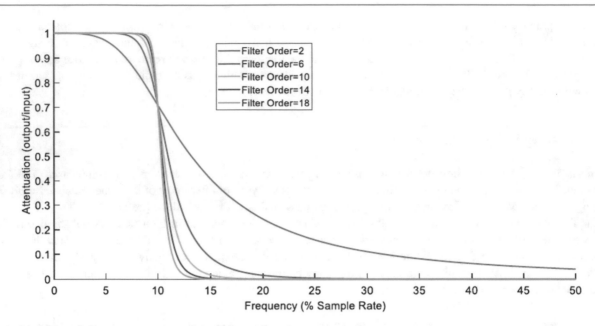

Fig. 4.6 For a Butterworth filter, the influence of filter order on the frequency magnitude response of the filter. Note that the cut-off for the filter was set at 10% of the signal sample rate

With the Butterworth filter as the data is filtered, it skews the data forward in time (phase lag). For some applications, this phase lag is not an issue, but for most biomechanical applications, the timings of events are an important part of an analysis. To remove the phase lag, the Butterworth filter can be applied in forward and reverse directions on a data set. The net effect of applying the Butterworth filter in forward and backward directions means the effective filter order is doubled, so if a second-order filter is used, the order after applying it in forward and backward directions is a fourth order. Therefore, a second-order filter applied in forward and reverse directions has the advantage of increasing the filter order, thus increasing the sharpness of the transition band. By using the filter twice, the frequency response is effectively squared; therefore:

$$H(f) = \frac{1}{1 + (f/f_0)^{2n}}$$

(4.15)

Rearrangement of this equation gives the filter effective cut-off frequency ($\frac{1}{\sqrt{2}}$; -3 dB) after m passes:

$$f = f_0\left(\sqrt{2} - 1\right)^{1/2m}$$

(4.16)

Therefore, if $m = 2$, $f = 0.802f_0$. So if a 10 Hz second-order Butterworth filter is used in forward and reverse directions, the order of the filter is doubled, and the cut-off frequency reduced by 0.802, that is to 8.02 Hz (see Fig. 4.7). Therefore, if the desired cut-off frequency is 10 Hz, the second-order filter should have a cut-off frequency of 12.27 Hz. Cut-off frequencies should be adjusted to allow for the influence of the double pass on the cut-off frequency.

Note in the figure that when the Butterworth filter is applied twice, the frequency magnitude response of the resulting filter is not the same as a fourth-order Butterworth filter; for comparison see Fig. 4.6. Therefore, the dual-pass second-order Butterworth filter is a fourth-order filter, but not strictly a Butterworth filter.

Remember the Butterworth filter has no previous data points at the start of the data set causing problems due to the recursive nature of the filter, but if to remove phase lag the filter is applied in reverse direction the same issues occur at the end of the data set. To circumvent these problems, the ends of a data set to be filtered can be padded. One way to do this is naturally, for example if the stance phase of walking is of interest but data are collected pre- and post-stance some of these data can be retained during data processing and only jettisoned once data processing is complete. Another option is to pad the data at either end of the data set; approaches for doing this include reflection of the data at the boundaries to generate 10–20 extra data points at either end of the data set (e.g., Smith 1989) or extrapolating 1 second of extra data at either end of the data

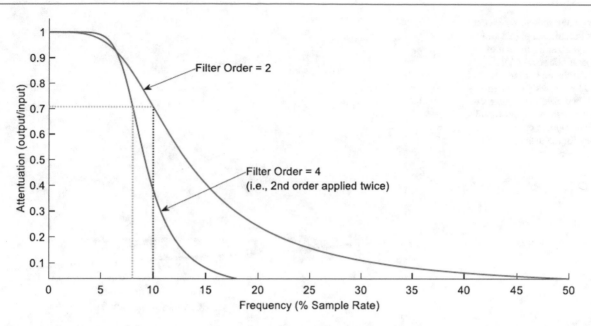

Fig. 4.7 For a Butterworth filter, the influence of single pass and double pass of the filter on the frequency magnitude response of the filter and therefore its cut-off frequency

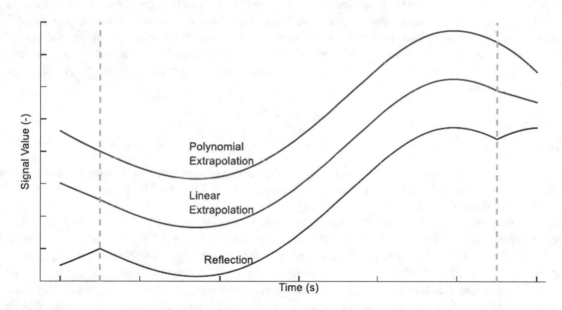

Fig. 4.8 Illustration of three data padding techniques, where from top to bottom there is second-order polynomial extrapolation (blue), linear polynomial extrapolation (red), and reflection (black). In all cases, the first and last 25 points of the original data set have been used to make the extrapolations. The vertical dashed lines indicate the end (first line) and start (second line) of the padding

set using a polynomial (e.g., Howarth and Callaghan 2009). The appropriate method to pad data and the amount of padding depend on the nature of the data to be processed, but to avoid problems at the ends of a data set, data padding is recommended. Figure 4.8 illustrates three methods: second-order polynomial extrapolation, linear polynomial extrapolation, and reflection. Notice how the data reflection does not work well for this data set, with a data window of 25 data points.

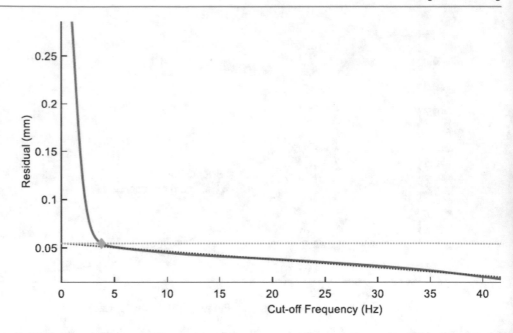

Fig. 4.9 The residual, difference between the original signal and filtered signal, as a function of Butterworth filter cut-off. The residuals at the higher cut-off values are fitted with a straight line and then used to estimate the noise level of the sampled signal and the appropriate cut-off frequency for the data set (marked with a diamond)

4.6.1   How Cut-Off Frequency is Determined

The impact of the Butterworth filter on the processed data will depend on the cut-off frequency used for the filter. In selecting the cut-off, some simply use the values adopted by others, analyzing the same or similar tasks. Some inspect their data pre- and post-processing to select a cut-off. The other option is to use an algorithm which automatically selects the cut-off based on some properties of the sampled data. Two of these automatic options will be described in the following paragraphs.

Winter (2009) proposed that the analysis of the residual could be used to select the Butterworth cut-off frequency. The residual is computed from:

$$R(f_0) = \sqrt{\frac{1}{n} \sum_{i=1}^{n} \left(f'(t_i) - f(t_i) \right)^2} \tag{4.17}$$

where $R(f_0)$ is the residual for a given cut-off frequency f_0, n is the number of data points in the time series, $f'(t_i)$ is the filtered output value at time instant t_i, and $f(t_i)$ is the unfiltered (raw) signal value. If the cut-off frequency is systematically varied, a plot can be made comparing the residual to cut-off (see Fig. 4.9).

At high cut-off frequencies, the presumption is that the true signal has been removed, so the residual reflects the remaining white noise. Therefore, once the filter cut-off exceeds the frequency where "movement" signal exists, the residual reflects the variance of the remaining noise in the signal, with the size of this residual reducing in a linear fashion with increasing cut-off frequency. The residual data points reflecting the white noise variance can be fitted to a line, and then the line interpolated back to the vertical axis, and this intersection represents the variance of the full-spectrum white noise. This noise variance value can then be used to select an appropriate signal cut-off frequency. This analysis assumes the key variable for setting a cut-off frequency is the noise variance, but really the cut-off should be selected at a frequency which reflects an appropriate signal to noise ratio.

Challis (1999) presented another automated procedure for computing the Butterworth cut-off frequency, where the cut-off frequency was estimated by exploiting the properties of the autocorrelation function of white noise. The autocorrelation of white noise has its maximum value at zero lag, with all other of values very small (see Chap. 3). To estimate the cut-off frequency, it is systematically varied until the difference between the filtered and unfiltered data is the best approximation to white noise as assessed using the autocorrelation function. Here the residual is a time series:

$$r(t) = f'(t) - f(t) \tag{4.18}$$

For each residual time series, the autocorrelation (R_{rr}) is computed, and the figure of merit is:

$$A = \sum_{i=1}^{m} (R_{rr}(i))^2 \qquad (4.19)$$

When this figure of merit is its smallest, then the optimum cut-off has been identified.

4.6.2 How to Compute Signal Derivatives

In many applications, signal derivatives (e.g., velocity, acceleration) are required. With the Butterworth filter once the data has been filtered, other methods must be used to determine signal derivatives. For example, the first-order finite difference equations can be used to determine signal derivatives (see Sect. 4.9 for more details).

4.7 Smoothing Splines

When all technical drawing had to be done by hand, some draftsman used a (mechanical) spline to try and fit a smooth line to a set of data points. The spline was often a flexible piece of wood whose shape could be adjusted by hanging weights from the spline. In 1967, Reinsch proposed a mathematical analog of the physical spline, extending the work of Schoenberg (1946a, b), designed to fit a smooth curve to a set of points. As the spline literature arose from math and statistics, it is typically referred to as a smoothing algorithm, although the spline behaves like a low-pass filter.

A spline is composed of a series of piecewise polynomials (Wahba 1990). These polynomials are joined at positions called knots; typically, the knots are the instants at which the data were sampled. The basic format for a polynomial in a spline is:

$$P(t) = c_1 + c_2 t + c_3 t^2 + c_4 t^3 \ldots + c_{nc} t^{nc-1} \qquad (4.20)$$

where $P(t)$ is the output of the polynomial of the spline for the knot at time t and $c_1, c_2, \ldots c_{nc}$ are the polynomial coefficients. For a spline, each time instant, or knot, might require a different set of polynomial coefficients. A spline is classified based on the order of the polynomials used; the most commonly used are described in Table 4.2.

In the fitting of the spline, the following function is minimized:

$$\text{Min} \Rightarrow \lambda \int_0^{t_{end}} (P^m(t))^2 dt + \sum_{i=1}^{n} (P(t_i) - y(t_i))^2 \qquad (4.21)$$

where λ is the parameter which adjusts the smoothness of fit, $P^m(t)$ is the mth derivative of the polynomial fitted at time t, m is the half order of the spline/polynomial (e.g., 2 for a cubic, 3 for a quintic, etc.), and $y(t_i)$ is the sampled data value at time t_i.

The first half of the equation specifies the smoothness of the function:

$$\lambda \int_0^{t_{end}} (P^m(t))^2 dt \qquad (4.22)$$

as the more the mth derivative of the polynomial is minimized, the smoother the signal, or inversely the roughness of the fit. For example, if m is 2, then this is similar to minimizing acceleration for a movement signal. The second half of the function specifies the tightness of the fit (its fidelity to the measured data):

Table 4.2 The order of the spline, the corresponding highest polynomial term, and the name assigned to a spline of that order

Order	Highest polynomial term	Name
4	$c_4.t^3$	Cubic spline
6	$C_6.t^5$	Quintic spline
8	$C_8.t^7$	Heptic spline

$$\sum\nolimits_{i=1}^{n} (P(t_i) - y(t_i))^2 \tag{4.23}$$

It measures how close polynomial predictions are to the sampled values of the points. If λ is zero, then the polynomial must go through each point, so none of the noise is removed; in this case, it would be an interpolating spline. As λ is increased, the polynomials appear smoother and tend to pass less close to each data point.

With m being the half order of the spline when the polynomials are fitted, there are certain conditions that must be met:

- The polynomials must be continuous at the knots up to and including the $(2\ m-2)$ derivative
- For natural splines, typically used in biomechanics, the end points are set to zero for the mth-order derivative and higher.

The constraints placed on the endpoints mean that, for example, for a cubic spline the endpoints of the second derivative are zero, and for a quintic spline the endpoints of the third derivative are zero. These constraints can be useful, for example, if for the task the body is stationary at the beginning and end of the task in which case using a cubic spline might be beneficial. But consideration should be given to the nature of the task when selecting the order of the spline; a quintic is most commonly used.

The performance of a spline with appropriate smoothness parameter (λ) behaves in a similar fashion to a low-pass Butterworth filter (Craven and Wahba 1979). Woltring et al. (1987) showed that the effective cut-off (f_0) for a spline fitted to noisy data equidistantly sampled ($\Delta t = \tau$) with an appropriate smoothness parameters is:

$$f_0 = \frac{1}{(\lambda\tau)^{2m}} \tag{4.24}$$

So the spline behaves like an mth-order Butterworth filter, but with the advantage that the spline does not introduce phase lag.

4.7.1 How to Tune Smoothing

The output from the spline at each time instant $\widehat{y}(t_i)$ is an estimate of the noiseless value of the signal ($y(t_i)$). If the variance of the noise (σ^2) corrupting the time series is known a priori, Reinsch (1967) proposed that the smoothness parameter λ should be set so that:

$$R(\lambda) = \frac{1}{n} \sum\nolimits_{i=1}^{n} (\widehat{y}_\lambda(t_i) - y(t_i))^2 \approx \sigma^2 \tag{4.25}$$

where $R(\lambda)$ is the mean-squared residual error for a given smoothing parameter and $\widehat{y}_\lambda(t_i)$ is the spline predicted output at time instant (t_i) with smoothing parameter value λ. Unfortunately, in many circumstances, the variance of the noise is not known in which case there are two options for selecting the degree of smoothing. The first is to use visual inspection of the curves produced by different smoothing parameters; while visual inspection of data is important, it is not a sufficiently objective method for selecting the degree of smoothing, making comparison of results between studies problematic. The second option is to use some properties of the data to be smoothed to determine the amount of smoothing.

With smoothing splines, a method called cross-validation can be used to estimate the appropriate smoothing parameter. In essence, a smoothing parameter value is selected, and then a spline fitted to the data, but with one of the data set values removed, then prediction error is computed as the difference between the actual value of the removed data point and its spline-predicted value. This is systematically repeated for all data set points, and the root mean square fitting value can be determined for the smoothing parameter value. This is repeated for multiple smoothing parameter values until the parameter is found which gives the best prediction value assessed using the root mean square fitting error. There are three broad categories of cross-validated spline: ordinary, generalized, and robust generalized.

The ordinary cross-validated (OCV) spline procedure was proposed by Wahba and Wold (1975). The idea is to minimize the following function:

$$\text{OCV}(\lambda) = \frac{1}{n} \sum_{i=1}^{n} \left(\widehat{y}_{\lambda,-i}(t_i) - y(t_i) \right)^2 \tag{4.26}$$

where $\widehat{y}_{\lambda,-i}(t_i)$ is the spline predicted output at time instant (t_i) for a spline fitted to the data set with data point i removed from the original data set and smoothing parameter value λ. With the OCV spline, each data point is systematically removed from the data set, and splines are fitted. The OCV spline produces results at least as good if not superior to other data filtering/smoothing methods (Challis and Kerwin 1988). In this form, the determination of the smoothing parameters is time consuming. The operation of a spline, for a given smoothing parameter, is contained in a matrix S_λ, and the OCV(λ) can be computed from:

$$\text{OCV}(\lambda) = \frac{1}{n} \sum_{i=1}^{n} \left(\frac{\widehat{y}_{\lambda,-i}(t_i) - y(t_i)}{1 - s_{\lambda,ii}} \right)^2 \tag{4.27}$$

where $s_{\lambda,\,ii}$ are the diagonal elements of the matrix S_λ which is the influence matrix that relates the vector of smoothed values $(\widehat{y}_\lambda(t_i))$ to the vector of measured values $(y(t_i))$. This form of the equations means that for each value of λ, we do not have to compute n splines, a significant time saving.

A variant on the OCV is the generalized cross-validation (GCV) spline proposed by Craven and Wahba (1979). For the GCV spline, the task is to minimize the following function:

$$\text{GCV}(\lambda) = \frac{1}{n} \sum_{i=1}^{n} \left(\frac{\widehat{y}_{\lambda,-i}(t_i) - y(t_i)}{1 - \text{tr}(S_\lambda)/n} \right)^2 \tag{4.28}$$

where $\text{tr}(S_\lambda)$ is the trace of matrix S_λ. This criterion function is computationally less demanding than the OCV criterion function, providing further computation time saving.

Wahba (1990) has highlighted that the GCV splines do not always provide reliable results when the number of samples is less than 30, and spurious results occur ~0.2% of the time for samples as large as 200. To circumvent this problem, Lukas (2006) and others have proposed the robust generalized cross-validation (RGCV). The criterion function to minimize is:

$$\text{RGCV}(\lambda) = \left(\gamma + (1 - \gamma) \left(\frac{\text{tr}(S_\lambda^2)}{n} \right) \right) \text{GCV}(\lambda) \tag{4.29}$$

where γ is the robustness parameters ($0 \leq \gamma \leq 1$). When $\gamma = 1$, then the RGCV is the GCV. As a general guide the lower value of γ the more robust the result, so in practical applications as the number of samples shrink the γ value should also decrease. The problem with the RGCV is that it introduces another parameter the user must specify; the rationale behind cross-validation was to make the determination of the required parameter automatic so this advantage is somewhat diminished.

4.7.2 How to Compute Derivatives

The basic format for a cubic polynomial, used in a cubic spline, is:

$$P(t) = c_1 + c_2\, t + c_3 t^2 + c_4 t^3 \tag{4.30}$$

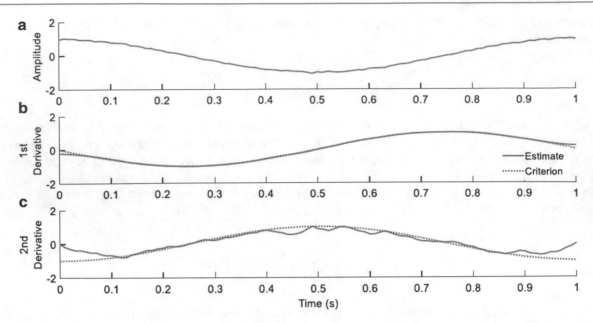

Fig. 4.10 Illustration of the influence of the order a spline on the estimate of signal derivatives. A noisy cosine wave has been fitted with a generalized cross-validated cubic spline, and then the first and second derivatives of the signal estimated. The figure comprises (**a**) the noisy cosine wave, (**b**) estimate of the first derivative, and (**c**) the estimate of the second derivative

If derivatives are required, the polynomial derivative can be used, so for a cubic polynomial, the polynomials are:

$$\dot{P}(t) = c_2 + 2\,c_3\,t + 3\,c_4 t^2 \tag{4.31}$$

$$\ddot{P}(t) = 2\,c_3 + 6\,c_4\,t \tag{4.32}$$

It is important to consider the order of the spline when planning to compute derivatives. With m as the half order of the spline, for natural splines, the end points are set to zero for the kth derivatives for $k = m, \ldots 2m-2$. This constraint means that, for example, unless you know that the acceleration at the ends of the task being examined is zero, a cubic spline should not be used for the determination of acceleration. Figure 4.10 illustrates this problem; a noisy cosine wave has been fitted with a generalized cross-validated cubic spline. The spline estimate of the first derivative matches the criterion well. The spline estimate of the second derivative does not match the criterion well, but in particular note that at either end of the signal, the estimates of the second derivative have been forced to a value of zero.

4.8   Truncated Fourier Series

A Fourier series can be fitted to data. The format for the discrete Fourier series is:

$$f(t) = a_0 + \sum_{k=1}^{N} a_k \, \cos\,(k\,\omega\,t) + b_k \, \sin\,(k\,\omega\,t) \tag{4.33}$$

where the fundamental frequency is:

$$\omega = \frac{2\pi}{T} \tag{4.34}$$

and T is the sample period. The coefficients for the Fourier series are computed from:

$$a_0 = \frac{1}{n} \sum_{i=1}^{n} f(\,t_i) \tag{4.35}$$

$$a_k = \frac{2}{n} \sum_{i=1}^{n} f(\,t_i) \, \cos\,((i-1)\omega\,t_i) \qquad k = 1 : N \tag{4.36}$$

Fig. 4.11 An example of signal processing using a truncated Fourier series where (**a**) is the noiseless signal, (**b**) is the noisy signal, and (**c**) is the reconstructed Fourier series, after the Fourier series has been fitted to the noisy signal and then the higher Fourier series terms set to zero

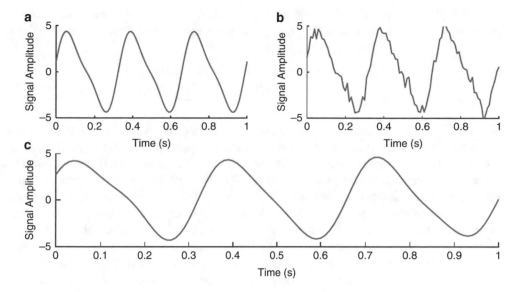

$$b_k = \frac{2}{n} \sum_{i=1}^{n} f(t_i) \, \sin\left((i-1)\omega \, t_i\right) \qquad k = 1 : N \qquad (4.37)$$

where N is the maximum number of coefficients:

$$N \leq \frac{n-1}{2} \qquad (4.38)$$

and n is the number of data points in the series to be modeled.

Imagine the sample period was 1 second and the data had been collected at 100 Hz; each coefficient would represent an integer frequency component from 1 to 50 Hz. If once a Fourier series is fitted to these data, only the first ten coefficients are used to reconstruct the signal; the new signal reflects the original signal with any signal components above 10 Hz removed. If the signal to be analyzed contained frequencies only in the lower part of the frequency spectrum, then the discarded higher coefficients would be associated with noise; therefore, if these coefficients were set to zero and the signal reconstructed from the remaining coefficients, then the truncated Fourier series acts like a low-pass filter. Figure 4.11 illustrates this, where a Fourier series has been fitted to noisy data, but by truncating the Fourier series, the reconstructed time series has removed the higher-frequency signal components, reflecting signal noise, and reproduced a good representation of the noise less signal.

There are two things to consider when fitting the Fourier series. The first is whether to de-trend the data before fitting the Fourier series; if this is done, so $f(t = 0) = f(t = T) = 0$ and then the cosine component of the series is not required. As the Fourier series assumes the data have repeating signal components, de-trending can improve results. The second thing to consider is whether to use the fast Fourier transformation, which provides the advantage of speed but does suffer from the need for the number of data points to conform to a certain number of points, which may not be feasible in certain applications.

4.8.1 How to Determine the Appropriate Number of Fourier Coefficients

The output from the reconstruction of a truncated Fourier series at each time instant $\widehat{y}(t_i)$ is an estimate of the noiseless value of the signal ($y(t_i)$). If the variance of the noise (σ^2) corrupting the time series is known a priori, then:

$$R(N_T) = \frac{1}{n} \sum_{i=1}^{N_T} \left(\widehat{y}_{N_T}(t_i) - y(t_i)\right)^2 \approx \sigma^2 \qquad (4.39)$$

where $R(N_T)$ is the mean-squared residual error for a given number (N_T) coefficients in the truncated Fourier series and $\widehat{y}_{N_T}(t_i)$ is the truncated Fourier series output at time instant (t_i). As is generally the case in data processing, the variance of the noise is not known a priori. There are a number of options for selecting the amount of Fourier series truncation.

Winter (2009) proposed that the analysis of the residual could be used to select the Butterworth cut-off frequency; this same procedure can be used to compute the appropriate number of Fourier terms. The residual is computed from:

$$R(N_T) = \sqrt{\frac{1}{n} \sum_{i=1}^{n} \left(f'(t_i) - f(t_i) \right)^2} \tag{4.40}$$

where $R(N_T)$ is the residual for N_T coefficients in the truncated Fourier series, n is the number of data points in the time series, $f'(t_i)$ is the Fourier series reconstructed time series at time instant t_i, and $f(t_i)$ is the unfiltered (raw) signal value. If the number of Fourier coefficients is systematically varied, then a plot can be made comparing the residual to the number of coefficients (see section on Butterworth filtering 4.6.1). Challis (1999) also used the residual of a processed time series to compute the optimal Butterworth filter cut-off; that same procedure can be used for the determination of the number of Fourier coefficients to use.

The determination of the appropriate order of the Fourier series to use is a classic example of an ill-posed problem (Tikhonov et al. 1995), where if additional constraints are not placed on the problem, the order cannot be selected. Anderssen and Bloomfield (1974) adopted a regularization procedure proposed by Cullum (1971) to fit Fourier series to noisy data. This procedure was subsequently adopted in biomechanics by Hatze (1981). The regularization procedure sought an order of the Fourier series which balanced the fit to the noisy data and the smoothness of the first derivative. Subsequent developments in this area have favored the generalized cross-validation procedures, previously presented for fitting splines, as a means to select Fourier series order (e.g., Amato and De Feis 1997; Angelini and Canditiis 2000).

4.8.2 How to Compute Derivatives

Once the truncated Fourier series to model the data has been selected, then the signal derivatives can be computed from:

$$\dot{f}(t) = \sum_{k=1}^{N_T} k\omega\, a_k\, \sin\left(k\,\omega\,t\right) - k\omega\, b_k\, \cos\left(k\,\omega\,t\right) \tag{4.41}$$

$$\ddot{f}(t) = -\sum_{k=1}^{N_T} (k\omega)^2\, a_k\, \cos\left(k\,\omega\,t\right) - (k\omega)^2\, b_k\, \sin\left(k\,\omega\,t\right) \tag{4.42}$$

4.9 Calculation of Derivatives

In a number of applications, it is necessary to compute signal derivatives, for example, to estimate velocity values after data filtering using a Butterworth filter. If a signal is represented by a Fourier series:

$$f(t) = a_0 + \sum_{k=1}^{N} a_k\, \cos\left(k\,\omega\,t\right) + b_k\, \sin\left(k\,\omega\,t\right) \tag{4.43}$$

Differentiation of this function indicates that for the first-order derivative, the signal magnitude increases with increasing frequency (see, e.g., Eq. 4.41), and second-order derivative in proportion to frequency squared (see, e.g., Eq. 4.42). Figure 4.12 illustrates the frequency content of a signal and its first derivative. Notice how for the signal there are large signal amplitudes up to 4 Hz; the remaining signal components reflect the amplitude of the noise. Once differentiated, the process of differentiation amplifies signal components, with greater amplification and with increasing frequency, which in turn amplifies the noise in the signal.

There are a number of options for computing derivatives; the most commonly used are the finite difference equations. These equations are simply derived from a Taylor series expansion:

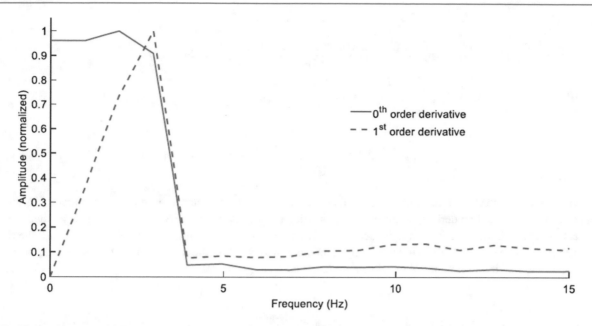

Fig. 4.12 The frequency-amplitude graphs of a signal and its derivative. This illustrates the action of differentiation and shows how high-frequency noise components are amplified by differentiation. Note that the amplitudes of the two signals have been normalized

$$f(t_i + \Delta t) = \sum_{j=0}^{\infty} \frac{f^j(t_i)}{j!} \Delta t^j \qquad (4.44)$$

where $f(t_i)$ is the value of function at time t_i, Δt is a small time increment, and $f^j(t_i)$ is the jth-order derivative of the function at time t_i. If we truncate the Taylor series at two terms and for now ignore the remaining terms of the equation:

$$f(t_i + \Delta t) = f(t_i) + \frac{\dot{f}(t_i)}{1} \Delta t = f(t_i) + \dot{f}(t_i) \Delta t \qquad (4.45)$$

Then make the derivative the result of the function, and an equation for derivative estimation is produced:

$$\dot{f}(t_i) = \frac{f(t_i + \Delta t) - f(t_i)}{\Delta t} \qquad (4.46)$$

This is the forward central difference equation, by a similar derivation central, and backward finite difference equations can be derived. In a similar vein and by a similar derivation different order finite difference equations can be derived. With a change in notation, the two term finite difference equations for the first derivative are:

$$\dot{x}(t_i) = \frac{x(t_i) - x(t_i - \Delta t)}{\Delta t} \qquad \text{(Backward)} \qquad (4.47)$$

$$\dot{x}(t_i) = \frac{x(t_i + \Delta t) - x(t_i - \Delta t)}{2\Delta t} \qquad \text{(Central)} \qquad (4.48)$$

$$\dot{x}(t_i) = \frac{x(t_i + \Delta t) - x(t_i)}{\Delta t} \qquad \text{(Forward)} \qquad (4.49)$$

At the start of a signal when there are no previous values, the forward difference equation is used, and at the end of the data set when there is no future value, the backward difference equation is used. The three term finite difference equations for the second derivative are:

$$\ddot{x}(t_i) = \frac{x(t_i) - 2\,x(t_i + \Delta t) + x(t_i + 2\Delta t)}{(\Delta t)^2} \quad \text{(Backward)} \tag{4.50}$$

$$\ddot{x}(t_i) = \frac{x(t_i + \Delta t) - 2\,x(t_i) + x(t_i - \Delta t)}{(\Delta t)^2} \quad \text{(Central)} \tag{4.51}$$

$$\ddot{x}(t_i) = \frac{x(t_i + 2\Delta t) - x(t_i + \Delta t) + x(t_i)}{(\Delta t)^2} \quad \text{(Forward)} \tag{4.52}$$

We can estimate the error associated with each of these formulae, due to the truncation of the Taylor series. For the central difference equation for the first derivative, the error is a function of $(\Delta t)^2$, and for the second derivative, is a function of $(\Delta t)^4$. Therefore, to improve the accuracy of an estimated derivative, it makes sense to make Δt as small as possible. If the time step size is too small, then the results will be influenced by computer rounding error; the smallest Δt is dictated by our measurement equipment and the maximum number of digits ($n$max) that a computer can use:

$$\Delta t_{\min} \approx 10^{n\text{max}/2} \tag{4.53}$$

The influence of reducing the Δt is illustrated in Table 4.3. As would be expected, the error reduces with decreasing sample interval and is greater for the second derivative compared with the first derivative.

There are finite difference equations with different number of terms (see Appendix F – Numerical Data Differentiation). Figure 4.13 presents the frequency magnitude response for central finite difference equations with increasing number of terms in the equations. An ideal differentiator has an increasing influence on the signal with increasing signal frequency component (see Fig. 4.13 for ideal response). None of the equations presented produce this ideal response, but at low frequencies, they provide a good approximation.

The ideal differentiator works well if the data is noiseless and if the signal has components up to the Nyquist frequency. Most biomechanical signals to be numerically differentiated have noise in the higher part of the frequency spectrum, so ideally these frequencies should be suppressed. There is some suppression with the finite difference equations, but not the triangular frequency magnitude response that we would desire. An option to address this is once the data has been differentiated, a low-pass filter is used to suppress the higher-frequency components that may have been magnified by the differentiation formula.

Custom filters can be designed by specifying the required frequency response of the filter, and then various algorithms try to produce the best approximation to the desired frequency response. The algorithms used to design the filter include the least-squares method (Sunder and Ramachandran 1994), the eigenfilter method (Vaidyanathan and Truong 1987), the Kaiser window method (Kaiser 1974), and the Remez exchange algorithm (Parks and McClellan 1972). Each approach has strengths and weaknesses. For example, Fig. 4.14 shows two frequency magnitude responses for a differentiator designed using the Parks-McClellan algorithm, which in turn uses the Remez exchange algorithm. It seeks an optimal fit between the desired and actual frequency responses. Given the finite precision of the calculations, the fit will rarely match precisely the required frequency response.

Table 4.3 The absolute error in estimating the first and second derivatives of $y = e^t$ when $t = 0$, using the finite difference equations

Δt (s)	Error first derivative	Error second derivative
0.1	0.0016675	0.000833611
0.01	1.66667×10^{-5}	8.33336×10^{-6}
0.001	1.66667×10^{-7}	8.34065×10^{-8}
0.0001	1.66689×10^{-9}	5.02476×10^{-9}
0.00001	1.21023×10^{-11}	1.02748×10^{-6}

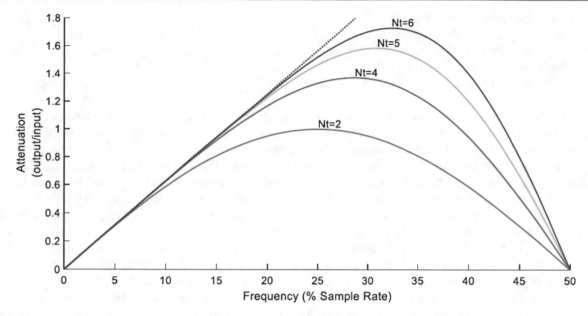

Fig. 4.13 The frequency magnitude response of central finite difference equations with an increasing number of terms, where Nt is the number of terms in the formula. The dotted line represents the theoretical ideal differentiator

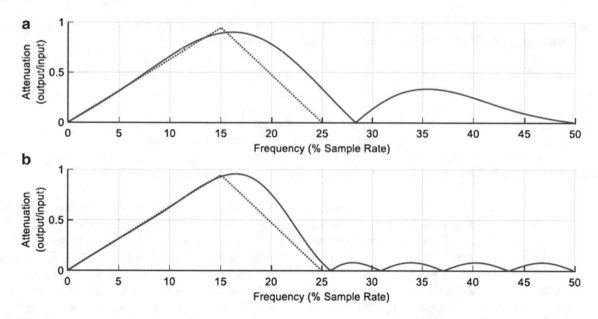

Fig. 4.14 The frequency magnitude response of a finite impulse response differentiating filter, realized using the Parks-McClellan algorithm. The design specifications were a passband edge of 15% of the sample rate and a stopband edge of 25% of the sample rate. Note the improved approximation to the desired frequency response with more coefficients in the filter, (**a**) 8 points and (**b**) 16 points. The dotted line represents the desired frequency response of the differentiator

4.10 Determining the Amount of Filtering

There are a variety of methods for the filtering (smoothing) of biomechanical data. For these methods, the question arises as to how the amount of filtering (smoothing) is selected. If the properties of the signal and noise were known then the amount of the filtering could easily be selected, but this is infrequently the case. There are a number of options for determining the amount of filtering:

- Use the previously published values.
- Tune to some property of the movement.
- Optimal regularization (e.g., Fourier series).
- Generalized cross-validation (e.g., splines).
- Autocorrelation-based procedure (e.g., Butterworth filter).

One approach is to simply adopt the cut-off frequency used by other researchers. This approach makes comparison of data with that of others easier but has the disadvantage that it does not account for any peculiarities of the collected data.

There are a number of ways in which the amount of filtering can be tuned to some property of the movement signal. For example, during the flight phase, the center of mass of the whole body should have a vertical acceleration equal to the local acceleration due to gravity; therefore, to exploit this information, the amount of filtering could be tuned to produce acceleration close to the known criterion. Similarly during flight, the whole body angular momentum remains constant and therefore provides another option for tuning of the amount of filtering. During many analyses, the motion and force plate data are sampled simultaneously, so it is feasible that the acceleration of the whole body center of mass can be determined from the force plate record and then the motion data processed to provide as close a match as feasible. Of course these approaches all rely on having good estimates of body segment inertial properties which will not always be the case.

The automatic procedures use information contained in the sampled signal to estimate the optimal amount of filtering. These approaches take potential subjectivity out of the process of selecting the amount of filtering. They do represent "black boxes" where the software takes care of selecting the amount of filtering; therefore, they should be used with some caution, and visual inspection of processed data is an important step. The automated procedures include optimally regularized Fourier series (e.g., Hatze 1981), generalized cross-validated splines (e.g., Craven and Wahba 1979; Woltring 1986), and an autocorrelation-based procedure (Challis 1999).

4.11 Assessment of Methods for Data Filtering/Smoothing

There are a variety of methods for filtering (smoothing) of biomechanical data. These methods all have their merits, for example:

- Splines do not require equidistantly sampled data, but Fourier series and Butterworth filters do.
- Fourier series and splines allow direct computation of derivatives, while Butterworth filters do not.
- The Butterworth filter and truncated Fourier series have direct meaning in the frequency domain, while the spline does not.
- Computationally, the Butterworth filter is fast compared with truncated Fourier series and splines.

There have been a variety of assessments of methods of filtering (smoothing) of biomechanical data, but these have not provided an obviously superior method. Most of these assessments rely on data sets with known derivatives as their criterion. For example, Pezzack et al. (1977) collected motion data using film analysis and criterion acceleration data from an accelerometer mounted on a mechanical arm. A subject grasped the mechanical arm and moved it, while the acceleration values were measured. The resulting data set comprised a set of noisy displacement data and criterion acceleration data. In theory, a good filtering and differentiation routine should provide good estimates of the acceleration data from the processing of the noisy displacement data. Lanshammar (1982) analyzed these data and assessed that the noise level was not reflective of typical biomechanical data so added more noise to the displacement data. The other approach is data sets generated from mathematical functions, where the first and second derivative values can be determined via differentiation of these functions. To test a filtering method noise is then added to the true "displacement" data. Appendix G presents a variety of data sets and functions which have been used for the assessment of the filtering of biomechanical data.

To illustrate some of the differences between different techniques for the processing of biomechanical data, the performance of a Butterworth filter, with the degree of cut-off selected using an autocorrelation-based procedure (ABP – Challis 1999), was compared with a quintic spline with the degree of smoothing selected using a generalized cross-validation procedure (GCVQS – Woltring 1986) (see Challis 1999 for more details and additional assessments). Dowling (1985) presented data where he simultaneously collected accelerometer data and motion analysis displacement data (see Fig. 4.15). The task was to process the noisy displacement data and evaluate methods by comparing estimated acceleration values with

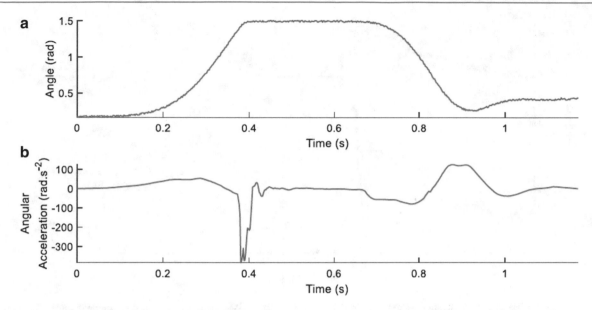

Fig. 4.15 The data of Dowling (1985) for the assessment of a filter and differentiation methods. The data comprise (**a**) the noisy angular displacement data and (**b**) the criterion acceleration data

the criterion values; these comparisons were made visually and numerically. Accuracy of acceleration estimates was quantified using the percentage root mean square difference (%RMSD) which is computed from:

$$\%\mathrm{RMSD} = 100\sqrt{\frac{\frac{1}{n}\sum_{i=1}^{n}(x_{ci} - x_i)^2}{\frac{1}{n}\sum_{i=1}^{n}(x_{ci})^2}} \qquad (4.54)$$

where n is the number of data points, x_{ci} is the ith value of the criterion signal, and x_i is the estimated ith value of the signal.

Figure 4.16 shows how well both the ABP and the GCVQS estimated the acceleration data. For the ABP, the %RMSD between the criterion and estimated acceleration values was 41.4%, and for the GCVQS was 38.4%. Both techniques assume that the signal to noise ratio is equal throughout the data set, but this is not the case as there are regions of high acceleration in contrast to regions of relatively low accelerations.

It is feasible to focus on regions of the acceleration data to examine performance of the two techniques for the estimation of acceleration values (see Fig. 4.17). The ABP was oscillatory in regions of relatively low acceleration and then underestimated the peak acceleration value. The GCVQS produced a closer estimate of the peak acceleration value, with the penalty that it was even more oscillatory than the ABP for the remainder of the data. Both techniques sought a compromise between overfiltering in the area of the peak acceleration value and not filtering sufficiently in the other sections of the curve. Both techniques have provided reasonable estimates of the acceleration values, but the analysis of these data has highlighted the problems which arise if the signal from which accelerations are to be estimated is not stationary. Unfortunately, most analyzed signals in biomechanics are not stationary, yet this is the common implicit assumption.

The results highlight some of the general issues associated with filtering of the noise associated with data collected for motion analysis, with implications for the processing of other types of data.

- Filtering cannot remove all of the noise.
- While reducing noise, some of the signal will also be reduced.
- Some approaches remove too much signal (e.g., ABP).
- Some approaches leave too much noise (e.g., GCVQS).
- Filtering is essential if reasonable estimates of derivatives are to be obtained.
- The higher order the derivative, the poorer the estimate.

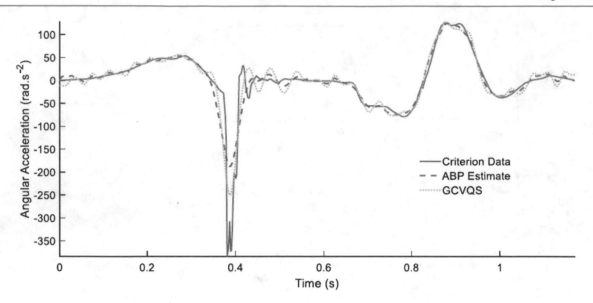

Fig. 4.16 The estimation of the acceleration values presented in Dowling (1985) based on noisy angular displacement values. The techniques evaluated were an autocorrelation-based procedure (ABP – Challis 1999) and a generalized cross-validation quintic spline (GCVQS – Woltring 1986)

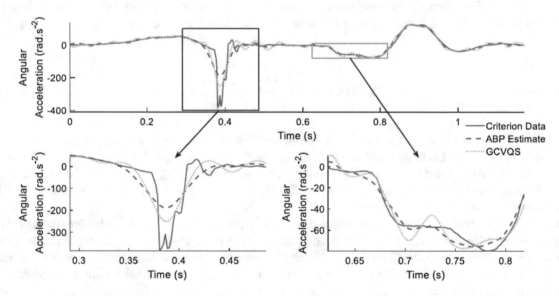

Fig. 4.17 The estimation of the acceleration values presented in Dowling (1985) based on noisy angular displacement values, for specific regions of the curve. The lower graphs show an area of high acceleration and relatively low accelerations. The techniques evaluated were an autocorrelation-based procedure (ABP – Challis 1999) and a generalized cross-validation quintic spline (GCVQS – Woltring 1986)

4.12 Conclusion

In this chapter, various data processing methods have been examined with a focus on the removal of key frequency components of a signal and the computation of derivatives.

4.13 Review Questions

1. With reference to digital filtering, what are the following:
   Passband

Cut-off frequency

Stopband frequency

Transition band

Attenuation

2. Describe the theoretical frequency magnitude response of an ideal low-pass, high-pass, band-pass, and band reject (notch) filters.

3. Suggests some candidate methods for removing (some) noise from motion analysis data. Explain the basic principles behind these methods.

4. How might you select the degree of filtering/smoothing for a data set? What options are available?

5. How can signal derivatives be computed?

References

Cited References

Amato, U., & De Feis, I. (1997). Convergence for the regularized inversion of Fourier series. *Journal of Computational and Applied Mathematics, 87*(2), 261–284.

Anderssen, R. S., & Bloomfield, P. (1974). Numerical differentiation procedures for non-exact data. *Numerische Mathematik, 22*, 157–182.

Angelini, C., & Canditiis, D. D. (2000). Fourier frequency adaptive regularization for smoothing data. *Journal of Computational and Applied Mathematics, 115*(1), 35–50.

Butterworth, S. (1930). On the theory of filter amplifiers. *Wireless Engineer, 7*, 536–541.

Challis, J. H. (1999). A procedure for the automatic determination of filter cutoff frequency of the processing of biomechanical data. *Journal of Applied Biomechanics, 15*(3), 303–317.

Challis, J. H., & Kerwin, D. G. (1988). An evaluation of splines in biomechanical data analysis. In G. De Groot, A. P. Hollander, P. A. Huijing, & G. J. Van Ingen Schenau (Eds.), *Biomechanics XI-B* (pp. 1057–1061). Amsterdam: Free University Press.

Craven, P., & Wahba, G. (1979). Smoothing noisy data with spline functions: Estimating the correct degree of smoothing by the method of generalised cross-validation. *Numerische Mathematik, 31*(4), 377–403.

Cullum, J. (1971). Numerical differentiation and regularization. *SIAM Journal on Numerical Analysis, 8*(2), 254–265.

Dowling, J. J. (1985). A modelling strategy for the smoothing of biomechanical data. In B. Jonsson (Ed.), *Biomechanics X-B* (pp. 1163–1167). Champaign: Human Kinetics Publishers.

Hatze, H. (1981). The use of optimally regularized Fourier series for estimating higher-order derivatives of noisy biomechanical data. *Journal of Biomechanics, 14*(1), 13–18.

Howarth, S. J., & Callaghan, J. P. (2009). The rule of 1 s for padding kinematic data prior to digital filtering: Influence of sampling and filter cutoff frequencies. *Journal of Electromyography and Kinesiology, 19*(5), 875–881.

Kaiser, J. F. (1974). Nonrecursive digital filter design using the I_0-sinh window function. In *Proceedings of the 1974 IEEE international symposium on circuits and systems*. San Francisco: IEEE.

Lanshammar, H. (1982). On practical evaluation of differentiation techniques for human gait analysis. *Journal of Biomechanics, 15*(2), 99–105.

Lukas, M. A. (2006). Robust generalized cross-validation for choosing the regularization parameter. *Inverse Problems, 22*(5), 1883.

Parks, T., & McClellan, J. (1972). Chebyshev approximation for nonrecursive digital filters with linear phase. *IEEE Transactions on Circuit Theory, 19*(2), 189–194.

Pezzack, J. C., Norman, R. W., & Winter, D. A. (1977). An assessment of derivative determining techniques used for motion analysis. *Journal of Biomechanics, 10*(5–6), 377–382.

Reinsch, C. H. (1967). Smoothing by spline functions. *Numerische Mathematik, 10*, 177–183.

Schoenberg, I. J. (1946a). Contributions to the problem of approximation of equidistant data by analytic functions: Part A. On the problem of smoothing or graduation. A first class of analytic approximation formulae. *Quarterly of Applied Mathematics, 4*(1), 45–99.

Schoenberg, I. J. (1946b). Contributions to the problem of approximation of equidistant data by analytic functions: Part B. On the problem of osculatory interpolation. A second class of analytic approximation formulae. *Quarterly of Applied Mathematics, 4*(2), 112–141.

Smith, G. (1989). Padding point extrapolation techniques for the Butterworth digital filter. *Journal of Biomechanics, 22*(8/9), 967–971.

Stigler, S. M. (1981). Gauss and the invention of least-squares. *Annals of Statistics, 9*(3), 465–474.

Sunder, S., & Ramachandran, V. (1994). Design of equiripple nonrecursive digital differentiators and Hilbert transformers using a weighted least-squares technique. *IEEE Transactions on Signal Processing, 42*(9), 2504–2509.

Tikhonov, A. N., Goncharsky, A. V., Stepanov, V. V., & Yagola, A. G. (1995). *Numerical methods for the solution of ill-posed problems*. New York: Springer.

Vaidyanathan, P., & Truong, N. (1987). Eigenfilters: A new approach to least-squares FIR filter design and applications including Nyquist filters. *IEEE Transactions on Circuits and Systems, 34*(1), 11–23.

Wahba, G. (1990). *Spline models for observational data*. Philadelphia: Society for Industrial and Applied Mathematics.

Wahba, G., & Wold, S. (1975). A completely automatic french curve: Fitting spline functions by cross validation. *Communications in Sattistics, 4*(1), 1–17.

Winter, D. A. (1995). Human balance and posture control during standing and walking. *Gait & Posture, 3*(4), 193–214.

Winter, D. A. (2009). Chapter 2 – Signal processing. In *Biomechanics and motor control of human movement*. Chichester: Wiley.

Woltring, H. J. (1986). A Fortran package for generalized, cross-validatory spline smoothing and differentiation. *Advances in Engineering Software*, *8*(2), 104–113.

Woltring, H. J., de Lange, A., Kauer, J. M. G., & Huiskes, R. (1987). Instantaneous helical axis estimation via natural cross-validated splines. In G. Bergmann, A. Kobler, & A. Rohlmann (Eds.), *Biomechanics: Basic and applied research* (pp. 121–128). Dordrecht: Martinus Nijhoff Publishers.

Useful References

Bloomfield, P. (2000). *Fourier analysis of time series*. New York: Wiley.

Stearns, S. D., & Hush, D. R. (2011). *Digital signal processing with examples in MATLAB* (2nd ed.). Boca Raton: CRC Press.

Electromyography

5

Overview

Electromyography (EMG) is the measurement of the electrical signal associated with muscle activity. The recorded signal reflects the electrical signal associated with a cascade of events related to the generation of muscle force. The actual recorded signal varies with various aspects of the measurement equipment, as well as with the amount of muscle activity. Compared with many other methods used in biomechanics, there are more variables to consider when appropriately collecting and processing an EMG signal.

That the production of muscular force is associated with electrical activity was first observed by the Italian physician Luigi Galvani (1737–1798). He examined the leg muscles of a frog and identified that "animal electricity" was important for the generation of muscular force (Galvani 1791). In 1849 the German physician Emil du Bois-Reymond (1818–1896) was the first to record the electrical signal associated with a voluntary muscular activity, examining the human upper limb (Du Bois-Reymond 1849). Our ability to measure muscle electricity activity greatly increased in the last century due to the development of microelectronics and the availability of computers; in addition our ability to process and analyze these signals was advanced by developments in signal processing.

The content of this chapter requires a working knowledge of signals and signal analysis (see Chaps. 3 and 4).

5.1 What Are the Electrical Events?

If electromyography is measuring the electrical signal associated with the activation of the muscle, what is the source and magnitude of this electrical activity?

The functional unit within a muscle responsible for responding to electrical activity and producing force is the motor unit. The motor unit comprises the cell body and dendrites of a motor neuron, the branches of its axon, and muscle fibers it directly innervates. For a single motor neuron the number of muscle fibers it innervates varies between muscles, for example, the platysma has a ratio of 1:25, the first dorsal interosseus 1:340, and the medial gastrocnemius 1:1600 (Feinstein et al. 1955). Similarly the number of motor units varies between muscles: the platysma has ~1000, the first dorsal interosseus ~120, and the medial gastrocnemius ~580 (Feinstein et al. 1955). The fibers associated with a motor unit are not necessarily located adjacent to one another; the fibers from different motor units are intermingled (Bodine et al. 1988). The stimulus from a motor neuron is an action potential (see Fig. 5.1).

The resting potential within a muscle fiber is about −90 mV, and the peak during an action potential is 40 mV (see Fig. 5.1 for typical time course). Because of the anatomy of the motor unit when a muscle fiber is stimulated, other muscle fibers will also be stimulated; therefore the recorded EMG signal represents the sum of multiple muscle fiber action potentials and results in the recording of a motor unit action potential. In many cases the EMG signal is the approximate sum of multiple motor unit action potentials (e.g., Keenan et al. 2005). The sampled waveform can be the result of the spatial-temporal summation of multiple motor unit action potentials. Figure 5.2 shows how a sampled surface EMG signal is comprised of the spatial-temporal summation of multiple motor unit action potentials. For surface EMG the signals conduct through the muscle tissue and adipose tissue; these tissues filter the signal which subsequently is sampled.

J. H. Challis, *Experimental Methods in Biomechanics*, https://doi.org/10.1007/978-3-030-52256-8_5

Fig. 5.1 Typical pattern of a muscle fiber action potential

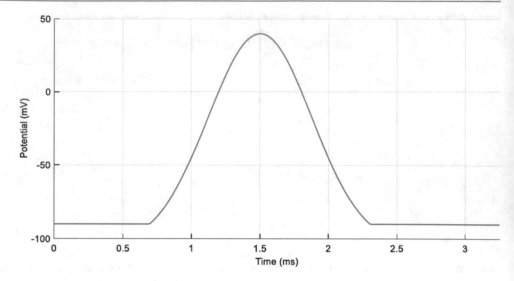

Fig. 5.2 A surface EMG signal comprised of the sum of the output of 100 motor unit action potentials generated by a model. Upper portion shows the motor unit action potential from ten representative motor units

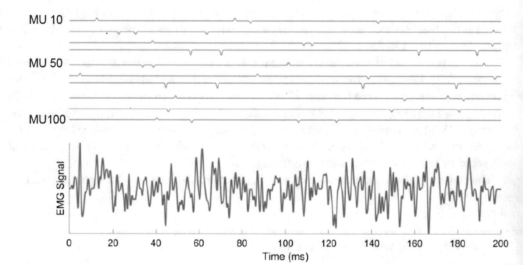

Table 5.1 The relative properties of the three types of motor units, *FF* fast fatigable, *FR* fast fatigue-resistant, and *S* slow

Property	Motor unit type		
	Type FF	Type FR	Type S
Neuron size	Large	Large	Small
Stimulation frequency	High	High	Low
Fiber diameter	Large	Large	Small
Maximum force	High	High	Small
Maximum speed	High	High	Small
Mitochondrial density	Low	High	High
Capillary density	Low	Medium	High

There are three broad types of motor unit: *Type FF* fast fatigable, *Type FR* fast fatigue-resistant, and *Type S* slow (Burke et al. 1973). The motor unit properties are more of a spectrum than these broad categories imply (Kernell 2006), but given these three motor unit types, their properties indicate how they may be used in movement and detected in EMG recordings. The properties of the different motor unit types are presented in Table 5.1 and therefore explain how each motor unit type acquired its name.

The force produced by a muscle depends on two factors related to the motor units: the number of motor units that is activated, and the rate of the action potential discharge for the motor units. These two factors are referred to as recruitment and rate coding, respectively. For the range of forces a muscle can produce both recruitment and rate coding contribute, but their relative contributions change with level of force. Recruitment is the dominant contributor to muscle force modulation at low forces, but at higher forces rate coding becomes more important (Milner-Brown et al. 1973). Motor units are recruited from smallest to largest in terms of their neuron size, Henneman's size principle (Henneman et al. 1965). This means that the motor units are recruited in the following order: slow, then fast fatigue-resistant, and finally fast fatigable. As the different motor unit types have different stimulation frequencies, these can sometimes be detected in EMG records permitting identification of how motor units are being used for a particular task.

5.2   Recording of the Signal

To record the EMG signal, electrodes are used to pick up the electrical signal; these will be discussed in a following sub-section. The placement of electrodes is a crucial factor influencing EMG signal quality and so will be reviewed followed by an examination of how the signal should be amplified and sampled.

5.2.1   Electrodes

There are two categories of electrodes: indwelling and surface. Indwelling electrodes are placed inside the muscle, while surface electrodes are placed on the skin over the muscle belly.

Irrespective of whether indwelling or surface electrodes are used, they can be used in monopolar or bipolar configurations. Under both conditions a reference electrode is also used which is placed on an electrically quiet location, for example, a bony landmark. The reference electrode should pick up some of the same ambient noise as the electrode(s) measuring the muscle signal; therefore some of this noise can be removed from signal of interest.

A monopolar electrode will pick up the signal near to the electrode but this can include an electrical signal not from the muscle of interest, for example, from electromagnetic signals from power cords and electrical outlets. Bipolar electrodes can circumvent the recording of unwanted electromagnetic signals. The signals from the two electrodes each record an electrical potential; these signals are passed to a differential amplifier which takes the difference between the signals from the pair of electrodes and amplifies this difference. The common mode noise between the two signals is eliminated (rejected); of course the rejection is not perfect but the differential signal from a bipolar electrode arrangement has a higher to signal to noise ratio than a monopolar electrode used on the same muscle.

For human studies indwelling electrodes are commonly needle electrodes much like a fine steel hypodermic needle with diameters as small as those of a human hair (see Fig. 5.3a). Due to the invasive nature of the indwelling electrodes, often a single electrode is used (monopolar) in humans; this occurs in particular if the aim is to measure a single motor unit. It can be hard to insert an electrode into a single motor unit; this is considered harder if bipolar electrodes are used. In animal studies the electrodes are often manufactured in the house and can be used in bipolar configurations. In animal studies chronic implantations are feasible.

Fig. 5.3 Typical (**a**) indwelling electrodes and (**b**) surface EMG electrodes

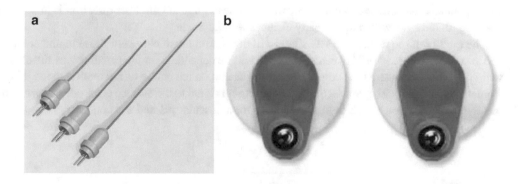

Surface electrodes consist of disks of metal of around 1 cm in diameter, most commonly made of silver/silver chloride (Fig. 5.3b). Surface preparation is crucial for reducing electrical impedance at the site of electrode placement. A typical protocol is to remove dead skin via abrasions (most often by shaving the site) and the removal of oils using rubbing alcohol. The electrodes are then attached using electrode gel and or some double-sided adhesive tape. The electrode gel aids in the pickup of the electrical signal. It is important to try and ensure that the electrode electrolyte junctions are equivalent for the pair of electrodes.

With the surface electrodes a crucial aspect is the distance between the electrodes; this must ideally be consistent between muscles and across subjects. The differential signal from the pair of electrodes acts like a band-pass filter. A larger interelectrode distance will lower the bandwidth and increase the chance of picking up signals from adjacent muscles (cross-talk). The typical configuration used in bipolar electrodes is an interelectrode distance of about 1 cm. Bipolar surface electrodes can be found fused into the same case meaning; the casing is fixed to the skin and therefore the electrodes. Such configurations control the interelectrode distance. Another option is rather than to have a pair of electrodes is to have an array of electrodes which span most of the length of the muscle (Merletti et al. 2001). Such an approach has the advantage that an electrode position is not crucial (see next section), and that given the array additional information is more readily obtained, for example, muscle electrical signal conduction velocity. They come with two principle drawbacks: that more sophisticated signal processing is required and that the array of electrodes can be a greater encumbrance to the subject than a pair of electrodes.

In many experimental settings indwelling electrodes are not practical but are more likely to provide information on motor unit action potential and aspects of motor unit firing rate and recruitment than surface electrodes. The surface electrodes still provide information about muscle activity including timing of that activity and its relative intensity for different phases of the movement. Some muscles may not be accessible to recording with surface electrodes in which case indwelling electrodes might be the only option.

Standard laboratory practice with electrodes includes ensuring all materials implanted are sterile, and while some surface electrodes are disposable if the electrodes are reusable their sterility should be ensured.

5.2.2 Electrode Placement

The placement of electrodes requires a good knowledge of the relevant muscular anatomy; for guidance with indwelling electrodes consult Perotto and Delagi (2011), and with surface electrodes consult Cram et al. (1998). With indwelling electrodes, the placement of the electrodes, in particular for deep lying muscles, can be guided using ultrasound imaging (Bianchi and Martinoli 2007). It is also important to be aware that there is no standard anatomy, for example, as many as 20% of people have a third head to their biceps brachii (Asvat et al. 1993), and the plantaris is absent in over 10% of a population (Freeman et al. 2008). The innervation zone is the location where the nerve and muscle fiber are connected. The EMG signal is not good close to the muscle-tendon junction or near the innervation zone (Gallina et al. 2013), so these regions should be avoided (see Barbero et al. 2012 for a guide).

The location of electrodes makes a difference to the quality of the recorded signal; therefore if subjects are to be measured on multiple days some type of semipermanent marking of electrode locations should be used (e.g., henna tattoos). The general recommended location of the electrodes should be in the midline of the muscle belly aligned with muscle fibers (Hermens et al. 2000). The midline positioning should minimize the influence of cross-talk. Of course some muscle have multiple fiber orientations (Infantolino and Challis 2014), and these fiber orientations change with fiber shortening (Chleboun et al. 2001), so for pennated muscles the maintenance of electrode alignment with fiber direction is impossible to maintain with the muscle fiber length changes which accompany changes in joint angle.

The quality of the recorded EMG signal will depend on the thickness of the skin and adipose tissue beneath the electrodes and so will therefore vary between subjects. It will be dependent on the proximity of the electrodes to the active motor units within the muscle, where the active motor units depend on the nature of the motion at the joints the muscle crosses (Ter Haar Romeny et al. 1982). The conduction of the electrical signal to the surface electrodes depends on the local blood flow and the metabolite production which in turn influences intramuscular pH and ion concentrations.

5.2.3 Signal Amplification

At some point close to the electrodes the signal is amplified. This signal amplification makes the signal larger so that any noise subsequently corrupting the signal will be small compared with the amplified signal. The amplification of the signal is typically of the order of 200 (gain); the hardware required to amplify the signal increases in size with amplification magnitude so gains greater than 200 are harder to achieve. Typically before the EMG signal is sampled, the signal is once again amplified to produce an appropriate voltage range for the analog to digital converter, as a consequence of this second amplification the first amplification is often referred to as preamplification.

With a bipolar electrode configuration, there is a differential amplifier, where the difference between the signals from the two electrodes is amplified. The act of the taking the difference attempts to eliminate common mode noise voltages. The ability of the electronics to remove common noise is quantified using the common mode rejection ratio (*CMRR*), which is the ratio between differential mode gain (A_D) and common mode gain (A_C). Therefore the *CMRR* is more commonly expressed in decibels by taking the log of the ratios:

$$\text{CMRR (dB)} = 20 \, \log_{10}\left(\frac{A_D}{A_C}\right) \tag{5.1}$$

Typical CMRR $\geq$ 100 dB for an EMG system, so an amplification of $\geq$100,000. For example, if the surface EMG signal has a magnitude of 3 mV but with (common) noise of magnitude of 200 mV and with a CMMR = 100 dB and an amplifier gain of 200, what is the change in signal to noise ratio? The output of the differential amplifier of the EMG signal would be:

$$\text{EMG}_{\text{Out}} = 200 \times 3 \text{ mV} = 600 \text{ mV} \tag{5.2}$$

The output of the differential amplifier of the noise will be:

$$\text{Noise}_{\text{Out}} = \frac{200 \times 200 \text{ mV}}{100,000} = 0.4 \text{ mV} \tag{5.3}$$

So the original signal to noise ratio was 0.015 but after differential amplification 1500.

5.2.4 Signal Sampling

The sample rate for an EMG signal is dictated by the sampling theorem which requires us to know the highest frequency present in the sampled signal (see Sect. 3.6). There are various estimates, for example, Winter (2009) reported the ranges of signal bandwidths to be 5–600 Hz for indwelling EMG electrodes and 5–500 Hz for surface EMG electrodes. Komi and Tesch (1979) report a highest frequency for surface EMG signals of 300 Hz, while De Luca et al. (2010) reported a bandwidth of 400 Hz.

5.3 Processing the EMG Signal

Once collected the raw EMG signal is sampled; it is processed to aid in the interpretation of the signal. The typical sequence is:

- High-pass filter
- Signal rectification
- Low-pass filter
- Signal normalization

In the following sub-sections, each of these steps will be outlined, with an example data set processed. The general sequence is illustrated in the Pseudo-Code 5.1.

Pseudo-Code 5.1: Basic processing of raw EMG signal

Purpose – To illustrate the computational process for raw EMG data	
Inputs	
EMG – Raw EMG data	
t – Time base, with n elements	
Output	
N_EMG – Normalized EMG signal after high-pass filtering, rectification, and low-pass filtering	
dt ← t(2) − t(1)	Sample interval
HP_EMG ← hpfilter(EMG, dt, 20)	20 Hz High-pass filter
R_EMG ← abs(HP_EMG)	Signal rectification (absolute value)
LP_EMG ← lpfilter(R_EMG, dt, 5)	5 Hz Low-pass filter
N_EMG ← LP_EMG/max(LP_EMG)	Signal normalization

5.3.1 High-Pass Filter

The EMG signal is broadband with frequency components up to at least 400 Hz. At the low-end of the frequency spectrum the noise magnitude is high relative to the signal magnitude. The major sources of this noise are cable motion and movement artifact which arises because the muscle under the electrodes moves or the skin to which the electrodes are affixed moves relative to the underlying muscles. To remove these low-frequency signal components the first processing step is to high-pass filter the raw EMG data (see Fig. 5.4). De Luca et al. (2010) examined a variety of filter cut-offs for the high-pass filtering of EMG signals and found the most appropriate was 20 Hz.

5.3.2 Rectification

The EMG signal is typically a sampling of multiple motor unit action potentials, which for surface EMG signals have been filtered through muscle and adipose tissue. As a consequence of the source of the EMG signal the averaging of the EMG signal in the time domain will yield a value close to zero. Therefore prior to further processing the signal is rectified. Older publications used half-wave rectification where all negative values were discarded, but now the full-wave rectification is the norm where all negative values are made positive (see Fig. 5.5). The previous processing of the EMG signal has removed some of the low-frequency components of the signal (high-pass filter), but the process of rectification reintroduces some low-frequency components into the signal.

The rectification of the EMG signal in most programming languages is straightforward as the making of negative values positive is easily achieved using the absolute function; typically the function has the syntax abs().

Fig. 5.4 The processing of an EMG sequence, (**a**) raw EMG signal, and (**b**) the high-pass filtered EMG signal (20 Hz cut-off)

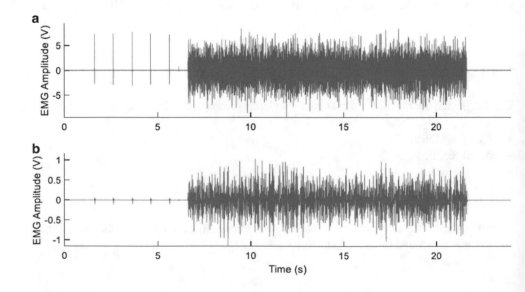

Fig. 5.5 The processing of an EMG sequence, (**a**) high-pass filtered EMG signal (20 Hz cut-off), and (**b**) the full-wave rectified signal

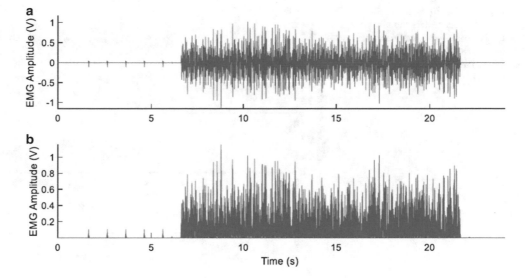

Fig. 5.6 The processing of an EMG sequence, (**a**) full-wave rectified high-pass filtered EMG signal (20 Hz cut-off), and (**b**) the 5 Hz low-pass filtered signal

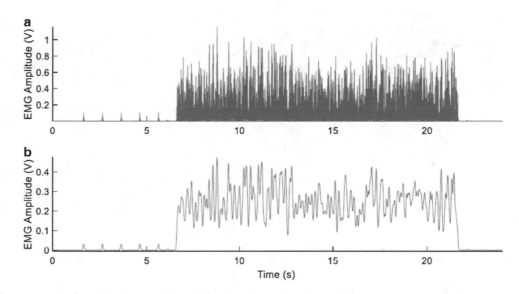

5.3.3   Low-Pass Filtering

The next stage is the elimination of the high-frequency components of the EMG signal by low-pass filtering. This stage is sometimes referred to as creating a linear envelope around the data. Figure 5.6 shows the effect of a 5 Hz low-pass filter on the processed EMG signal.

In the literature, a variety of filter cut-offs have been used for the low-pass filtering of the EMG signal. Figure 5.7 illustrates the influence of different cut-off frequencies. For example, De Luca (1997) proposed a cut-off frequency of 20 Hz, but others have used much lower frequencies cut-off from 3 to 10 Hz (e.g., Buchanan et al. 2004).

5.3.4   Normalization

The EMG signal reflects some aspects of the electrical activity of the muscle of interest. But there are many variables which influence the EMG signal, so to relate the signal to the relative intensity of the activity of the muscles, the EMG data is often normalized. There are two ways in which this is performed. The first is to normalize all of the EMG signals with respect to the highest voltage in the sampled signal. The second approach is to normalize all of the EMG signals with respect to the highest voltage in another signal. This second signal is typically the EMG signal sampled during a maximum isometric muscle action.

Fig. 5.7 The influence of the cut-off frequency for the low-pass filtering of a full-wave rectified high-pass filtered EMG signal, (**a**) 20 Hz cut-off, (**b**) 10 Hz cut-off, (**c**) 5 Hz cut-off, and (**d**) 2 Hz cut-off

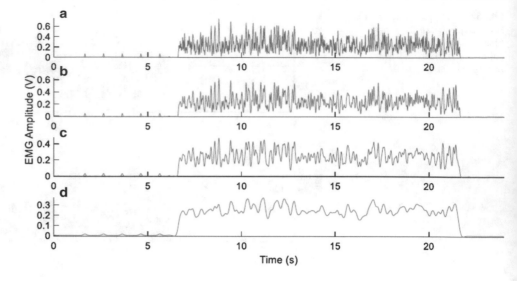

Fig. 5.8 The processing of an EMG sequence, (**a**) 5 Hz low-pass filtered full-wave rectified high-pass filtered EMG signal (20 Hz cut-off), and (**b**) the signal normalized with respect to the maximum value in the signal

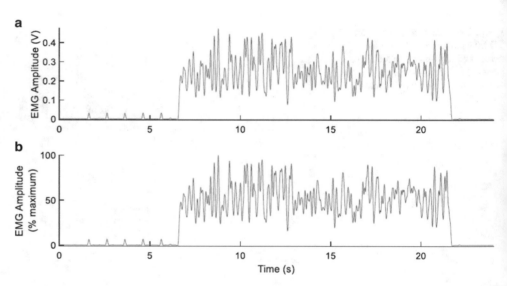

The second approach is more attractive, and it is useful for comparing the relative muscle activity across muscles of the same subject, unfortunately due to the problems with measuring EMG, the signals normalized in this way can have relative activity values greater than 1 or 100% depending on how the normalization is expressed. For convenience and to reduce a subject's time commitment the first method of normalization is commonly used (Fig. 5.8).

The surface EMG signal is typically a sampling of multiple motor unit action potentials, but in sampling motor unit action potentials from different motor units these can be out of phase in which case as they are averaged there can be some signal cancellation. Keenan et al. (2005) demonstrated that if EMG activity is being related to muscle activity that the relationship is poor if the data are not first normalized. Therefore normalization provides a better approximation to muscle activity than the non-normalized signal as the normalization helps account for reduction in EMG signal amplitude as a consequence of motor unit action potential cancellation.

5.4 Analyzing the EMG Signal

Once the sampled EMG signal has been processed as described in the previous section, the signal can be analyzed to provide insight into the relationship of muscle activity to human movements.

5.4.1   Timing of Muscle Activity

In many applications it useful to know when the muscle became active and then ceased to be active. In these applications the problem becomes identifying when the muscle became active and ceased to be active in an EMG signal. Given the noisy nature of the EMG signal, it is not simply a case of identifying when the EMG signal is first, or last, greater than zero. Munro and Steele (2000) used a signal threshold and time duration to identify the start and end of muscle activity. They used 7% of the maximum EMG signal amplitude as the threshold and identified the start of muscle activity when this threshold had been exceeded for 14 consecutive samples, and at the end of the activity when the EMG signal had been below this threshold for 14 consecutive samples (their data was sampled at 1000 Hz). In a similar fashion Hodges and Bui (1996) used an EMG threshold magnitude and time duration as their criteria, but here the EMG magnitude was multiples of the standard deviation of the baseline value (noise). Bonato et al. (1998) presented a more sophisticated threshold identification algorithm for EMG signals, but to date there is no generally accepted method.

A useful way of comparing two time series is to compute the cross-correlation of the two signals (see Sect. 3.7). Such an approach can help indicate the relative coordination between the timings of different muscles (Wren et al. 2006).

When evaluating muscle activity timing, as indicated by EMG, with movement kinematics and kinetics, it is important to consider electromechanical delay. This delay is the time between an electrical signal being registered and a muscle force being generated (see Fig. 5.9). Similarly there is a delay between the cessation of electrical activity and cessation of muscular force. These time delays are important when attempting, for example, to correlate EMG activity with the timing of movements. While some have queried the existence of electromechanical delay (e.g., Corcos et al. 1992), others have measured this time delay (e.g., Komi and Cavanagh 1977; Vos et al. 1991; Winter and Brookes 1991). The nature of the electromechanical delay appears to vary between muscles, with values reported as low as ~40 ms (Winter and Brookes 1991) and as great as ~120 ms (Vos et al. 1991). The duration of the electromechanical delay depends on two factors: the propagation of the action potential and the excitation-contraction coupling processes, and the properties and current state of the tendon experiencing and therefore transmitting force. Nordez et al. (2009) examined aspects of this timing by electrically stimulating the medial gastrocnemius and using high-speed ultrasound imaging to measure tendon motion (indicative of an applied force); they identified time delays associated with the mechanical events of around 22 ms.

5.4.2   EMG and Muscle Force

The EMG signal is a sampling of the signal sent to the muscle to produce muscle forces; therefore research has attempted to find and exploit the relationship between the EMG signal and muscle force. There have been a number of studies which have examined the relationship between the EMG signal and isometric muscle force production. These studies have typically used an integrated processed EMG signal (*IEMG*):

$$\text{IEMG} = \int_{t}^{t+\Delta t} \text{EMG}(t)\, dt \tag{5.4}$$

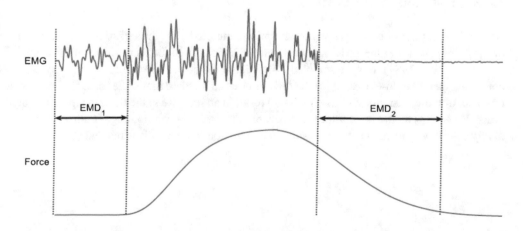

Fig. 5.9 The simultaneous time course of the EMG signal and the resulting muscle force, with the electromechanical display indicated. EMD_1 is the time delay between the EMG signal and the generation of muscle force, and EMD_2 is the delay between the end of the EMG signal and the cessation of muscle force

Fig. 5.10 The integration of an EMG signal, (**a**) 5 Hz low-pass filtered full-wave rectified high-pass filtered EMG signal (20 Hz cut-off), (**b**) the integrated signal with integration interval is the duration of EMG activity, and (**c**) the integrated signal with integration interval of 80 ms

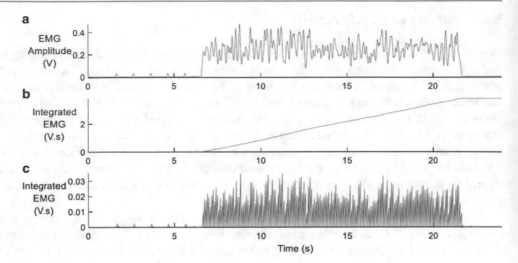

Where the IEMG is taken over the interval of Δt and EMG(t) is the value of the previously processed EMG signal at time t. The question arises as to the time interval to use. If the whole of the time period of data collection is used the IEMG signal simply increases with increasing time. The more common approach is to integrate over a small time window (e.g., 80 ms), throughout the data set (see Fig. 5.10). Given an IEMG signal, from surface electrodes for an isometric muscle action with increasing force production, Lippold (1952) reported a linear relationship between force and IEMG signal, while Vredenbregt and Rau (1973) reported a non-linear relationship. Given the many variables at play when collecting, sampling, and processing EMG signals, the lack of uniformity in establishing a relationship between EMG signals and muscle force is not surprising.

Using surface EMG electrodes, Lawrence and De Luca (1983) showed that the root mean square value of the EMG signal was approximately linearly related to isometric muscle force. The root mean square (RMS) value of the EMG signal was computed on EMG data which had been band-passed filtered but had no other processing. The RMS was computed from:

$$\text{RMS} = \left(\frac{1}{\Delta t} \int_{t}^{t+\Delta t} \text{EMG}^2(t) \, dt \right)^{1/2} \tag{5.5}$$

While a number of studies have related EMG signals to isometric muscle force, these are normally under constrained experimental conditions. Another approach has been to use a processed EMG signal as a measure of muscle activation and then use this to help drive a muscle model. In these models the EMG signal follows the typical sequence of EMG signal processing with the low-pass filtering using a relatively low cut-off frequency. For example, Li et al. (1999) used a 5 Hz cut-off for the low-pass filter and Buchanan et al. (2005) 8 Hz. This modeling approach requires models of the muscles involved in the activity and therefore the associated model parameters to be able to estimate muscle forces.

5.4.3 EMG and Motor Units

As the different motor unit types which comprise a muscle have different stimulation frequencies, the presence, or otherwise, of these frequencies in the EMG signal should be indicative of the recruitment of these units for the production of muscular action (De Luca 1984). For the sampled EMG signal the power spectrum is computed, from which the median frequency is computed (see Pseudo-Code 5.2). If the task requires an increase in muscle force this may be accompanied by the recruitment of some, or more, fast motor units which will result in an increase in the median signal frequency. In contrast during fatiguing muscular actions as the fast motor units become fatigued and can no longer contribute they will drop out leaving predominantly the slow units active which results in a reduction in the median frequency.

Pseudo-Code 5.2: Computation of median EMG signal power

Purpose – To illustrate the computation of the median EMG signal power	
Inputs	
Freq – Frequencies from the Fourier analysis of a time series of EMG data	
Power – Power at each frequency for Fourier analysis of time series of EMG data	
Output	
medianPwr – Frequency at which median power value occurs	
cumPwr ← cumtrapz(Freq, Power)	Compute area under frequency-power curve
totPwr ← cumPwr(end)	Compute total power
All50Pwr ← find(cumPwr >= 0.5 *totPwr)	Find 50% of total power
medianPwr ← Freq(All50Pwr(1))	Find frequency when 50% total power occurs

5.5 Conclusions

Compared to many other experimental methods used in biomechanics there are many decisions the experimenter makes in collecting and processing EMG data which strongly influence the final signal to be analyzed. There are useful guidelines for the collection and processing of EMG data, but the experimenter is still left with key decisions; it is perhaps these issues which instigated the following statement:

Electromyography is a seductive muse because it provides easy access to physiological processes that cause the muscle to generate force, produce movement, and accomplish the countless functions that allow us to interact with the world around us... To its detriment, electromyography is too easy to use and consequently too easy to abuse. (De Luca 1997)

The experimenter using EMG must be able to rationally justify each of their decisions in the EMG data collection and processing sequence.

5.6 Review Questions

1. List the measures to consider when collecting data using an EMG system.
2. Describe a sequence for the processing of a raw EMG signal.
3. Explain what useful information can be obtained from an EMG signal.
4. What is the theoretical relationship between EMG activity and muscle force? What is obtained in actuality?
5. What is electromechanical delay? What implications does this have for the analysis of human movement?
6. How does the EMG signal change with fatigue? Why is this?

References

Cited References

Asvat, R., Candler, P., & Sarmiento, E. E. (1993). High incidence of the third head of biceps brachii in South African populations. *Journal of Anatomy, 182,* 101–104.

Barbero, M., Merletti, R., & Rainoldi, A. (2012). *Atlas of muscle innervation zones: Understanding surface electromyography and its applications.* New York: Milan/Springer.

Bianchi, S., & Martinoli, C. (2007). *Ultrasound of the musculoskeletal system.* New York: Springer.

Bodine, S. C., Garfinkel, A., Roy, R. R., & Edgerton, V. R. (1988). Spatial distribution of motor unit fibers in the cat soleus and tibialis anterior muscles: Local interactions. *Journal of Neuroscience, 8*(6), 2142–2152.

Bonato, P., Alessio, T. D., & Knaflitz, M. (1998). A statistical method for the measurement of muscle activation intervals from surface myoelectric signal during gait. *IEEE Transactions on Biomedical Engineering, 45*(3), 287–299.

Buchanan, T. S., Lloyd, D. G., Manal, K., & Besier, T. F. (2004). Neuromusculoskeletal modeling: Estimation of muscle forces and joint moments and movements from measurements of neural command. *Journal of Applied Biomechanics, 20*(4), 367–395.

Buchanan, T. S., Lloyd, D. G., Manal, K., & Besier, T. F. (2005). Estimation of muscle forces and joint moments using a forward-inverse dynamics model. *Medicine and Science in Sports and Exercise, 37*(11), 1911–1916.

Burke, R. E., Levine, D. N., Tsaris, P., & Zajac, F. E. (1973). Physiological types and histochemical profiles in motor units of cat gastrocnemius. *Journal of Physiology, 234*, 723–748.

Chleboun, G. S., France, A. R., Crill, M. T., Braddock, H. K., & Howell, J. N. (2001). In vivo measurement of fascicle length and pennation angle of the human biceps femoris muscle. *Cells Tissues Organs in vivo, in vitro, 169*(4), 401–409.

Corcos, D. M., Gottlieb, G. L., Latash, M. L., Almeida, G. L., & Agarwal, G. C. (1992). Electromechanical delay: An experimental artifact. *Journal of Electromyography and Kinesiology, 2*(2), 59–68.

Cram, J. R., Kasman, G. R., & Holtz, J. (1998). *Introduction to surface electromyography*. Gaithersburg: Aspen.

De Luca, C. J. (1997). The use of surface electromyography in biomechanics. *Journal of Applied Biomechanics, 13*(2), 135–163.

De Luca, C. J., Donald Gilmore, L., Kuznetsov, M., & Roy, S. H. (2010). Filtering the surface EMG signal: Movement artifact and baseline noise contamination. *Journal of Biomechanics, 43*(8), 1573–1579.

Du Bois-Reymond, E. (1849). *Untersuchungen über Thierische Elektricität*. Berlin: Georg Reimer.

Feinstein, B., Lindegard, B., Nyman, E., & Wohlfart, G. (1955). Morphologic studies of motor units in normal human muscles. *Acta Anatomica, 23* (2), 127–142.

Freeman, A. J., Jacobson, N. A., & Fogg, Q. A. (2008). Anatomical variations of the plantaris muscle and a potential role in patellofemoral pain syndrome. *Clinical Anatomy, 21*(2), 178–181.

Gallina, A., Merletti, R., & Gazzoni, M. (2013). Innervation zone of the vastus medialis muscle: Position and effect on surface EMG variables. *Physiological Measurement, 34*(11), 1411.

Galvani, L. (1791). *De Viribus Electricitatis in Motu Musculari Commentarius*. Bologna: Accademia delle Scienze.

Henneman, E., Somjen, G., & Carpenter, D. O. (1965). Functional significance of cell size in spinal motoneurons. *Journal of Neurophysiology, 28* (3), 560–580.

Hermens, H. J., Freriks, B., Disselhorst-Klug, C., & Rau, G. (2000). Development of recommendations for SEMG sensors and sensor placement procedures. *Journal of Electromyography and Kinesiology, 10*(5), 361–374.

Hodges, P. W., & Bui, B. H. (1996). A comparison of computer-based methods for the determination of onset of muscle contraction using electromyography. *Electroencephalography and Clinical Neurophysiology/Electromyography and Motor Control, 101*(6), 511–519.

Infantolino, B. W., & Challis, J. H. (2014). Pennation angle variability in human muscle. *Journal of Applied Biomechanics, 30*(5), 663–667.

Keenan, K. G., Farina, D., Maluf, K. S., Merletti, R., & Enoka, R. M. (2005). Influence of amplitude cancellation on the simulated surface electromyogram. *Journal of Applied Physiology, 98*(1), 120–131.

Kernell, D. (2006). *The motoneurone and its muscle fibres*. Oxford: Oxford University Press.

Komi, P. V., & Cavanagh, P. R. (1977). Electromechanical delay in human skeletal muscle. *Medicine and Science in Sports and Exercise, 9*(1), 49.

Komi, P. V., & Tesch, P. (1979). EMG frequency spectrum, muscle structure, and fatigue during dynamic contractions in man. *European Journal of Applied Physiology, 42*(1), 41–50.

Lawrence, J. H., & De Luca, C. J. (1983). Myoelectric signal versus force relationship in different human muscles. *Journal of Applied Physiology, 54*(6), 1653–1659.

Li, M. M., Herzog, W., & Savelberg, H. H. C. M. (1999). Dynamic muscle force predictions from EMG: An artificial neural network approach. *Journal of Electromyography and Kinesiology, 9*(6), 391–400.

Lippold, O. C. J. (1952). The relation between integrated action potentials in a human muscle and its isometric tension. *Journal of Physiology, 117* (4), 492–499.

Merletti, R., Rainoldi, A., & Farina, D. (2001). Surface electromyography for noninvasive characterization of muscle. *Exercise and Sport Sciences Reviews, 29*(1), 20–25.

Milner-Brown, H. S., Stein, R. B., & Yemm, R. (1973). Changes in firing rate of human motor units during linearly changing voluntary contractions. *Journal of Physiology, 230*(2), 371–390.

Munro, B. J., & Steele, J. R. (2000). Does using an ejector chair affect muscle activation patterns in rheumatoid arthritic patients? A preliminary investigation. *Journal of Electromyography and Kinesiology, 10*(1), 25–32.

Nordez, A., Gallot, T., Catheline, S., Guével, A., Cornu, C., & Hug, F. (2009). Electromechanical delay revisited using very high frame rate ultrasound. *Journal of Applied Physiology, 106*(6), 1970–1975.

Perotto, A., & Delagi, E. F. (2011). *Anatomical guide for the electromyographer: The limbs and trunk* (5th ed.). Springfield: Charles C. Thomas.

Ter Haar Romeny, B. M., Denier van Der Gon, J. J., & Gielen, C. C. A. M. (1982). Changes in recruitment order of motor units in the human biceps muscle. *Experimental Neurology, 78*(2), 360–368.

Vos, E. J., Harlaar, J., & Van Ingen Schenau, G. J. (1991). Electromechanical delay during knee extensor contractions. *Medicine and Science in Sports and Exercise, 23*(10), 1187–1193.

Vredenbregt, J., & Rau, G. (1973). Surface electromyography in relation to force, muscle length and endurance. In J. E. Desmedt (Ed.), *New developments in electromyography and clinical neurophsiology* (Vol. 1, pp. 607–621). Basel: Karger.

Winter, D. A. (2009). *Biomechanics and motor control of human movement*. Chichester: Wiley.

Winter, E. M., & Brookes, F. B. C. (1991). Electromechanical response times and muscle elasticity in men and women. *European Journal of Applied Physiology, 63*(2), 124–128.

Wren, T. A. L., Patrick Do, K., Rethlefsen, S. A., & Healy, B. (2006). Cross-correlation as a method for comparing dynamic electromyography signals during gait. *Journal of Biomechanics, 39*(14), 2714–2718.

Useful References

Basmajian, J. V., & de Luca, C. J. (1985). *Muscles alive: Their functions revealed by electromyography*. Baltimore: Williams & Wilkins.

Cram, J. R., Kasman, G. R., & Holtz, J. (1998b). *Introduction to surface electromyography*. Gaithersburg: Aspen.

Loeb, G. E., & Gans, C. (1986). *Electromyography for experimentalists*. Chicago: University of Chicago Press.

http://www.seniam.org/. Provides useful guidelines for surface EMG, in particular electrode placement.

Motion Analysis

6

Overview

On October 19, 1878, Eadweard Muybridge's images of a galloping horse, Abe Edgington, dominated the cover of *Scientific American*. The ability to capture the features of animal locomotion was considered a significant technological achievement and immediately appreciated for the important information it provided for scientists and artists. Today, we have a variety of methods for making recordings of animal motion at high sample rates. The broad area of study is referred to as motion analysis and is simply the measurement of the motion of a body. Typical methods include goniometers, accelerometers, and image-based methods. Image-based motion analysis is the measurement of body position, and consequently its motion, based on recording images of the movement and making measurements from the images. Historically, as was the case with the horse Abe Edgington, images were recorded on film, but other technology has replaced film as the preferred medium for capturing images.

Edward Muggeridge (1830–1904) is considered one of the forefathers of modern biomechanics. He changed his name twice, first to Eadweard Muygridge and then second to Eadweard Muybridge. His contribution from a biomechanical perspective was in the area of motion analysis. He did not take measurements, but was one of the first to capture motion and did so for many species and movements. In 1884, Muybridge started his work at the University of Pennsylvania attempting to photographically catalog human and animal movements. These series of images are particularly noteworthy for the range of activities examined (e.g., walking, running, ascending stairs, swinging a pick axe, ironing, and climbing a ladder). His photographic sequences typically consisted of images taken from two views, with the subjects most often naked. Muybridge (1907) wrote:

> The figures illustrating the various movements are reproduced from the original negatives by the photo-gelatine process of printing, without any attempt having been made to improve their pictorial effect, either in outline or detail; or to conceal their imperfections.

Study briefly the following sequence of a woman "Dropping and lifting a handkerchief" (see Fig. 6.1). Note the position of the handkerchief in the third image of the top row, and compare it with its rearview counterpart in the bottom row. In the bottom row, the third image is actually a repeat of the first image. Today, Muybridge might be accused of data massage. There are a number of incidences of such anomalies in his work. These errors stated, the technological innovations of the work of Muybridge were important.

Image-based motion analysis techniques will be presented followed by other methods for measuring aspects of human movement (goniometers, accelerometers, and inertial measurement units). The content of this chapter requires a working knowledge of matrix algebra (see Appendix A).

6.1 Image-Based Motion Analysis

For image-based motion analysis, the basic structure is that a camera records images of an object of interest. Data is collected from these images, typically the locations of key body landmarks or locations of markers placed on a body of interest. The image coordinates of the key body landmarks are then transformed into real-world (object space) coordinates using methods from photogrammetry; this process is referred to as reconstruction. The next section explains the important aspects of

J. H. Challis, *Experimental Methods in Biomechanics*, https://doi.org/10.1007/978-3-030-52256-8_6

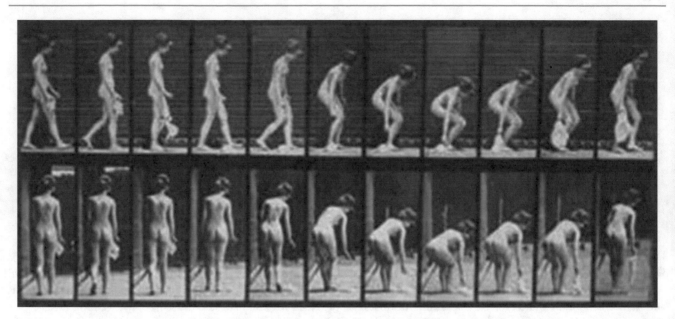

Fig. 6.1 Images of a woman "Dropping and lifting a handkerchief" from the work of Muybridge

photogrammetry for biomechanical analyses, while the following sub-sections outline options for recording the images and the locations of key landmarks on the image.

Most image-based motion analysis systems rely on a camera of some kind to record images. Cameras are designed so that when a shutter is opened, light is focused on a sensing unit by a lens. When the shutter is closed, no light reaches the sensing unit. The sensing unit is either a film or a charge-coupled device which records the pattern of light that falls onto them. Once the image has been registered the process of digitizing occurs, which in this context is the process of obtaining image coordinates of points in the image, typically key body landmarks. In the following sub-sections, methods for recording images and digitizing those images are described.

6.1.1 Film

The original medium for recording images was film, as used by Muybridge in the nineteenth century. As a method of recording images it was used in biomechanics almost exclusively until it was superseded by video-based technology. Much of the terminology used for recording images on film are used with more recent methods for recording and analyzing images for biomechanical purposes.

In biomechanics film has been used to record a sequence of images, frames, of the task at hand. Film is a plastic medium which on one side is a coating of a gelatin emulsion. The emulsion contains small silver halide crystals. When exposed to light the crystals react, with the amount of reaction proportional to the intensity and wavelength of the light. Each frame of a strip of film is moved through the camera and sequentially exposed to the light at different instants in time. Once the film has been exposed, images only become visible after chemical processing (development). Unlike contemporary methods for recording images the images on film are not immediately available to the user.

Film comes in different sizes, and varies in the distribution of the silver halide crystals that influence the resolution and sensitivity of the film. Movies typically use 35 mm film, but biomechanics has preferentially used 16 mm film (see Fig. 6.2). The oft-used 16 mm film has 40 frames per foot, meaning for 2 seconds trials sampled at 100 frames per second for 8 subjects would use 40 feet (12.192 m) of film.

The cameras have variable frame rates, effectively giving the user a choice of sample rate, with typical cameras having sample rates up to 500 frames per second but rates up to 5000 Hz are available. Users of film have a choice in the sensitivity of the film to use and can adjust the aperture on the camera to alter the amount of light reaching the film. For high-speed motions to ensure sufficient light is reaching the film the experimenter often illuminates the activity area with lights. These choices can make the recording of an activity challenging, in particular because the actual images cannot be viewed until the film had been processed.

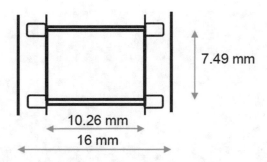

Fig. 6.2 The dimensions of a single frame of 16 mm film

Once the film has been processed the locations of key body landmarks have to be measured on the image. The film is projected onto a surface on which the points can be measured, often a digitizing tablet and then the locations of landmarks are measured. The resolution of the digitizer influences the accuracy of reconstructed landmarks. With film the user can place markers on the subject but these are not required. The process of digitizing can be time-consuming, for example, imagine measuring the locations of 20 landmarks on 100 frames of film for 3 trials per subject for 8 subjects. If a landmark is obscured in the film sequence for a period of time then the operator performing the digitizing can estimate its location using their knowledge of anatomy and the nature of the task.

6.1.2   Video-Based

Video technology relies on the use of charge-coupled devices (CCD) to record images, which were first developed by Boyle and Smith in 1970. In a CCD an image is projected through a lens onto an array of capacitors, and the light causes each capacitor to accumulate an electric charge that is proportional to the nature of the light falling on that capacitor. The color signal from each capacitor comprising the CCD represents one pixel in the complete set of pixels which represent the recorded image. The information from the CCD can be recorded; originally this was on videotape, but now digital formats are used (e.g., avi, mpeg). Video cameras vary in the number of sensors, capacitors, and CCDs, for example, 1125 pixels vertically by 2436 pixels horizontally in this case a total of 2,740,500 pixels (or 2.7 megapixels).

CCDs have now almost completely superseded film as the medium for recording movement. This shift has arisen for a number of reasons:

- The CCD images are available (almost) immediately.
- The CCDs are more sensitive to light than film.
- The data is stored digitally, so image enhancement and analysis are easily implemented in a digital environment.

Standard video cameras operate at the same frequency as the mains frequency (50 or 60 Hz), but specialized cameras will give higher rates. For example, many smartphones can operate at multiples of a 60 Hz sample rate. The required sample rate depends on the nature of the task being analyzed, but rates greater than 60 Hz are the norm and for high-speed movements are essential. For example, for the analysis of ball-foot interactions in soccer kicking video was sampled at 4000 Hz (Tsaousidis and Zatsiorsky 1996). In other domains even greater rates are used, for example, sample rate of 50 kHz for analyzing the behavior of lasers (Zhao et al. 2017) with even greater rates feasible (in excess of 200 kHz).

Once video data has been collected the locations of key body landmarks have to be measured on the image. This can be achieved by manual digitization of the images, but this can be a time-consuming process, although once a few frames have been processed the digitization software can predict where the marker will be in the next frame and move the cursor appropriately giving the user only minor movements to make to locate the point of interest. With video the user can place markers on the subject but these are not required. If markers are used, then it is feasible that the markers once initially identified can be automatically tracked. This is normally achieved by covering the markers in retroreflective paint or tape which appears as bright white in the images with computer analysis of the images allows each marker to be identified in each frame of the video (e.g., Krishnan et al. 2015). Recently, there has been some success with tracking body segments without the need for markers. For example, using machine learning methods body segments have been successfully tracked in video sequences (e.g., Pishchulin et al. 2016; Mathis et al. 2018).

Fig. 6.3 Elements of a passive marker-based image-based motion analysis system, (**a**) a set of typical markers, (**b**) the marker illuminated, and (**c**) a camera. (Note the ring of array of infrared light-emitting diodes mounted around the camera lens)

The technology of these video systems is also the basis of passive marker automatic digitizing systems. A number of these systems are commercially available, but some have been described in the scientific literature (e.g., Taylor et al. 1982; Furnee 1990), with the earliest commercial system, VICON, available in 1981 (MacLeod et al. 1990). The general principle of these systems is that the image coordinates of markers are measured automatically at the camera, before these data are combined for all cameras to allow reconstruction of marker locations in three-dimensional space. The markers are covered in retroreflective material and so appear very bright in the images; therefore, the scan of an image looks for the bright edges of the spherical markers from which the center of the sphere can be computed (see Fig. 6.3). Spherical markers are used because whatever the viewing angle from different cameras the center of the marker is the same. Marker illumination is enhanced by an array of infrared light-emitting diodes (LEDs), mounted around the camera lens (see Fig. 6.3). The LEDs are strobed to synchronize with camera shutter opening to further enhance marker image intensity.

The automated systems have sample rates that vary with processing power, for both the processor in the camera which locates marker edges and the computer which collates and processes the data from multiple cameras. With increasing processor power, in part due to Moore's Law (Golio 2015), the number of cameras which can be simultaneously used and at what sample rates has been increasing steadily.

6.1.3   Other Image-Based

Another automatic digitization motion analysis system exploits the properties of lateral photodetectors (Woltring 1975). Light falling on the surface, which is a photodetector, generates a photocurrent that can be converted into a position-dependent signal; thus the image coordinates are automatically obtained. These systems rely on infrared light-emitting diodes (LED) to produce light, so LEDs are placed on the body of interest. To measure multiple points in the field of view the LEDs are multiplexed, this temporal sequencing means the LEDs are automatically identified. The spatial resolution of these systems is high, while the temporal resolution varies with the number of markers. The "cameras" based on the lateral photodetectors are stable, compared to CCD-based cameras, so they often come factory calibrated and so can be used without system calibration before every operation. A disadvantage is the use of active markers which means the subjects have to carry a power supply with wires leading to each LED or have bulky markers each with their own power supply. In 1975, SELSPOT (SElective Light SPOT recognition) was the first commercially available system exploiting this technology (Woltring 1977).

Roentgen stereophotogrammetric analysis uses two X-ray sources to produce the images, but there are no lenses, but the perspective centers are the Roentgen sources. The configuration for Roentgen stereophotogrammetry puts the object of interest in between the Roentgen source and the photographic plate effectively within the camera. Even though the geometry of Roentgen stereophotogrammetry is different from traditional cameras, the same camera calibration and object space reconstruction can be used (Selvik et al. 1983). Identification of the positions of the bones can be enhanced by implanting tantalum balls, 0.5 to 1 mm in diameter in the bones (Choo and Oxland 2003). One advantage of using tantalum balls is that useful images can be obtained using a lower radiation exposure than would be required to image the bones without the radiopaque balls.

6.2 Photogrammetry

Photogrammetry is the science of making measurements of objects in the physical world from recordings of their images; specifically, it focuses on the recording, measurement, and interpretation of images. There are various subdivisions of photogrammetry including satellite, aerial, and close-range photogrammetry. In biomechanics, the most common are close-range photogrammetry (e.g., Abdel-Aziz and Karara 1971), and Roentgen stereophotogrammetry (RSA; e.g., Choo and Oxland 2003).

6.2.1 Central Projectivity

The basic principle for a camera is that all points within the field of view are projected along straight lines which all intersect at the perspective center (in theory the middle of the lens) and are projected onto a plane behind the lens which is the image plane (Fig. 6.4). The image plane might be the plane in which the film lies or the plane of the CCD. This is the law of central projectivity.

There is a basic mathematical relationship between the object space coordinates of a point and the image coordinates of that point. It can be expressed in equation form (note for simplicity the subscript referring to a specific point has been dropped from this equation):

$$\begin{bmatrix} X - X_C \\ Y - Y_C \\ Z - Z_C \end{bmatrix} = \lambda R \begin{bmatrix} x - x_p \\ y - y_p \\ 0 - C \end{bmatrix} \tag{6.1}$$

where X, Y, and Z are the object space coordinates of a point; Xc, Yc, and Zc are the coordinates of perspective center (lens); λ is a scale factor; R is a 3 x 3 matrix which describes orientation of camera coordinate system relative to object space coordinate system; x and y are the image coordinates of that point; x_p and y_p are the image coordinates of the principal point;

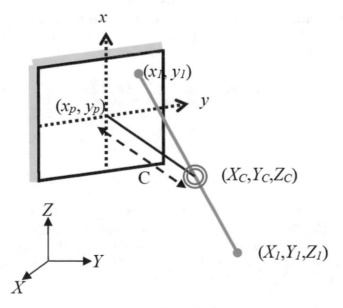

Fig. 6.4 Illustration of the law of central projectivity, showing a point in the object space and in the image plane. Here, X_1, Y_1, and Z_1 are the object space coordinates of a point; Xc, Yc, and Zc are the object space coordinates of perspective center (lens); x_1 and y_1 are the image coordinates of the point; x_p and y_p are the image coordinates of principal point; and C is the distance from the perspective center to the principal point

and C is the distance from perspective center to the principal point. The matrix R while being a nine-element matrix can be described by three parameters (see Chap. 9 for details on deconstructing attitude matrices down to three parameters).

The following are important definitions:

Perspective Center – the common point on the optical axis at which the light rays are considered to intersect (theoretically the center of the lens).

Principal Point – the projection of the perspective center onto the image plane.

Principal Distance – distance measured along optical axis from perspective center to the principal point. This distance is sometimes referred to as the focal length or calibrated focal length.

To be able to model a camera, there are therefore six external camera parameters:

- Camera position (3)
- Camera orientation (3)

and at least three internal camera parameters:

- Principal point (2)
- Principal distance/focal length (1)

If lens distortion is also accounted for, in modeling the cameras, there can also be additional internal parameters (see Sub-sect. 6.2.4 on modeling lens distortion).

The process of determining the camera model coefficients is referred to as calibration, and the process of using the camera model along with the calibration coefficients to determine object space coordinates of points measured on images is referred to as reconstruction. One of the more popular methods for camera calibration and reconstruction is the direct linear transformation (DLT). To illustrate the process of calibration and reconstruction, the DLT will be described in the following sub-section and other methods for calibration and reconstruction outlined in a subsequent sub-section.

6.2.2 The Direct Linear Transform

The interior and exterior camera parameters can be determined explicitly to allow reconstruction or determined implicitly with a model (mathematical) representation of the camera. A common method is the direct linear transformation (DLT, Abdel-Aziz and Karara 1971). This paper is sufficiently important that it was reprinted 40 years after it was first published (Abdel-Aziz et al. 2015). In the DLT, the basic mathematical relationship between the image coordinates and the object space coordinates of a point has been cast in a form that is easily addressed by methods of linear algebra (see Appendix A). The basic forms of the DLT equations for a given camera are:

$$x + \Delta x = \frac{L_1 X + L_2 Y + L_3 Z + L_4}{L_9 X + L_{10} Y + L_{11} Z + 1} \tag{6.2}$$

$$y + \Delta y = \frac{L_5 X + L_6 Y + L_7 Z + L_8}{L_9 X + L_{10} Y + L_{11} Z + 1} \tag{6.3}$$

where Δx and Δy are the errors which corrupt the measured image coordinates and L_1 to L_{11} are the DLT calibration coefficients.

For one camera if we had one point in the field of view whose 3D coordinates we know, then we have 5 knowns (x, y, X, Y, Z) but 11 unknowns (L_1 to L_{11}). Therefore, more calibration points are required; given six, we have sufficient to determine the DLT calibration coefficients. There are advantages for more calibration points (see later Sub-sect. 6.2.6). The following equation represents the rearranged DLT equations which permit estimation of the DLT calibration coefficients, n is the number of calibration points, and the subscripts refer to each point:

$$
\begin{bmatrix}
X_1 & Y_1 & Z_1 & 1 & 0 & 0 & 0 & 0 & -x_1X_1 & -x_1Y_1 & -x_1Z_1 \\
0 & 0 & 0 & 0 & X_1 & Y_1 & Z_1 & 1 & -y_1X_1 & -y_1Y_1 & -y_1Z_1 \\
X_2 & Y_2 & Z_2 & 1 & 0 & 0 & 0 & 0 & -x_2X_2 & -x_2Y_2 & -x_2Z_2 \\
0 & 0 & 0 & 0 & X_2 & Y_2 & Z_2 & 1 & -y_2X_2 & -y_2Y_2 & -y_2Z_2 \\
\vdots & \vdots & \vdots & \vdots & \vdots & \vdots & \vdots & \vdots \\
\vdots & \vdots & \vdots & \vdots & \vdots & \vdots & \vdots & \vdots \\
X_n & Y_n & Z_n & 1 & 0 & 0 & 0 & 0 & -x_nX_n & -x_nY_n & -x_nZ_n \\
0 & 0 & 0 & 0 & X_n & Y_n & Z_n & 1 & -y_nX_n & -y_nY_n & -y_nX_n
\end{bmatrix}
\begin{bmatrix} L_1 \\ L_2 \\ L_3 \\ L_4 \\ L_5 \\ L_6 \\ L_7 \\ L_8 \\ L_9 \\ L_{10} \\ L_{11} \end{bmatrix}
=
\begin{bmatrix} x_1 \\ y_1 \\ x_2 \\ y_2 \\ \vdots \\ \vdots \\ x_n \\ y_n \end{bmatrix}
\tag{6.4}
$$

The matrices and vectors can be represented in summary as:

$$
A L = B \tag{6.5}
$$

where A is a $2n \times 11$ matrix, L is an 11×1 vector, and B is a $2n \times 1$ vector. To compute the vector of calibration coefficients, the Moore-Penrose inverse can be used (see Appendix A) and very efficiently computed using the singular value decomposition (see Appendix H):

$$
L = \left(A^T A \right)^{-1} A^T B \tag{6.6}
$$

Given the calibration coefficients, reconstruction can now be performed; to do this, a landmark must be viewable in at least two cameras simultaneously (see Fig. 6.5). Equation 6.7 is the relevant equation, which presumes the landmark is visible in all nc cameras ($nc \geq 2$):

$$
\begin{bmatrix}
L_{1,1} - L_{1,9}x_1 & L_{1,2} - L_{1,10}x_1 & L_{1,3} - L_{1,11}x_1 \\
L_{1,1} - L_{1,9}y_1 & L_{1,1} - L_{1,9}y_1 & L_{1,3} - L_{1,11}y_1 \\
\vdots & \vdots & \vdots & \vdots & \vdots \\
\vdots & \vdots & \vdots & \vdots & \vdots \\
L_{nc,1} - L_{nc,9}x_{nc} & L_{nc,1} - L_{nc,9}x_{nc} & L_{nc,3} - L_{nc,11}x_{nc} \\
L_{nc,1} - L_{nc,9}y_{nc} & L_{nc,1} - L_{nc,9}y_{nc} & L_{nc,3} - L_{nc,11}y_{nc1}
\end{bmatrix}
\begin{bmatrix} X \\ Y \\ Z \end{bmatrix}
=
\begin{bmatrix} x_1 - L_{1,4} \\ y_1 - L_{1,8} \\ \vdots \\ \vdots \\ x_{nc} - L_{nc,4} \\ x_{nc} - L_{nc,4} \end{bmatrix}
\tag{6.7}
$$

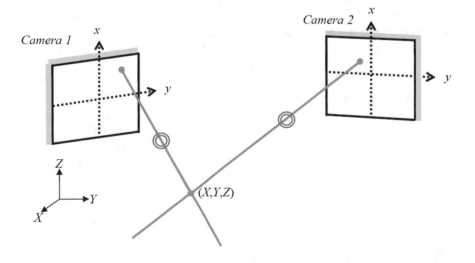

Fig. 6.5 Illustration of a landmark in three-dimensional space, with its image recorded in two cameras simultaneously

The matrices and vectors can be represented in summary as:

$$D E = F \tag{6.8}$$

where D is an $2nc \times 3$ matrix, E is an 3×1 vector, and F is an $2nc \times 1$ vector. To compute the vector of object space coordinates of a point, once again, the Moore-Penrose inverse can be used:

$$E = \left(D^T D\right)^{-1} D^T F \tag{6.9}$$

The calibration coefficients as well as reconstructing image coordinates in three-dimensional space can also be used to determine the external and internal camera parameters. These parameters can be useful to determine if there is an error in the experimental setup, for example, if camera positions are determined but the distance between cameras does relate to measures made during experimental setup. The coordinates of the principal point (x_P, y_P) are computed from:

$$x_P = \frac{L_1 L_9 + L_2 L_{10} + L_3 L_{11}}{L_9^2 + L_{10}^2 + L_{11}^2} \tag{6.10}$$

$$y_P = \frac{L_5 L_9 + L_6 L_{10} + L_7 L_{11}}{L_9^2 + L_{10}^2 + L_{11}^2} \tag{6.11}$$

Then the principal distance (C) is computed from:

$$C_x = \frac{L_1^2 + L_2^2 + L_3^2}{L_9^2 + L_{10}^2 + L_{11}^2 - x_P^2} \tag{6.12}$$

$$C_y = \frac{L_5^2 + L_6^2 + L_7^2}{L_9^2 + L_{10}^2 + L_{11}^2 - y_P^2} \tag{6.13}$$

$$C = \frac{C_x + C_y}{2} \tag{6.14}$$

Camera position (X_C, Y_C, Z_C) can be computed from:

$$\begin{bmatrix} X_C \\ Y_C \\ Z_C \end{bmatrix} = \begin{bmatrix} L_1 & L_2 & L_3 \\ L_5 & L_6 & L_7 \\ L_9 & L_{10} & L_{11} \end{bmatrix}^{-1} \begin{bmatrix} -L_4 \\ -L_8 \\ -1 \end{bmatrix} \tag{6.15}$$

The camera orientation is defined by an attitude matrix from which three orientation angles can be extracted (see Chap. 9 for description of how orientation can be defined). Given a rotation sequence about X, Y, and Z axes, respectively, with corresponding angles of α, β, and γ, these can be determined using:

$$\alpha = \tan^{-1}\left(\frac{-L_{10}}{L_{11}}\right) \tag{6.16}$$

$$\beta = \sin^{-1}\left(\frac{-L_9}{\sqrt{L_9^2 + L_{10}^2 + L_{11}^2}}\right) \tag{6.17}$$

$$\gamma = \cos^{-1}\left(\frac{L_1 - x_P\,L_9}{C\,\cos(\beta)\sqrt{L_9^2 + L_{10}^2 + L_{11}^2}}\right) \qquad (6.18)$$

There are 11 DLT calibration coefficients, but notice there are only 10 camera parameters (Xc, Yc, Zc, α, β, γ, x_P, y_P, C, λ); therefore, there is a redundancy in the DLT coefficients. Bopp and Krauss (1978) demonstrated that this redundancy can be exploited to improve reconstruction accuracy, and it requires an iterative computational process.

There is also a two-dimensional version of the DLT where the base equations are:

$$x + \Delta x = \frac{L_1 X + L_2 Y + L_3}{L_7 X + L_8 Y + 1} \qquad (6.19)$$

$$y + \Delta y = \frac{L_4 X + L_5 Y + L_6}{L_7 X + L_8 Y + 1} \qquad (6.20)$$

A typical approach for two-dimensional analysis has been to use one camera where the camera image plane is assumed to be parallel to the plane of motion. Under these conditions, image plane and the object space plane are parallel, and then the denominator in the equations reduces to 1 ($L_7 = L_8 = 0$). Of course, these equations have the advantage that if a single camera is used, then the image plane does not have to be precisely parallel to the object space plane. These equations also allow information from multiple cameras to be used to measure motion in some predefined plane. The unknown calibration coefficients can be determined by measuring the locations of a planar distribution of control points. At least four control points are required, and these should form a quadrangle.

6.2.3 Other Calibration and Reconstruction Methods

The DLT works well to illustrate the procedure for using photogrammetry to determine the three-dimensional locations of points in the field of view of the cameras. There are other techniques which offer greater flexibility, but which are based on the same principles. Some of these methods will be outlined in the following paragraphs with a focus on those with applications in biomechanics.

To perform the DLT requires calibration points which encompass the volume of interest. For some applications, this necessitates a large calibration object to be manufactured, measured, and stored. To provide more flexibility, Challis (1995) presented a method where cameras could be calibrated using a calibration object placed in multiple positions in the field of view.

Woltring (1980) presented a method, Simultaneous Multiframe Analytical Calibration (SMAC), which obviated the need to have an object containing three-dimensional distribution of control points. Control points were provided by a calibration plane that contained a two-dimensional distribution of points. The calibration plane was moved to various positions throughout the activity space. The advantages of this method are that a cumbersome three-dimensional calibration object does not need to be manufactured and that a planar calibration object is easier to manufacture.

Dapena et al. (1982) presented a method, non-linear transformation (NLT), which required the recording of images from two crosses with known dimensions and a control of object of known length (e.g., a rod with markers on its ends). The internal camera parameters are calculated from the measurements of the crosses and the external camera parameters from the control object of known length. This method once again removes the need for a complex calibration object. Dapena (1985) eventually discovered an error in his equations but found the original incorrect equations produced superior reconstruction accuracy suggesting the incorrect equations had inadvertently corrected for not deliberately modeled aspects of camera-lens system distortion.

Most commercially available systems use wand calibration. With these methods, a three-point, or more, reference is placed in the middle of the volume of interest to give axes directions and the origin of the reference system. Then a wand with a pair of markers at a known distance apart is moved throughout the volume of interest. With commercial systems, the precise details of the algorithm are not typically provided. But in 2001 Borghese et al. (2001) outlined a wand calibration method, principal points indirect estimate (PIE). Effectively by moving the wand throughout the volume of interest a redundant data set of the image coordinates of the pair of markers is obtained, and the camera parameters can then be estimated in a least-squares sense using an optimization algorithm.

It should be mentioned that there are one-camera methods that permit determination of the three-dimensional coordinates of a point (e.g., Miller et al. 1980; Eian and Poppele 2002). With these methods there are additional constraints placed on the reconstruction method. For example, Eian and Poppele (2002) required that pairs of markers were a known distant apart. These methods while attractive, due to the requirement of less equipment, have the drawback that the accuracy of reconstruction is poor for the depth axis (axis perpendicular to the camera plane).

6.2.4   Lens

With many image motion analysis systems, there is a lens in the system. The lens can cause a distortion of the image, which in turn influences the accuracy of the reconstruction of points (Conrady 1919). There are two categories of lens distortion: tangential and radial. Tangential distortion occurs when the lens and image plane are not parallel to one another. Radial distortion is the most common type of lens distortion, it is manifested by the amount of distortion changing with increasing radial distance from the perspective center of the lens, and therefore it is radially symmetric. Radial lens distortions can usually be classified as either barrel distortions or pincushion distortions (see Fig. 6.6). With barrel distortion, the magnification of the image decreases as distance from the center of lens increases, while for pincushion distortion, the magnification of image increases as distance from the center of lens increases.

Correction of lens distortion is often referred to as lens linearization and can improve accuracy of reconstruction, and often determination of the lens model parameters is part of the calibration process. The following equations represent typical corrections for lens distortion (Brown 1966):

$$x' = x + \bar{x}\left(k_1 r^2 + k_2 r^4 + k_3 r^6 + \ldots\right) \tag{6.21}$$

$$+ \left\{c_1\left(r^2 + 2\bar{x}^2\right) + 2c_2\bar{x}\bar{y}\left(r^2 + 2\bar{x}^2\right)\right\}\left(1 + c_3 r^2 + \ldots\right)$$

$$y' = y + \bar{y}\left(k_1 r^2 + k_2 r^4 + k_3 r^6 + \ldots\right) \tag{6.22}$$

$$+ \left\{c_2\left(r^2 + 2\bar{y}^2\right) + 2c_1\bar{x}\bar{y}\left(r^2 + 2\bar{x}^2\right)\right\}\left(1 + c_3 r^2 + \ldots\right)$$

where x' and y' are the corrected image coordinates; x and y are the (distorted) image coordinates; $\bar{x} = x - x_p, \bar{y} = y - y_p, xp$, and yp are the image coordinates of the principal point; $r = \sqrt{x^2 + y^2}$ and k_i are the coefficients associated with the modeling of radial distortion; and c_i are the coefficients associated with the modeling of tangential distortion.

In most analyses, only radial distortion is modeled. The appropriate lens calibration coefficients can be determined by recording an image of a planar grid which fills the image plane.

Fig. 6.6 Illustration of (**a**) undistorted image, (**b**) barrel distortions, and (**c**) pincushion distortion

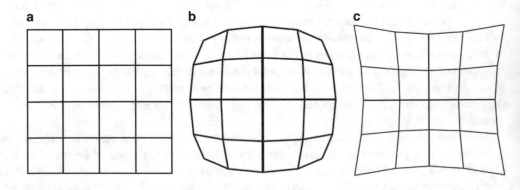

6.2.5 Marker Identification and Tracking

Marker identification, or marker sorting, is the process of identifying markers from one time instant to the next and in the same way from multiple views (cameras). For example, if a marker is placed on the lateral malleolus then in each camera and at every sampled instant that marker must be accurately identified as the marker placed on the lateral malleolus. If there is only one marker used identification is not problematic, but this is rarely the case. With active marker systems marker identification is typically performed in hardware as these systems control the sequencing of the illumination of the markers thus making marker identification just a temporal sequencing problem. With passive marker systems, the images from multiple camera views are collected and then markers identified. Figure 6.7 illustrates the starting point for marker identification, so the task is to match pairs of markers in the two camera views and to do this consistently across frames of data.

From each camera, there is a ray (line) from the image plane through the camera perspective center to each marker. With 2 cameras and just 6 markers, there are 36 possible combination of rays, but they must be accurately matched; more cameras and more markers make the situation more complicated. Ideally pairs of rays intersect at a marker allowing easy pairing (Khan and Shah 2003), but this may not uniquely be the case (see Fig. 6.8). Measurement inaccuracy means the rays are unlikely to perfectly intersect but hopefully do so within a small tolerance. Indeed increased accuracy of motion analysis can reach a level where the accuracy is greater than that required for biomechanical assessment but has the advantage that it permits better marker identification. The use of neural networks for the identification of markers has shown promise (e.g., Holzreiter 2005).

The problem of marker identification is compounded by phantom markers and lost markers. Phantom markers occur when more markers appear in the camera images than were actually placed there by the experimenter and due to objects in the field of view causing a bright reflection. For example, reflective strips placed on some sports shoes might produce a phantom marker. Lost markers occur where the view of a marker by a camera is in some way blocked. If there are multiple cameras losing a marker, this may not be an issue as long as the marker is visible in at least two camera views. Markers may also be lost if one marker covers another, and this creates a problem, because which of the markers is missing, and which is which when the lost marker reappears. If a marker position is tracked this can have an advantage with marker identification as when

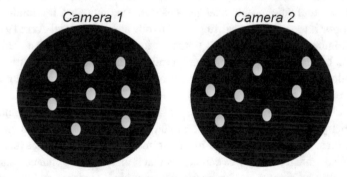

Fig. 6.7 The images of markers in two camera views. Note in most applications there are more than two cameras

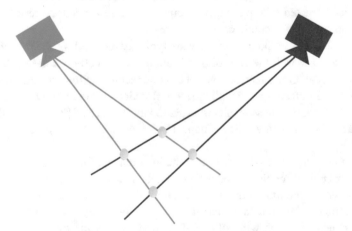

Fig. 6.8 The rays from two cameras intersecting at markers, with the correct pairing of rays not clear

analyzing a time instant tracking can provide a region in which to look for a marker, and of course marker tracking is also useful when a marker cannot be viewed in any of the cameras for a period of time. Companies producing passive marker automated motion analysis systems have propriety software which produces marker identification, exploiting the intersecting rays from markers to cameras. Normally with these systems the user has to identify which marker is which in the first few collected frames of data.

Marker tracking is using information about the path of a marker to predict its future path. Marker tracking has two purposes: it is used as part of marker identification as a prediction of markers' position for a future time instant helps identify which marker is which, and it helps fill in gaps in a markers' path if the image of that marker is lost for a period of time. To track and predict marker path, a polynomial of some form, typically a spline, can be fit a markers' known path (e.g., Akbari-Shandiz et al. 2018). More sophistication methods include the use of a Kalman filter (Yeasin and Chaudhuri 2000), principal component analysis (Liu and McMillan 2006), or convolutional neural networks (Pischchulin et al. 2016). Companies producing passive marker automated motion analysis systems have propriety software which track markers, exploiting the intersecting rays from markers to cameras, but they are based on the methods outlined and may be task-specific. For example, the system may be designed to track markers during gait and therefore may not work so successfully for other tasks.

6.2.6 Other Considerations

If multiple cameras are employed then ideally the cameras should have their data collection synchronized. This is normally achieved by cabling between the cameras that ensures synchronized collection, but if this is not feasible post hoc synchronization is feasible. With film and video if a timing device is visible by each camera image coordinates can be interpolated to a common time base (Rome 1995). Another approach can be based on identifying common events in the task which if sufficient events are feasible to identify should permit the establishment of a common time base.

There are two competing concerns when positioning cameras: the relative orientation of the cameras for maximum accuracy and the positioning of cameras to ensure the landmarks of interest can be seen in multiple images. The second concern is crucially important for automatic motion analysis systems. Theoretically, if a landmark is visible in two cameras, reconstruction accuracy is superior if the cameras are orthogonal to one another (Abdel-Aziz 1974), but for markers placed on the human body with this configuration, there is a high potential for a marker to be obscured from one camera view. The optimal positioning of cameras has to balance camera configurations that ensure maximum reconstruction accuracy while ensuring visibility of body landmarks. These camera configurations are task-specific, consider walking gait compared with a simple upper limb horizontal reaching task.

To determine the location of landmarks in three-dimensional space typically requires the landmark to be viewed in at least two cameras. In photogrammetry it is recognized that redundancy, a landmark viewed in multiple cameras, increases reconstruction accuracy (e.g., Fraser et al. 2005). With most automatic motion analysis systems the number of cameras is typically greater than four. One advantage of multiple cameras is that it increases the chances that a landmark is viewable in at least two cameras. Accuracy would be highest if a landmark was seen in all cameras and in most cases lowest when viewed only in a pair of cameras. This means that the accuracy of the reconstruction of a landmark can differ throughout the movement being analyzed. For example, Pribanić et al. (2007) used a nine-camera system and then assessed reconstruction accuracy when points were reconstructed with pairs of the cameras. The least accurate result for a camera pair was over an order of magnitude greater than the most accurate camera pair result.

To perform calibration control points in the field of view are typically required. In the model fitting process some residual error is often provided, reflecting how well the control points are predicted based on the calibration fit. Challis and Kerwin (1992) demonstrated that this residual error underestimates the actual reconstruction error. Using an independent set of points, the reconstruction accuracy was 2.3 mm compared with the reconstruction accuracy of the control points of 1.7 mm, while for another case, it was 4.7 mm and 1.4 mm. These results indicate the importance of having an independent means of assessing 3D photogrammetry reconstruction accuracy. (See Chap. 11 for description of methods for assessing reconstruction accuracy.)

With the DLT, Wood and Marshall (1986) demonstrated the importance of the control point distribution. Reconstruction accuracy is superior if the control points are distributed throughout the field of view. Challis and Kerwin (1992) confirmed this finding and demonstrated that reconstruction accuracy was superior the more control points used. These results indicate that the control points must encompass the area in which the tasks to be analyzed occurs. If this recommendation is not followed the method used for reconstruction will not be calibrated for all of the movement space, which means there will be high inaccuracy for points reconstructed outside of the calibrated volume relative to points falling within the calibration volume.

6.2.7 Image-Based Motion Analysis Perspectives

Automatic motion analysis systems have their roots in human biomechanics (MacLeod et al. 1990) but have now been adopted for animation purposes in the movie and gaming industries. The advantage of this is that the market for these systems has been increased and therefore made this equipment cheaper for those working in biomechanics. The technological developments in this domain have been significant over the last three decades, and the question arises: What the future may hold?

Lab-based three-dimensional image-based motion analysis typically calibrates a volume in which the activity is analyzed. In gait studies, the subject typically walks several steps before and after the volume in which the data is collected. An increased number of steps could be recorded with more cameras or a cheaper option if the cameras could pan and possibly tilt to track the motion. Pan and tilt cameras have been used for analyzing ski jumping (Yeadon 1989), but it is feasible that future systems will actively track the subject and then adjust the camera calibration to allow for the changes in camera orientation.

What else might future marker-based motion analysis systems provide: potentially better marker identification algorithms, marker tracking algorithms, higher sample rates, and increased reconstruction accuracy. Increased reconstruction accuracy will assist with better marker identification. Of course, there are developments in marker-less motion tracking (e.g., Scott et al. 2017; Mathis et al. 2018), which may have an important impact on the ability of biomechanists to make ecologically valid data collection.

6.3 Goniometers

A goniometer is an instrument which allows the measurement of a joint angle. One limb of the goniometer is aligned with body segment on one side of the joint, and the other goniometer limb aligned with the body segment on the other side of the joint. Goniometers are often used in clinical settings for measuring static joint angles (see Fig. 6.9).

In biomechanics goniometers have been developed which allow the dynamic recording of joint angles, sometimes referred to as electrogoniometers. The simplest in concept are much like the static goniometer but with a potentiometer at the axis of rotation (e.g., Kettelkamp et al. 1970). As the limbs of the goniometer move, the potentiometer is rotated, and its resistance to electricity changes. Suitable calibration permits relating the resistance change to the angle. Other sensors have been used to determine joint angles. Nicol (1987) used a bendable beam mounted with strain gauges, and the beam was positioned to cross the joint of interest with the strain in the strain gauges varying with changes in joint angles. Subsequently, this method was extended to permit the measurement of the three-dimensional angles at a joint (Nicol 1988). The calibration of these goniometers can be complex (Legnani et al. 2000). Donno et al. (2008) presented a goniometer which used a laser beam propagating in a single-mode optical fiber, and changes in the light being registered were related to the change in joint angle. Mohamed et al. (2012) compared strain gauge and fiber optic goniometers with data from a passive marker-based motion analysis system. They evaluated knee angles during activities of daily living, reporting errors of up to 10 degrees, but with very similar performances between the two types of goniometers.

Fig. 6.9 A goniometer

One advantage of goniometers is that it is feasible to make them yourself (e.g., Wang et al. 2011), but there are drawbacks. Goniometers based on potentiometers can be quite cumbersome if intended to measure three angular degrees of freedom at a joint (e.g., Chao 1980). With potentiometer-based goniometers, the axis of the potentiometer should be appropriately aligned with the joint axis. Alignment of the goniometers to appropriately measure a joint angle can be problematic if cross talk is to be avoided. While the mounting of a goniometer can be problematic, an additional problem can be the cabling associated with measuring the signal from the goniometer(s). Finally, goniometers only provide angular data but for many analyses, additional kinematic data are required.

With goniometers and some other measurement systems (e.g., accelerometers, electromyographic system), the signal has to be recorded by some means. The simplest approach is that the signal from the sensor can be connected via wires directly to an analog-to-digital converter. Such a configuration can present problems as the wires could constrain subject motion but also increase noise in the sampled data. Another option is to use a data logger on the subject and at the end of the task transmit the signal to a computer. These devices have become small so that they now do not add significantly to the size of the sensor. The final approach is a telemetry system, where the sensor data is transmitted via a radio signal (e.g., Bluetooth) to a receiver adjacent to computer with an analog-to-digital converter.

6.4   Accelerometers

Accelerometers are sensors that have almost become ubiquitous with developments in microelectronics. Accelerometers can be found in smartphones, various gaming systems, and subject-worn health trackers. In biomechanics they can be used for several purposes, one to simply measure accelerations (e.g., Wosk and Voloshin 1981); another to determine another parameter from the acceleration signal, for example, foot strike during gait (e.g., Giandolini et al. 2014); and/or in combination with other measures (fusion) to determine other variables, for example, segmental kinematics (e.g., To and Mahfouz 2013). The latter purpose is common with inertial measurement units (IMU) (see the following section).

6.4.1   Basic Design

Accelerometers are used for measuring acceleration, and they are typically either strain gauge or piezoelectric accelerometers (see Chap. 7 for details about strain gauges and piezoelectric crystals). For a typical strain gauge accelerometer, a cantilevered mass is attached to a base (see Fig. 6.10). As the base is accelerated, the seismic mass is "left behind" due to its inertia and causes a change in strain in the strain gauges, and these strains can be related to the acceleration.

There are accelerometers that exploit fiber optic technology (Brown and Garrett 1991). A typical design has coils of fiber optics mounted on a plate or disk, which flexes when the system is accelerated (see Fig. 6.11). As the plate or disk flexes, the fiber optic is stressed and its ability to transmit light changes. These accelerometers have the advantage that they can be manufactured without metal components.

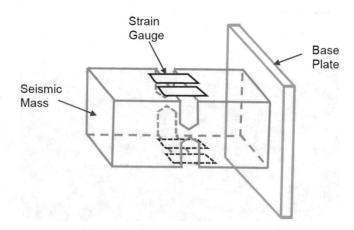

Fig. 6.10 Basic design of a cantilever mass accelerometer. In this design, there are four strain gauges

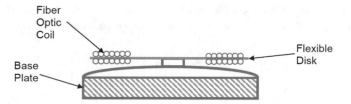

Fig. 6.11 Basic design of a fiber optic accelerometer

6.4.2   Key Properties

Accelerometers come with a range of capabilities, and these must be matched to nature of the task from which data is to be collected. The following are key capabilities that are often provided in the data sheets which accompany the accelerometer.

Mass and Size The size of the accelerometer is typically proportional to its mass. For the mounting on the human body the mass should not be too great; otherwise, its inertia could influence the movement being analyzed, and it could be more susceptible to displacement relative to the bony landmark whose acceleration is supposed to be measured.

Amplitude Range This is the range of accelerations which can be measured with a given accelerometer, often expressed in terms of multiples of the acceleration due to gravity (e.g., 5 g, 10 g, etc.). For the analysis of impacts during running an accelerometer with a range of 50 g is appropriate (e.g., Mercer et al. 2003), but a greater range, greater than 180 g, is needed to examine the accelerations associated with head impacts in American football (e.g., Rowson et al. 2009).

Shock Limit This is the limit acceleration that the accelerometer can tolerate. Once again expressed in terms of multiples of the acceleration due to gravity, it is typical multiples of the amplitude range giving a safety margin for accelerometer operation.

Temperature Range This is the viable operating range of the accelerometer. The sensors used to determine the acceleration can be temperature-sensitive, so an accelerometer should be used in its specified temperature range. There can be temperature-specific calibration adjustments.

Frequency Range This reports the range of frequencies that the accelerometer can accurately measure. Typically, the frequency response is up to -3 dB attenuation. While accelerometers have a frequency range, it is a good practice to only collect data that uses a fraction of this frequency range to safeguard against unusual frequency components in the sampled signal.

Amplitude Linearity This reflects any non-linearity up to the amplitude range limit. For example, if the full-range amplitude is 50 g and the linearity is reported to be 1%, then at full scale the deviation is potentially 0.5 g.

Sensitivity This indicates the response, voltage output, for a given acceleration. For example, if the sensitivity is 10 mV/g for a 50 g acceleration the output voltage would be 0.5 volts.

Resolution This reflects the smallest sensitivity of the accelerometers and represents a lower limit of acceleration values which can be interpreted as meaningful; below that value, the signal is noise.

6.4.3   Calibration

Given a known input and recorded output, accelerometers can be calibrated. For example, for uniaxial accelerometers:

$$a = c(v - b) \tag{6.23}$$

where a is the acceleration, c is a calibration coefficient, v is the accelerometer output, and b is an offset (bias). Many accelerometers are triaxial, so Eq. 6.23 becomes a series of vectors and a calibration matrix:

$$\begin{bmatrix} a_x \\ a_y \\ a_z \end{bmatrix} = \begin{bmatrix} c_x & 0 & 0 \\ 0 & c_y & 0 \\ 0 & 0 & c_z \end{bmatrix} \left(\begin{bmatrix} v_x \\ v_y \\ v_z \end{bmatrix} - \begin{bmatrix} b_x \\ b_y \\ b_z \end{bmatrix} \right) \tag{6.24}$$

In reality, each sensor is likely sensitive to orthogonal accelerations, that is, some cross talk is present. To account for this, the calibration matrix is more complex:

$$C = \begin{bmatrix} c_{xx} & c_{xy} & c_{xz} \\ c_{yx} & c_{yy} & c_{yz} \\ c_{zx} & c_{zy} & c_{zz} \end{bmatrix} \tag{6.25}$$

In this form, the matrix accounts for cross-axis sensitives. For some triaxial accelerometers, the accelerometers are not necessarily orthogonal to one another (e.g., Crisco et al. 2004), which means for useful output the matrix C is multiplied by an attitude matrix to ensure the resulting output accelerations represent accelerations associated with three orthogonal coordinate axes (see Chap. 9 for details on attitude matrices).

For a triaxial accelerometer there are three desired outputs that are dependent on 12 parameters (9 matrix elements and 3 bias terms). Therefore, given a redundant set of known accelerations and sensor outputs, the 12 calibration coefficients can be determined. The equation for calibration can now be recast in matrix form:

$$a = C(v - b) \tag{6.26}$$

$$C^{-1}a = v - b \tag{6.27}$$

$$v = C^{-1}a + b \tag{6.28}$$

$$v = \overline{C}\overline{a} \tag{6.29}$$

where
$$\overline{C} = \begin{bmatrix} C^{-1} b \end{bmatrix} \qquad\qquad \overline{a} = \begin{bmatrix} b \\ 1 \end{bmatrix}$$

Given a set of n accelerations and associated sensor outputs, the vectors become matrices, therefore:

$$V = \overline{C}\,\overline{A} \tag{6.30}$$

where $V = \begin{bmatrix} v_{x1} & v_{x2} & \ldots & v_{xn} \\ v_{y1} & v_{y2} & \ldots & v_{yn} \\ v_{z1} & v_{z2} & \ldots & v_{zn} \end{bmatrix}$

$$\overline{A} = \begin{bmatrix} a_{x1} & a_{x2} & \ldots & a_{xn} \\ a_{y1} & a_{y2} & \ldots & a_{yn} \\ a_{z1} & a_{z2} & \ldots & a_{zn} \\ 1 & 1 & \ldots & 1 \end{bmatrix}$$

To compute the calibration matrix, the Moore-Penrose inverse can be used (see Appendix A) very efficiently using the singular value decomposition (see Appendix H). If $n \geq 6$, then the calibration matrix can be determined from:

$$\overline{C} = V\overline{A}^T \left(\overline{A}\,\overline{A}^T \right)^{-1} \tag{6.31}$$

If the accelerometer is encased, for example, in a cube, then the cube can be placed on a horizontal surface on each of its six sides, this produces a matrix of known accelerations, and therefore, the set of criterion accelerations are:

$$\bar{A} = \begin{bmatrix} g & -g & 0 & 0 & 0 & 0 \\ 0 & 0 & g & -g & 0 & 0 \\ 0 & 0 & 0 & 0 & g & -g \\ 1 & 1 & 1 & 1 & 1 & 1 \end{bmatrix} \tag{6.32}$$

and the set of outputs are:

$$V = \begin{bmatrix} v_{x1} & v_{x2} & v_{x3} & v_{x4} & v_{x5} & v_{x6} \\ v_{y1} & v_{y2} & v_{y3} & v_{y4} & v_{y5} & v_{y6} \\ v_{z1} & v_{z2} & v_{z3} & v_{z4} & v_{z5} & v_{z6} \end{bmatrix} \tag{6.33}$$

Such an approach to calibration, using static measures, is common (e.g., Lötters et al. 1998; Won and Golnaraghi 2010). Of course, this does not calibrate for a full range of accelerations, but is a simple but effective method for performing a spot-check on the factory calibration of an accelerometer. A fuller spectrum of accelerations can be obtained using a motion rate table, which can provide a highly redundant set of measures.

6.4.4 Determining Kinematics

An accelerometer cannot differentiate between the acceleration caused by motion and the acceleration due to gravity; therefore, some compensation for the orientation of the accelerometer with respect to the gravity vector can be appropriate. The accuracy of motion measurements is therefore dependent on how well the gravity component can be estimated and filtered from the measured signal. For a simple uniaxial accelerometer, the measured acceleration is:

$$a_m = a + g \cos(\theta) \tag{6.34}$$

where a_m is the measured acceleration, a is the actual acceleration, g is the acceleration due to gravity, and θ is the angle between vertical (gravity vector) and the line of action of the acceleration. Imagine a uniaxial accelerometer on the tibial tuberosity trying to measure the acceleration during footstrike of running. If the angle in the sagittal plane of the tibia is not too great, errors can be small (e.g., $\cos\left(\frac{\pi}{9}\right) \approx 0.94$), which has experimentally been shown to be the case (Lafortune and Hennig 1991). In contrast, movements that comprise a lot of angular motion would require some correction for gravity. In inertial measurement units, a magnetometer can provide information on the gravity vector, and in other studies, a motion analysis system can be used to determine accelerometer orientation (e.g., Ladin and Wu 1991). It is also feasible if the initial orientation of the accelerometer is known to correct for the influence of gravity by using the accelerometer information to determine other kinematic variables.

Clearly accelerometers provide information about acceleration, but it is feasible to determine other kinematic quantities from accelerometer data. The acceleration of a point A on a rigid body can be described by:

$$a_A = a_0 + \omega \times (\omega \times r) + \alpha \times r \tag{6.35}$$

where a_A is the acceleration of point A on the rigid body, a_0 is the acceleration of point 0 on the rigid body (for analysis purposes, often the segment's center of mass), ω is the angular acceleration of the body, r is the vector describing the relative position of point A with respect to point 0, and α is the angular acceleration of the body. Notice that the acceleration of point A has two components: a normal component ($\omega \times (\omega \times r)$) and a tangential component ($\alpha \times r$). If the rigid body is rotating, point A will have a circular motion relative to point 0; therefore, the acceleration of point A has one component due to the acceleration of point 0 and another due to its acceleration with respect to point 0 (see Fig. 6.12).

If there is no rotation of the rigid body, then $a_A = a_0$. With rigid body rotation, it is feasible to compute angular kinematics from the linear acceleration data. The acceleration a_A is measured from an accelerometer (or group of accelerometers). Equation 6.35 can be rewritten so:

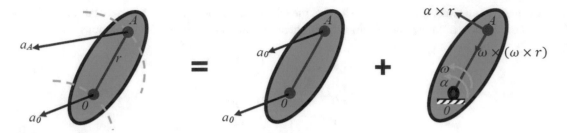

Fig. 6.12 The acceleration of a point A in terms of point 0, given the angular kinematics of the segments

$$a_A - a_0 = \ A\,r + B\,r \tag{6.36}$$

where:

$$A = \begin{bmatrix} -\omega_y^2 - \omega_z^2 & \omega_x\omega_y & \omega_x\omega_z \\ \omega_x\omega_y & -\omega_z^2 - \omega_x^2 & \omega_y\omega_z \\ \omega_x\omega_z & \omega_y\omega_z & -\omega_x^2 - \omega_y^2 \end{bmatrix}$$

$$B = \begin{bmatrix} 0 & -\alpha_z & \alpha_y \\ \alpha_z & 0 & -\alpha_x \\ -\alpha_y & \alpha_x & 0 \end{bmatrix}$$

which in turn can be expressed as:

$$\begin{bmatrix} a_x \\ a_y \\ a_z \end{bmatrix} - \begin{bmatrix} a_{0x} \\ a_{0y} \\ a_{0z} \end{bmatrix} = \begin{bmatrix} -\omega_y^2 - \omega_z^2 & \omega_x\omega_y - \alpha_z & \omega_x\omega_z + \alpha_y \\ \omega_x\omega_y + \alpha_z & -\omega_z^2 - \omega_x^2 & \omega_y\omega_z - \alpha_x \\ \omega_x\omega_z - \alpha_y & \omega_y\omega_z + \alpha_x & -\omega_x^2 - \omega_y^2 \end{bmatrix} \begin{bmatrix} x \\ y \\ z \end{bmatrix} \tag{6.37}$$

where:$r = \begin{bmatrix} x \\ y \\ z \end{bmatrix}$

    In this format, the equations have 9 unknowns (a_{0x}, a_{0x}, a_{0x}, ω_x, ω_y, ω_z, ω_y, α_x, α_y, α_z), so in theory, these unknowns can be determined if there is information from 9 accelerometers (e.g., Padgaonkar et al. 1975), although accuracy can be better with 12 accelerometers (Zappa et al. 2001).

    When acceleration and/or velocity data are available, integration can be used to determine velocities and displacements. For example, given linear accelerations:

$$\int_{t1}^{t2} a(t)\,dt = v(t) + c \tag{6.38}$$

where $t1$ and $t2$ are two time instants, a is the current acceleration, v is the current velocity, and c is the constant of integration. The same process can be used to determine displacement information from velocity data. Once integration has been performed, there is a constant in Eq. 6.38, which must be determined; otherwise, there is a constant bias in any kinematic quantities. Typically, these constants can be determined if some initial or final condition is known, for example, that the limb is stationary at one of these points. The integrations are typically performed numerically for small time steps, where the results from one time step provide initial conditions for the subsequent time step. The areas under sub-sections of the curve, for example, the time-acceleration curve, can be determined by modeling each section of the curve as a trapezoid (see Fig. 6.13).

    These approaches are attractive, but assessments of the ability of a set of accelerometers to provide segmental kinematics have highlighted problems. For example, Giansanti et al. (2003) examined the feasibility of collecting accelerometer data from 6 to 9 accelerometers connected to one segment and computing segment kinematics. Errors grew over the sample period

Fig. 6.13 Dimensions of a trapezoid and the equation used to compute its area

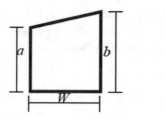

$$area = \frac{a+b}{2}W$$

and were quite significant even for a period as short as 4 seconds. For example, segment angles had errors of 10 degrees in one of the better cases. Clearly caution should be taken when using this approach or interpreting such data.

6.4.5   Mounting Accelerometers

The target in most studies is to use accelerometers to obtain a measure of the acceleration of a bone. One option is to drill pins into the bones and attach accelerometers to the pins (e.g., Lafortune and Hennig 1991). In most studies, less invasive methods are used. There are some good options, for example, a bite bar with an accelerometer is a good indicator of skull acceleration (e.g., Muir et al. 2013). In a similar fashion, accelerometers can be glued or, via wax, attached to the nails. A common approach is to place the accelerometer over a bony landmark and bind it tightly to that spot using tape of some kind. Ziegert and Lewis (1979) compared this approach with accelerometers mounted to bone pins; if the binding of the surface-mounted accelerometers was tight enough and the accelerometer of low mass (1.5 grams), they gave values similar to those obtained from the bone-mounted accelerometers.

6.5   Inertial Measurement Units

Inertial measurement units (IMU) consist of a number of devices for measuring aspects of the units' kinematics. Common components include triaxial accelerometer, gyroscope (to measure orientation), and magnetometer (to measure orientation with respect to a magnetic field). On occasions, they also contain a Global Positioning System (GPS). Outside of biomechanics these devices have been used to make measurements on unmanned vehicles, satellites, and spacecraft and in some gaming systems to highlight just a few examples. They are now used in biomechanics with one unit attached to a segment potentially providing information on the segment's angular and linear kinematics.

  Many IMU use propriety algorithms for combining the data from the various sensors in the IMU, and this combination is referred to as data fusion. Typically, specific details of the propriety algorithms are not available, but most are based on a Kalman filter (1960). In a general sense the Kalman filter is a set of mathematical equations, recursive in formulation, that estimate the state of a system or process that minimizes the mean of the squared error (Grewal and Andrews 2011). These filters are specified by the equations used which can vary between users modeling the same system. There are instances where authors have published details of their algorithm, for example, Caron et al. (2006) describe the data fusion methods implemented within a Kalman filter to determine IMU kinematics. In biomechanics, a variety of Kalman filter implementations have been examined, but Caruso et al. (2018) did not find one clearly superior and recommended that task-specific tuning of the Kalman filter is required to provide good filter kinematic output.

6.6   Review Questions

1. What are the intrinsic and extrinsic camera parameters?
2. What is calibration and reconstruction in image-based motion analysis?
3. List the advantages and disadvantages of the following classes of motion analysis systems: cine film, automatic passive marker motion analysis system, and automatic active marker motion analysis system.
4. Explain the role of the lens in image-based motion analysis.
5. Explain the relationship between a point measured in both image and object space reference frames.
6. What are marker tracking and marker identification? Why are they important?
7. Briefly outline how goniometers and accelerometers monitor motion.

8. Briefly outline the limitations of goniometers and accelerometers for the analysis of human movement.
9. Suggest activities for which each of the following is particularly suited: goniometers and accelerometers.
10. Why is integration typically necessary when processing accelerometer data?

References

Cited References

Abdel-Aziz, Y. I. (1974). Expected accuracy of convergent photos. *Photogrammetric Engineering, 40*, 1341–1346.

Abdel-Aziz, Y. I., & Karara, H. M. (1971). Direct linear transformation from comparator co-ordinates into object space co-ordinates in close range photogrammetry. In *ASP symposium on close range photogrammetry* (pp. 1–18). Falls, Church: American Society of Photogrammetry.

Abdel-Aziz, Y. I., Karara, H. M., & Hauck, M. (2015). Direct Linear Transformation from comparator coordinates into object space coordinates in close-range photogrammetry. *Photogrammetric Engineering & Remote Sensing, 81*(2), 103–107.

Akbari-Shandiz, M., Mozingo, J. D., Holmes, D. R., III, & Zhao, K. D. (2018). An interpolation technique to enable accurate three-dimensional joint kinematic analyses using asynchronous biplane fluoroscopy. *Medical Engineering & Physics, 60*, 109–116.

Bopp, H. P., & Krauss, H. (1978). An orientation and calibration method for non-topographic application. *Photogrammetric Engineering and Remote Sensing, 44*(9), 1191–1196.

Borghese, N. A., Cerveri, P., & Rigiroli, P. (2001). A fast method for calibrating video-based motion analysers using only a rigid bar. *Medical & Biological Engineering & Computing, 39*(1), 76–81.

Boyle, W. S., & Smith, G. E. (1970). Charge coupled semiconductor devices. *The Bell System Technical Journal, 49*(4), 587–593.

Brown, D. A., & Garrett, S. L. (1991). *Interferometric fiber optic accelerometer.* Paper presented at the SPIE Microelectronic Interconnect and Integrated Processing Symposium.

Brown, D. C. (1966). Decentering distortion of lenses. *Photogrammetric Engineering, 32*(3), 444–462.

Caron, F., Duflos, E., Pomorski, D., & Vanheeghe, P. (2006). GPS/IMU data fusion using multisensor Kalman filtering: introduction of contextual aspects. *Information Fusion, 7*(2), 221–230.

Caruso, M., Bonci, T., Knaflitz, M., Croce, U. D., & Cereatti, A. (2018). A comparative accuracy analysis of five sensor fusion algorithms for orientation estimation using magnetic and inertial sensors. *Gait & Posture, 66*, S9–S10.

Challis, J. H. (1995). A multiphase calibration procedure for the Direct Linear Transformation. *Journal of Applied Biomechanics, 11*, 351–358.

Challis, J. H., & Kerwin, D. G. (1992). Accuracy assessment and control point configuration when using the DLT for cine-photogrammetry. *Journal of Biomechanics, 25*(9), 1053–1058.

Chao, E. Y. S. (1980). Justification of triaxial goniometer for the measurement of joint rotation. *Journal of Biomechanics, 13*, 989–1006.

Choo, A. M., & Oxland, T. R. (2003). Improved RSA accuracy with DLT and balanced calibration marker distributions with an assessment of initial-calibration. *Journal of Biomechanics, 36*(2), 259–264.

Conrady, A. E. (1919). Decentred lens-systems. *Monthly Notices of the Royal Astronomical Society, 79*(5), 384–390.

Crisco, J. J., Chu, J. J., & Greenwald, R. M. (2004). An algorithm for estimating acceleration magnitude and impact location using multiple nonorthogonal single-axis accelerometers. *Journal of Biomechanical Engineering, 126*(6), 849–854.

Dapena, J. (1985). Correction for 'Three-Dimensional cinematography with control object of unknown shape. *Journal of Biomechanics, 18*, 163.

Dapena, J., Harman, E. A., & Miller, J. A. (1982). Three-Dimensional cinematography with control object of unknown shape. *Journal of Biomechanics, 15*(1), 11–19.

Donno, M., Palange, E., Di Nicola, F., Bucci, G., & Ciancetta, F. (2008). A new flexible optical fiber goniometer for dynamic angular measurements: Application to human joint movement monitoring. *IEEE Transactions on Instrumentation and Measurement, 57*(8), 1614–1620.

Eian, J., & Poppele, R. (2002). A single-camera method for three-dimensional video imaging. *Journal of Neuroscience Methods, 120*(1), 65–85.

Fraser, C. S., Woods, A., & Brizzi, D. (2005). Hyper redundancy for accuracy enhancement in automated close range photogrammetry. *The Photogrammetric Record, 20*(111), 205–217.

Furnee, E. H. (1990). PRIMAS: Real-time image-based motion measurement system. In J. S. Walton (Ed.), *Proceedings of mini-symposium on image based motion measurement* (Vol. 1356, pp. 56–62). Washington: SPIE.

Giandolini, M., Poupard, T., Gimenez, P., Horvais, N., Millet, G. Y., Morin, J.-B., & Samozino, P. (2014). A simple field method to identify foot strike pattern during running. *Journal of Biomechanics, 47*(7), 1588–1593.

Giansanti, D., Macellari, V., Maccioni, G., & Cappozzo, A. (2003). Is it feasible to reconstruct body segment 3-D position and orientation using accelerometric data? *IEEE Transactions on Biomedical Engineering, 50*(4), 476–483.

Golio, M. (2015). Fifty years of Moore's Law. *Proceedings of the IEEE, 103*(10), 1932–1937.

Grewal, M. S., & Andrews, A. P. (2011). *Kalman filtering: theory and practice using MATLAB.* New York: John Wiley & Sons.

Holzreiter, S. (2005). Autolabeling 3D tracks using neural networks. *Clinical Biomechanics, 20*(1), 1–8.

Kalman, R. E. (1960). A new approach to linear filtering and prediction problems. *Journal of Basic Engineering, 82*(1), 35–45.

Kettelkamp, D. B., Johnson, R. J., Smidt, G. L., Chao, E. Y., & Walker, M. (1970). An electrogoniometric study of knee motion in normal gait. *Journal of Bone and Joint Surgery (America), 52*(4), 775–790.

Khan, S., & Shah, M. (2003). Consistent labeling of tracked objects in multiple cameras with overlapping fields of view. *IEEE Transactions on Pattern Analysis and Machine Intelligence, 25*(10), 1355–1360.

Krishnan, C., Washabaugh, E. P., & Seetharaman, Y. (2015). A low cost real-time motion tracking approach using webcam technology. *Journal of Biomechanics, 48*(3), 544–548.

Ladin, Z., & Wu, G. (1991). Combining position and acceleration measurements for joint force estimation. *Journal of Biomechanics, 24*(12), 1171–1187.

Lafortune, M. A., & Hennig, E. M. (1991). Contribution of angular motion and gravity to tibial acceleration. *Medicine & Science in Sports & Exercise, 23*(3), 360–363.

Legnani, G., Zappa, B., Casolo, F., Adamini, R., & Magnani, P. L. (2000). A model of an electro-goniometer and its calibration for biomechanical applications. *Medical Engineering & Physics, 22*, 711–722.

Liu, G., & McMillan, L. (2006). Estimation of missing markers in human motion capture. *The Visual Computer, 22*(9), 721–728.

Lötters, J. C., Schipper, J., Veltink, P. H., Olthuis, W., & Bergveld, P. (1998). Procedure for in-use calibration of triaxial accelerometers in medical applications. *Sensors and Actuators A: Physical, 68*(1), 221–228.

MacLeod, A., Morris, J. R. W., & Lyster, M. (1990). Highly accurate video coordinate generation for automatic 3D trajectory calculation. In J. S. Walton (Ed.), *Image-based motion measurement* (Vol. 1356, pp. 12–18). SPIE.

Mathis, A., Mamidanna, P., Abe, T., Cury, K., & Murthy, V. (2018). Markerless tracking of user-defined features with deep learning. *arXiv, arXiv:1804*, 03142.

Mercer, J. A., Devita, P., Derrick, T. R., & Bates, B. T. (2003). Individual effects of stride length and frequency on shock attenuation during running. *Medicine & Science in Sports & Exercise, 35*(2), 307–313.

Miller, N. R., Shapiro, R., & McLaughin, T. M. (1980). A technique for obtaining spatial kinematic parameters of segments of biomechanical systems from cinematographic data. *Journal of Biomechanics, 13*, 535–547.

Mohamed, A. A., Baba, J., Beyea, J., Landry, J., Sexton, A., & McGibbon, C. A. (2012). Comparison of strain-gage and fiber-optic goniometry for measuring knee kinematics during activities of daily living and exercise. *Journal of Biomechanical Engineering-Transactions of the ASME, 134*(8), 084502.

Muir, J., Kiel, D. P., & Rubin, C. T. (2013). Safety and severity of accelerations delivered from whole body vibration exercise devices to standing adults. *Journal of Science and Medicine in Sport, 16*(6), 526–531.

Muybridge, E. (1907). *Animals in motion: an electro-photographic investigation of consecutive phases of muscular actions* (3rd ed.). London: Chapman & Hall, Ld.

Nicol, A. C. (1987). A new flexible electrogoniometer with widespread applications. In B. Jonsson (Ed.), *Biomechanics XB* (pp. 1029–1033). Champaign: Human Kinetics.

Nicol, A. C. (1988). A triaxial flexible electrogoniometer. In G. de Groot, A. P. Hollander, P. A. Huijing, & G. J. van Ingen Schenau (Eds.), *Biomechanics XI* (Vol. B, pp. 964–967). Amsterdam: Free University Press.

Padgaonkar, A. J., Krieger, K. W., & King, A. I. (1975). Measurement of angular acceleration of a rigid body using linear accelerometers. *Journal of Applied Mechanics-Transactions of the ASME, 42*(3), 552–556.

Pishchulin, L., Insafutdinov, E., Tang, S., Andres, B., Andriluka, M., Gehler, P., & Schiele, B. (2016). *DeepCut: Joint subset partition and labeling for multi person pose estimation*. Paper presented at the 2016 IEEE Conference on Computer Vision and Pattern Recognition (CVPR).

Pribanić, T., Sturm, P., & Cifrek, M. (2007). Calibration of 3D kinematic systems using orthogonality constraints. *Machine Vision and Applications, 18*(6), 367–381.

Rome, L. C. (1995). A device for synchronising biomechanical data with cine film. *Journal of Biomechanics, 28*(3), 333–345.

Rowson, S., Brolinson, G., Goforth, M., Dietter, D., & Duma, S. (2009). Linear and angular head acceleration measurements in collegiate football. *Journal of Biomechanical Engineering, 131*(6), 061016.

Scott, J., Collins, R. T., Funk, C., & Liu, Y. (2017). 4D model-based spatiotemporal alignment of scripted Taiji Quan sequences. In *Paper presented at the ICCV 2017 PeopleCap Workshop*. Venice, Italy.

Selvik, G., Alberius, P., & Aronson, A. S. (1983). A Roentgen stereophotogrammetric system. Construction, calibration and technical accuracy. *Acta Radiologica Diagnosis, 24*(4), 343–352.

Taylor, K. D., Mottier, F. M., Simmons, D. W., Cohen, W., Pavlak, R., Cornell, D. P., & Hankins, G. B. (1982). An automated motion measurement system for clinical gait analysis. *Journal of Biomechanics, 15*(7), 505–516.

To, G., & Mahfouz, M. R. (2013). Quaternionic attitude estimation for robotic and human motion tracking using sequential monte carlo methods with von Mises-Fisher and nonuniform densities simulations. *IEEE Transactions on Biomedical Engineering, 60*(11), 3046–3059.

Tsaousidis, N., & Zatsiorsky, V. (1996). Two types of ball-effector interaction and their relative contribution to soccer kicking. *Human Movement Science, 15*(6), 861–876.

Wang, P. T., King, C. E., Do, A. H., & Nenadic, Z. (2011). A durable, low-cost electrogoniometer for dynamic measurement of joint trajectories. *Medical Engineering & Physics, 33*(5), 546–552.

Woltring, H. J. (1975). Single- and dual-axis lateral photodetectors of rectangular shape. *IEEE Transactions on Electron Devices, 22*(8), 581–590.

Woltring, H. J. (1977). Bilinear approximation by orthogonal triangularization. *Photogrammetria, 33*(3), 77–93.

Woltring, H. J. (1980). Planar control in multi-camera calibration for three-dimensional gait studies. *Journal of Biomechanics, 13*(1), 39–48.

Won, S. P., & Golnaraghi, F. (2010). A triaxial accelerometer calibration method using a mathematical model. *IEEE Transactions on Instrumentation and Measurement, 59*(8), 2144–2153.

Wood, G. A., & Marshall, R. N. (1986). The accuracy of DLT extrapolation in three-dimensional film analysis. *Journal of Biomechanics, 19*(9), 781–785.

Wosk, J., & Voloshin, A. (1981). Wave attenuation in skeletons of young healthy persons. *Journal of Biomechanics, 14*(4), 261–267.

Yeadon, M. R. (1989). A method for obtaining three-dimensional data on ski jumping using pan and tilt cameras. *International Journal of Sport Biomechanics, 5*, 238–247.

Yeasin, M., & Chaudhuri, S. (2000). Development of an automated image processing system for kinematic analysis of human gait. *Real-Time Imaging, 6*(1), 55–67.

Zappa, B., Legnani, G., van den Bogert, A. J., & Adamini, R. (2001). On the number and placement of accelerometers for angular velocity and acceleration determination. *Journal of Dynamic Systems, Measurement, and Control, 123*(3), 552–554.

Zhao, C., Fezzaa, K., Cunningham, R. W., Wen, H., De Carlo, F., Chen, L., Rollett, A. D., & Sun, T. (2017). Real-time monitoring of laser powder bed fusion process using high-speed X-ray imaging and diffraction. *Scientific Reports, 7*(1), 3602.

Ziegert, J. C., & Lewis, J. L. (1979). Effect of soft-tissue on measurements of vibrational bone motion by skin-mounted accelerometers. *Journal of Biomechanical Engineering-Transactions of the ASME, 101*(3), 218–220.

Useful References

Ghosh, S. K. (1979). *Analytical photogrammetry. U.S.A.* New York: Pergamon Press Inc..

Hartley, R., & Zisserman, A. (2000). *Multiple view geometry in computer vision.* Cambridge University Press.

Slama, C. C. (1980). *Manual of photogrammetry* (4th ed.). Falls Church: American Society of Photogrammetry.

Overview

In 1687 Isaac Newton published his Philosophiæ Naturalis Principia Mathematica (Mathematical Principles of Natural Philosophy), revealing to the world, among other things, his three laws of motion. These laws laid the foundation for classical mechanics. Importantly these three laws explain the relationship between the motion of a body and the forces applied to that body. If interested in human movement or lack of movement, then measuring the forces applied to the human body is important. Giovanni Alfonso Borelli (1608–1679) tried to relate the forces applied to the human body and the corresponding muscle forces in his posthumously published work (Borelli 1685). Borelli was unable to measure any forces directly and applied some of the mechanics incorrectly, but his work highlights the importance of being able to measure forces.

Étienne-Jules Marey (1830–1904) pioneered many ways of measuring human movement including some of the earliest efforts to measure the pressures and forces between the foot and the ground during human locomotion (Marey and Demenÿ 1885). In contemporary biomechanics, a common piece of apparatus for the measurement of ground reaction forces is the force plate. Here the focus will be on the force plate as an example of force measurement, but many of the principles which guide the design, use, and analysis of force plates can be generalized to other ways of measuring forces.

The content of this chapter requires a working knowledge of matrix algebra (see Appendix A).

7.1 Force Measurement

To measure a force requires a transducer. Generally a transducer can be defined as a device which converts energy from one form to another. Therefore to measure a force the transducer must respond in some way to the force applied to it. There are a variety of ways of measuring forces and a number of different transducers; the three major ways will be reviewed in the following sub-sections.

7.1.1 Springs

If a force is applied to a spring, the length of that spring changes. Many bathroom scales use springs to measure forces, and Elftman (1938) designed a force plate to measure ground reaction forces, which relied on the measurement of spring displacements to estimate forces. Ideally a spring would follow Hooke's Law:

$$F = k \, \Delta L \tag{7.1}$$

where F is the force applied to the spring, k is the spring constant, and ΔL is the change in length of the spring. Figure 7.1 illustrates the law.

According to Hooke's Law when twice the load is applied to the spring, the amount of stretch is doubled. Therefore if we know the spring constant and measure of the change in length the applied force can be determined. In reality springs may

J. H. Challis, *Experimental Methods in Biomechanics*, https://doi.org/10.1007/978-3-030-52256-8_7

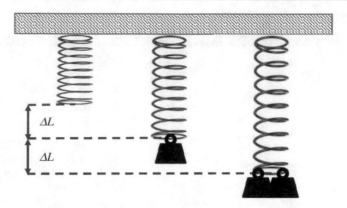

Fig. 7.1 Illustration of Hooke's Law, showing twice the stretch (Δ*L*) in a spring when the weight on the spring is doubled

Fig. 7.2 The force-extension curve for (**a**) a material behaving linearly and (**b**) a material behaving non-linearly with hysteresis

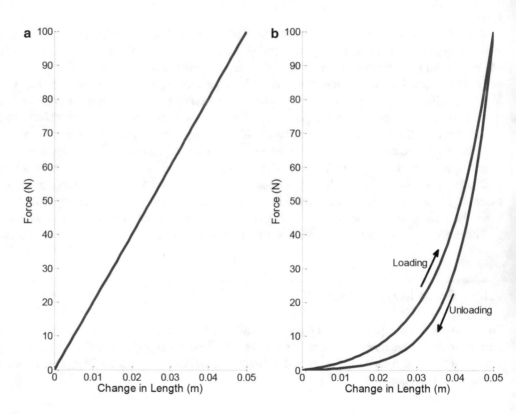

approximately have a linear relationship between the applied force and change in spring length, but they are not strictly linear. For springs where the relationship between spring extension and the applied force are not linearly related to one another, the following equation is more appropriate:

$$F = k_1\left(e^{k_2 \Delta L} - 1\right) \tag{7.2}$$

Therefore, to use this relationship, it would be necessary to know two coefficients (k_1, k_2); suitable calibration should provide these coefficients.

One other property that should be considered for a spring, which also applies to other force-measuring transducers, is hysteresis. Hysteresis occurs when the loading and unloading curves do not match (see Fig. 7.2b). For the same load during loading and unloading, the extension of the spring is greater during unloading. If the force-extension curve is plotted the hysteresis can be seen, with the area between the two curves representing the energy dissipated, typically due to material internal friction. For force measurement, ideally the transducer used should have low hysteresis.

Fig. 7.3 Illustration of strain as the ratio of the change in length of a material relative to its original length

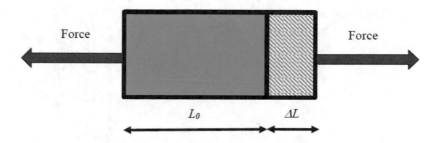

7.1.2 Strain Gauges

Lord Kelvin noticed that the electrical resistance of a wire changes as the wire is stretched. The electrical resistance (R) of wire of given length (L) is given by:

$$R = \frac{\rho L}{A} \tag{7.3}$$

where ρ is the resistivity of the wire and A is the cross-sectional area of the wire. Therefore as the strain in the wire increases, the wire length increases, so resistance increases; similarly the wire's cross-sectional area decreases, so resistance increases. In fact for most materials as strain increases so does their resistivity. Therefore, these factors mean as the strain in the wire increases the resistance increases.

A strain gauge is a sensor whose resistance to an electrical signal changes with the strain in the sensor. Therefore, after suitable calibration measuring the change in resistance of the strain gauge permits determination of the force being applied to the strain gauge. The strain experienced by the strain gauge can be positive (tensile) therefore there is elongation, or negative (compressive) therefore there is contraction. Strain can be computed from:

$$\varepsilon = \frac{\Delta L}{L_0} \tag{7.4}$$

where ε is the strain, ΔL is the change in length due to the applied force, and L_0 is the unloaded length. The basic nomenclature is illustrated in Fig. 7.3.

A typical strain gauge is a thin metal wire or metal foil. This metal can be attached directly to the object of interest (unbounded strain gauge), or the metal is arranged in a grid which is bonded to a carrier which is then attached to the object of interest (bonded metal wire strain gauges and bonded metal foil strain gauges). For bonded strain gauges, the strain experienced by the object of interest is transferred to the carrier and therefore to the metal of the sensor (see Fig. 7.4).

Each strain gauge has a gauge factor (GF), which is the ratio of the relative change in resistance to the relative change in length:

$$GF = \frac{\left(\Delta R / R\right)}{\left(\Delta L / L_0\right)} = \frac{\left(\Delta R / R\right)}{\varepsilon} \tag{7.5}$$

where ΔR is the change in resistance of the strain gauge relative to the resistance in the unstrained strain gauge (R). A metal foil strain gauge typically has a gauge factor between 2 and 5 and an unstrained resistance of 120 to 350 Ω. The measured strains are often very small $10^{-6} < \varepsilon < 10^{-3}$. Given a typical range of strains, a gauge factor of 2, and an unstrained resistance of 120 Ω, the range of changes in resistance is 0.00024 Ω to 0.24 Ω. The system will use a bridge to measure the change in resistance.

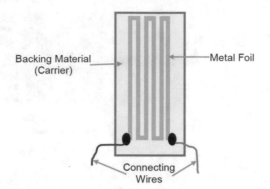

Fig. 7.4 Illustration of bonded strain gauge with metal foil etched into carrier. Some strain gauges have the foil orientated in multiple directions, each with an independent output, for the simultaneous measurement of the forces in multiple directions

Fig. 7.5 The arrangement of three sets of components of a three component piezoelectric load cell, (a) relative orientation of crystals, and (b) the electrodes (represented as red lines) used to measure electric charge

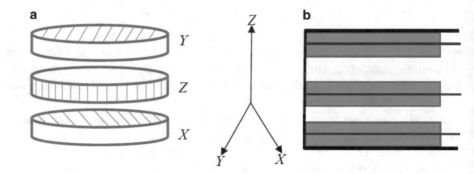

7.1.3 Piezoelectric Crystals

Piezoelectric crystals when squeezed produce an electric charge; this phenomenon was first identified by Pierre Curie (1859–1906) and his brother Paul-Jacques Curie (1856–1941). Strictly these crystals due to a mechanical force generate a surface charge, with the magnitude of the charge being proportional to the magnitude of the applied force. Piezoelectric crystals grow naturally, but most crystals used as sensors now are specifically grown and processed quartz.

The crystals only produce force when loaded from one direction; therefore a three-dimensional force sensor can be manufactured by stacking crystals each with a different orientation (see Fig. 7.5). Electrodes inserted in the crystals permit measurement of the electric charge. When packed into a load washer, the system is very stiff; typically under a load of 10 kN, the transducer compresses by only 0.001 mm. For most force-sensing equipment, this high stiffness is useful. Typical output is 2.3 pC/N (Stefanescu, 2011). Piezoelectric crystals dissipate the charge over time due to leakage; therefore the interface to the crystals is designed to minimize this problem, but it does mean that piezoelectric force transducers are more sensitive to drift with time than strain gauges.

7.1.4 Calibration

Force sensors are calibrated by applying known loads and measuring output (Δs in subsequent equations), for example, length change, change in resistance, change in electric charge, and then modeling the relationship between output and load. The modeling of this relationship can be a linear equation via Hooke's Law:

$$F = k\,\Delta s \tag{7.6}$$

or a non-linear representation:

$$F = k_1\left(e^{k_2 \Delta s} - 1\right) \tag{7.7}$$

or fitted with a polynomial:

$$F = c_1 + c_2 \Delta s + c_3 \Delta s^2 + c_4 \Delta s^3 \ldots + c_n \Delta s^{n-1} \tag{7.8}$$

Equation coefficients are typically fitted in a least-squares sense, with an algorithm such as the Levenberg-Marquardt algorithm used for non-linear equations (More 1977). The appropriate equation and coefficients influence the accuracy of the model fit and therefore force predictions.

In the process of calibration, a number of factors must be considered. For example, is the output of the measurement system influenced by moisture or temperature? If this is the case then either the operating conditions should be specified or the equation can allow for adjustment, for example, for variations in temperature. Commercial force-measuring devices often come with specifications such as maximum force, working force range, accuracy, resolution, drift, desired operating conditions (e.g., temperature and humidity), and the influence of static and dynamic loading.

7.1.5   Applications

In the following the use of force sensors for measuring ground reaction forces will be discussed, but these force sensors have broader applications. For example Bergmann et al. (1993) instrumented artificial hip joints in two subjects, so that the loads applied to these joints could be quantified in vivo. Gregor et al. (1991) placed a buckle transducer on the Achilles tendon so the forces produced by the triceps surae could be measured during walking and running. The buckle transducer looks like a belt buckle, with strain gauges attached so that changes in the buckle shape due to the force applied to it by the tendon can be quantified. Arndt et al. (2002) used strain gauges to measure the loading patterns in the second metatarsal during barefoot walking; an instrumented staple was placed in the second metatarsal of eight subjects.

7.2   Force Plates

To complement his analysis of human locomotion, Herbert Elftman made a force plate; its sensors were springs which he could film to measure their deformations (Elftman 1938). At the time of his work, Elftman did not have access to other force-measuring transducers. In 1938 Edward Simmons Jr. (1911–2004) working at the California Institute of Technology and Arthur Ruge (1905–2000) working at the Massachusetts Institute of Technology, both independently developed the modern bonded wire resistance strain gauge. It was not until the 1950s due to the work of Walter Kistler (1918–2015) that the properties of piezoelectric crystals were exploited for the measurement of force. Modern force plates rely on either strain gauges or piezoelectric crystals, analog-to-digital converters, and computers to measure the ground reaction forces.

7.2.1   Reaction Forces

Force plates are typically used to measure ground reaction forces, although they have been used to measure forces in other situations, for example, at the shoulders during rugby scrummaging (Milburn 1990). The ground reaction force is the force applied by the ground which is equal in magnitude but in the opposite direction to the forces applied by the subject to the ground. The measured forces reflect the acceleration of the center of mass of the human body:

$$F_x = m\, a_x \tag{7.9}$$

$$F_y = m\, a_y \tag{7.10}$$

$$F_z = m\, (a_z + g) \tag{7.11}$$

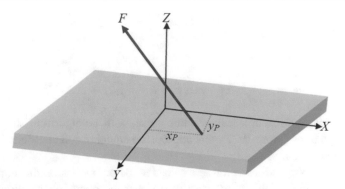

Fig. 7.6 The coordinate directions for a force plate, along with the force vector (F) and the origin of the force – the center of pressure (x_P, y_p)

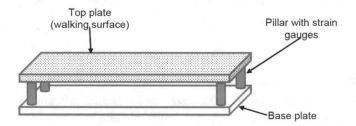

Fig. 7.7 Basic configuration of a strain gauge-based force plate

where F_x, F_y, and F_z are the ground reaction forces in the three-coordinate directions; m is the mass of the body; a_x, a_y, and a_z are the center of mass acceleration in the three-coordinate directions; and g is the acceleration due to gravity.

While force is a vector quantity (has both magnitude and direction), the ground reaction force is technically a fixed vector because the origin of the vector must be specified (see Fig. 7.6). In biomechanics the interception of the vertical ground reaction force with the force plate (support surface) is called the center of pressure.

7.2.2 Design Criteria

There have been a number of papers describing how to build a force plate (e.g., Payne 1974; Bonde-Petersen 1975; Gola 1980; Heglund 1981; Hsieh 2006). While it is feasible to build a force plate most researchers use commercially produced plates; in these plates the typical design is to have force sensors toward the four corners of the top plate. A force plate using piezoelectric crystals will have a washer of crystals at each corner, with each washer consisting of crystals orientated to measure forces in the three-coordinate directions. A force plate with strain gauges sits on four pillars which support the top plate (see Fig. 7.7). The base plate is anchored, while the top plate is flush with the walking surface. The four pillars are instrumented with strain gauges.

As force plates either use strain gauges or piezoelectric crystals, what are the relative merits of force plates using either technology? In many senses they perform similarly (see Table 7.1); the main differences are in hysteresis and drift. Strain gauge-based force plates have greater hysteresis than piezoelectric crystal-based force plates, but the piezoelectric plates have greater drift. To some extent these deficiencies can be corrected with appropriate software, but depending on the application, these deficiencies might influence the selection of a force plate for a study.

The force plate should have these key properties: rigid, strong, imperceptible displacement, high natural frequency, no cross-talk between axes, easy and stable calibration, and light. The following will consider each of these.

Rigid

Most surfaces we walk or run across have negligible deformation; the force plate top surface should mirror these properties.

Table 7.1 Performance of strain gauge and piezoelectric force plates

Characteristic	Strain gauge	Piezoelectric
Range	Unlimited	Unlimited
Linearity	>96%	>99.5%
Cross-talk	Low	Low
Hysteresis	<4%	<0.5%
Threshold	50 mN to 100 mN	<5 mN
Accuracy	<5%	<5%
Drift	None	~0.01 N.s^{-1}

Strong

The force plate must be sufficiently strong to tolerate the loads placed on it. For example, they might need to be capable of tolerating loads applied during horse locomotion (e.g., Biknevicius et al. 2006) and yet sensitive enough to measure the forces when frogs jump (e.g., Nauwelaerts and Aerts 2006).

Imperceptible Displacement

For the strain gauges and piezoelectric crystals to register a force, there has to be some deformation; therefore the force sensor must be stiff yet sufficiently sensitive so that the subject interacting with the force plate does not notice the deformation.

High Natural Frequency

Objects have a natural frequency, a frequency at which they vibrate when struck. For example, a tuning fork when struck will vibrate at its natural frequency which produces a characteristic sound. For a mass hanging from a (massless) spring, the natural frequency (f_n) is:

$$f_n = \frac{1}{2\pi}\sqrt{\frac{k}{m}} \tag{7.12}$$

where k is the stiffness of the spring and m is the mass of the mass suspended from the spring. For most objects, the natural (resonant) frequency is a function of the mass and stiffness of the materials comprising the object but will also depend on how the materials are organized and interact. For a force plate, its natural frequency can be assessed by striking it and then measuring its frequency of vibration. Most force plates have different natural frequencies in the different axis directions. The general recommendation is that the force plate is mounted on a concrete block which is independent of the building in which the force plate is installed. By being mounted in such a way, the force plate can be considered independent of the building which means it is not picking up vibrations of the building, for example, due to doors slamming shut or elevator use. Of course in some situations, the ideal mounting is not feasible in which case the natural frequency of the plate will also incorporate frequencies arising from the surroundings. Force plates are often placed on a mounting frame which is then bolted on the ground. Kerwin and Chapman (1988) have shown that these frames also influence the natural frequency of the force plate, and an appropriately designed mounting frame can increase the natural frequency.

The importance of the natural frequency is that this frequency must be greater than the frequencies of the signals to be collected. If the natural frequency is sufficiently high, then basic signal processing (e.g., low-pass filtering) can separate the frequencies of the signal of interest from the vibrations of the force plate. Most ground reaction forces from human movement have frequency content below 50 Hz, so if, for example, the natural frequency of the force plate is 200 Hz, it is easy to separate one from the other. Commercially available force plates appropriately mounted have natural frequencies typically above 300 Hz.

No Cross-Talk Between Axes

Some of the early designs of force plates would display cross-talk, for example, some horizontal forces were registered when only a vertical force was applied: there was cross-talk between axes. Suitable design and calibration will remove this problem.

Stable and Easy Calibration

If the calibration of a force plate is stable that means the calibration does not have to be performed on a frequent basis. Most commercial force plates are very stable, without any recalibration required for years (although spot checks on the calibration are recommended). Considering the loads that might be applied to a force plate and the corresponding accuracy that is required in measuring the forces and in particular the center of pressure calibration can be problematic. For example, imagine loading a force plate with the equivalent of 3000 N of force but with submillimeter positional accuracy of load placement. Under these conditions, calibration is not necessarily that easy; a subsequent sub-section outlines the process for calibration and spot-checks of a force plate.

Light
For the force plate to be light is a desirable quality if it is to be moved to different locations for use in a variety of experiments.

7.2.3 Problems with Force Plates

In gait analysis a typical requirement is that the subjects' foot makes a full contact with a force plate, given that a typical force plate size of 0.60 m by 0.40 m this can be problematic. There are two options to ensure the foot makes full contact with the force plate. One is to have the subject take as many trials as necessary until they naturally make full contact with the force plate; with certain subjects the number of trials they can complete might be limited in which case such a strategy might not be feasible. The alternative is to ask the subject to plan their approach to ensure the force plate is struck. Grabiner et al. (1995) have suggested that visual targeting does not influence ground reaction forces in walking, but in running Challis (2001) has shown that force plate targeting does influence the ground reaction forces. One solution to the problem with striking the force plate is to use a large plate so that targeted striking can be avoided. While this approach is attractive during walking both feet cannot be in contact with the force plate simultaneously which might occur with a large force plate, for running this is not an issue but there is a second problem. As the force plate becomes larger its natural frequency decreases, so very large force plates are not a viable option.

The top of a force plate is metal, commonly aluminum, but that is not a normal walking or running surface. It is feasible to place another surface on top of the force plate, for example, carpet tile or artificial turf, but this means the force measured by the plate is not the one applied by the subject to the ground by their foot (maybe via their footwear), but the force transmitted through the material of the surface placed on top of the force plate.

Some of the problems with force plates can be circumvented using a treadmill system that has force plates beneath the treadmill belt (e.g., Bundle et al. 2015a; Dierick et al. 2004). With such devices, the natural frequency of the plate is reduced and of course susceptible to the vibrations of the treadmill. In addition the computation of the center of pressure requires detailed knowledge of the treadmill belt kinematics.

There are free standing force plates designed for postural studies, which are portable and can be placed on suitable surfaces providing some flexibility for postural studies. There are drawbacks to these portable plates in terms of their natural frequency and sensitivity to the local environment.

7.2.4 Calibration

Imagine a single pillar to the top of which a force is applied (see Fig. 7.8). The response of six strain gauges attached to the pillar is measured; their output would be in volts (s_i, $i = 1, 6$). Each strain gauge is designed to measure a different aspect of the loading on the pillar (force: Fx, Fy, Fz, moments: Mx, My, Mz).

If no cross-talk exists, then:

$$F_x = a_1 s_1 \qquad F_y = a_2 s_2 \qquad F_z = a_3 s_3 \qquad (7.13, 7.14, 7.15)$$
$$M_x = a_4 s_4 \qquad M_y = a_5 s_5 \qquad M_z = a_6 s_6 \qquad (7.16, 7.17, 7.18)$$

Unfortunately cross-talk does exist; therefore the calibration is more complex, but if the equations are expanded, then cross-talk can be accounted for. The resulting equations are:

$$F_x = a_{11} s_1 + a_{12} s_2 + a_{13} s_3 + a_{14} s_4 + a_{15} s_5 + a_{16} s_6 \qquad (7.19)$$

$$F_y = a_{21} s_1 + a_{22} s_2 + a_{23} s_3 + a_{24} s_4 + a_{25} s_5 + a_{26} s_6 \qquad (7.20)$$

Fig. 7.8 Loads applied to a
pillar, as illustration of calibration

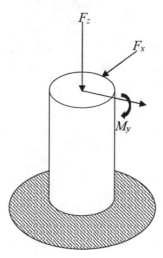

$$F_z = a_{31}\,s_1 + a_{32}\,s_2 + a_{33}\,s_3 + a_{34}\,s_4 + a_{35}\,s_5 + a_{36}\,s_6 \tag{7.21}$$

$$M_x = a_{41}\,s_1 + a_{42}\,s_2 + a_{43}\,s_3 + a_{44}\,s_4 + a_{45}\,s_5 + a_{46}\,s_6 \tag{7.22}$$

$$M_y = a_{51}\,s_1 + a_{52}\,s_2 + a_{53}\,s_3 + a_{54}\,s_4 + a_{55}\,s_5 + a_{56}\,s_6 \tag{7.23}$$

$$M_z = a_{61}\,s_1 + a_{62}\,s_2 + a_{63}\,s_3 + a_{64}\,s_4 + a_{65}\,s_5 + a_{66}\,s_6 \tag{7.24}$$

where a_{ij} are the calibration coefficients. This sequence of equations can then be placed in matrix form:

$$
\begin{bmatrix} F_x \\ F_y \\ F_z \\ M_x \\ M_y \\ M_z \end{bmatrix}
=
\begin{bmatrix}
a_{11} & a_{12} & a_{13} & a_{14} & a_{15} & a_{16} \\
a_{21} & a_{22} & a_{23} & a_{24} & a_{25} & a_{26} \\
a_{31} & a_{32} & a_{33} & a_{34} & a_{35} & a_{36} \\
a_{41} & a_{42} & a_{43} & a_{44} & a_{45} & a_{46} \\
a_{51} & a_{52} & a_{53} & a_{54} & a_{55} & a_{56} \\
a_{61} & a_{62} & a_{63} & a_{64} & a_{65} & a_{66}
\end{bmatrix}
\begin{bmatrix} s_1 \\ s_2 \\ s_3 \\ s_4 \\ s_5 \\ s_6 \end{bmatrix}
\tag{7.25}
$$

This equation given the output from the load cells and the calibration matrix allows determination of the loads applied. If no cross-talk exists, all non-diagonal elements of the calibration matrix will be zero; this is very rarely the case. For calibration known loads are applied and load cell outputs measured; given the number of knowns and unknowns at least six sets of loads are required. Therefore in equation form:

$$L = C\,S \tag{7.26}$$

where L is a $6 \times n$ matrix containing n different applied loads, C is a 6×6 matrix containing the calibration coefficients, and S is a $6 \times n$ matrix containing the sensor outputs to each of the n different applied loads. To compute the matrix of calibration coefficients, the Moore-Penrose inverse can be used (see Appendix A) and very efficiently using the singular value decomposition (see Appendix H):

$$L\,S^T \left(S\,S^T \right)^{-1} = C \tag{7.27}$$

In the process of calibration a range of loads representing the range of loads expected in the task(s) to be examined is important, and while six sets of loads are sufficient some redundancy will increase accuracy.

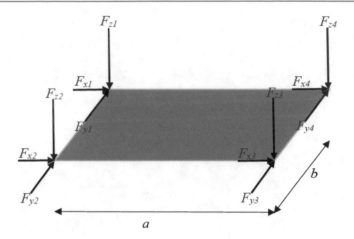

Fig. 7.9 The outputs from a piezoelectric force plate

The process for calibrating a single pillar with strain gauges mounted on it can be generalized to the calibration of strain gauge force plates which typically have four pillars. For a piezoelectric force plate the sensors register forces so additional computations are required to determine the moments. From each set of sensors three sets of forces are measured (see Fig. 7.9); from these forces the resultant force and moments can be determined.

Given the output from four sets of sensors under the support surface of a piezoelectric force plate, the ground reaction forces and moments (M_x, M_y, M_z) can be computed from:

$$F_x = F_{x1} + F_{x2} + F_{x3} + F_{x4} \tag{7.28}$$

$$F_y = F_{y1} + F_{y2} + F_{y3} + F_{y4} \tag{7.29}$$

$$F_z = F_{z1} + F_{z2} + F_{z3} + F_{z4} \tag{7.30}$$

$$M_x = \frac{b}{2}\left(-F_{z1} + F_{z2} + F_{z3} - F_{z4}\right) \tag{7.31}$$

$$M_y = \frac{a}{2}\left(-F_{z1} + F_{z2} + F_{z3} + F_{z4}\right) \tag{7.32}$$

$$M_z = \frac{b}{2}\left(F_{x3} + F_{x4} - F_{x1} - F_{x2}\right) + \frac{a}{2}\left(F_{y1} + F_{y4} - F_{y2} - F_{y3}\right) \tag{7.33}$$

For a force plate system, given the forces and moments applied to the surface of the force plate, the location of the center of pressure can be determined:

$$COP_x = \frac{-h\,F_x - M_y}{F_z} \tag{7.34}$$

$$COP_y = \frac{-h\,F_y + M_x}{F_z} \tag{7.35}$$

where COP_x and COP_y are the center of pressure location in the two horizontal directions and h is the distance from the sensors to the top surface of the plate.

In addition the free moment (T_Z) can be computed, there are no reaction moments about the other two axes. The free moment of the ground reaction is a moment about a vertical axis which is due to shear forces between the foot and ground; it is computed from:

$$T_Z = M_Z - \mathrm{COP}_x F_y + \mathrm{COP}_y F_x \qquad (7.36)$$

The magnitude of the free moment has been linked to the amount of ankle pronation in running (Holden and Cavanagh 1991), and to lower limb running injuries (Milner et al. 2006).

Hall et al. (1996) presented a procedure which can be used for the confirmation of force plate calibration. The basis of the system was a series of calibrated weights of known masses. Using a pulley system, these weights were used to apply known horizontal forces to a force plate, and a vertical loading device where the force was applied to the top plate of the force plate was via a 15 mm diameter ball bearing. Collins et al. (2009) presented a simpler process for confirming force plate calibration, which required the manual production of forces. It consists of a rod with a point at one end to apply forces to the force plate surface and a single axis load cell to measure those forces. Attached to the rod are motion analysis system markers which permit determination of the direction of the load and point of force application. In use this system is much simpler than the process used by Hall et al. (1996) but has the draw back that it requires more equipment. Both of these approaches, Hall et al. (1996) and Collins et al. (2009), are static in nature; an alternative to this is the method proposed by Fairburn et al. (2000). They used an oscillating pendulum mounted in a frame placed on top of a force plate to apply dynamic loads.

Bobbert and Schamhardt (1990) examined the accuracy with which a force plate provided the center of pressure location. By applying loads across the surface of a force plate, they had known center of pressure locations and force plate system output. They noted that center of pressure measurement accuracy was worst toward the edges of the force plate, particularly beyond the sensors. They improved the accuracy of center of pressure measures using a polynomial fit to their records of force plate output and known center of pressure values.

7.3 Force Plate Data Analysis

Force plate data can be useful on its own for biomechanical analysis, and various aspects of the analysis of force plate signals will be described in the following sub-sections. Often force plate data are combined with other data, for example, to compute resultant joint moments; these will be discussed later (see Chap. 10).

7.3.1 Signal Peaks

It is relatively easy in any time series to find the peak or minimum value in the series, and ground reaction forces for a variety of tasks have been quantified using this approach. A time series potentially contains a lot of information, and characterizing it using one point in the series, or two with a peak and a minimum, may not be sufficient to adequately characterize the sequence even if such an approach lends itself to easy application of statistical tests. For some ground reaction forces, there are events in the time series which are considered to be important. For example, for heel strikers running Nigg (2001) has identified that the vertical ground reaction force profile has two distinct peaks (see Fig. 7.10); the first is usually referred to as the impact peak and the second as the active peak.

7.3.2 Frequency Components

In the analysis of normal walking Antonsson and Mann (1985) demonstrated that of the power in the vertical ground reaction force signal, 98% was below 10 Hz. In a similar analysis Stergiou et al. (2002) reported that for the ground reaction forces in walking, 99% of the signal power was below 23 Hz, 15 Hz, and 13 Hz in the medial-lateral, anterior-posterior, and vertical directions, respectively, for their younger subjects, with similar values for their older subjects.

During running the frequency spectrum is broader, for example, Gruber et al. (2017) used wavelet analysis to compute the mean frequency in a group of rear foot strikers and non-rear foot strikers. The mean frequency was 26.4 Hz in the rear foot striking group and 16.5 Hz for the non-rear foot strikers for the vertical ground reaction force. Shorten and Mientjes (2011) demonstrated that in heel strikers there is a much broader frequency range around the time of the impact peak compared with the remainder of the support phase.

In quiet upright stance, the frequency spectrum is, as would be expected, narrower than in walking or running. Prieto et al. (1996) had subjects stand still on a force plate for 30 seconds; the body does not stay perfectly still it moves in part because of the nature of the body (inverted pendulum) and due to physiological processes. For example, young subjects had 95% of their

Fig. 7.10 The four variables
identified for the vertical ground
reaction forces for the stance
phase of running: impact peak
magnitude, active peak
magnitude, time from initial
contact to impact peak, and time
from initial contact to active peak

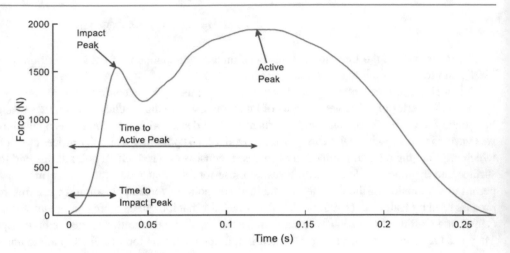

signal power of their center of pressure motion below 0.93 Hz and 1.05 Hz in the anterior-posterior directions with the eyes open and closed, respectively. In older subjects, these frequency ranges were increased (0.99 Hz and 1.10 Hz).

When collecting force plate data, it is tempting to set the sample rate considering sampling theorem and the frequency content of the task being analyzed. This approach ignores the sampled signal components due to the vibrations of the force plate (at its natural frequency). Therefore the data should be sampled at high enough a sample frequency to capture the vibrations of the force plate, and then the collected data can be low-pass filtered to remove the vibrations of the force plate. The alternative approach is to have an analog filter so that before the signal from the force plate is sampled, it is low-pass filtered to remove force plate high-frequency vibrations. With such an analog filter the sample rate can be tuned to the frequency content of the tasks being analyzed.

7.3.3 Impulse (Integration)

In mechanics impulse (symbolized by J) is the integral of a force, $F(t)$, over the time interval t_1 to t_2, for which it acts, therefore:

$$J = \int_{t_1}^{t_2} F(t)\, dt \qquad (7.37)$$

Impulse is a vector quantity, which is equivalent to the change in linear momentum of the body to which the force is applied:

$$J = \int_{t_1}^{t_2} F(t)\, dt = mv(t_2) - mv(t_1) \qquad (7.38)$$

where m is the mass of the body, v is the velocity of the center of mass of the body, and the mass of the body remains constant over the time period.

If the mass of the body remains constant for the duration of the application of the force, then the impulse is proportional to the change in velocity of the center of mass of the body. Therefore computation of impulse from force-time records for humans in motion indicates the change in velocity of the center of mass. For example, during steady-state running, the ground reaction forces in the anterior-posterior direction should have zero net impulse for a foot fall because if the running is steady state the center of mass horizontal velocity at foot strike should be the same as the center of mass horizontal velocity at foot off. During constant speed running, the impulse in the anterior-posterior direction is negative during the first half of stance representing a reduction in center of mass velocity and positive for the second half of stance representing an increase in velocity.

In a vertical jump if the subject is stationary at the start, then the computed vertical impulse from the start to takeoff divided by subject mass gives the vertical velocity at takeoff. Given the takeoff velocity using the equations of uniformly accelerated

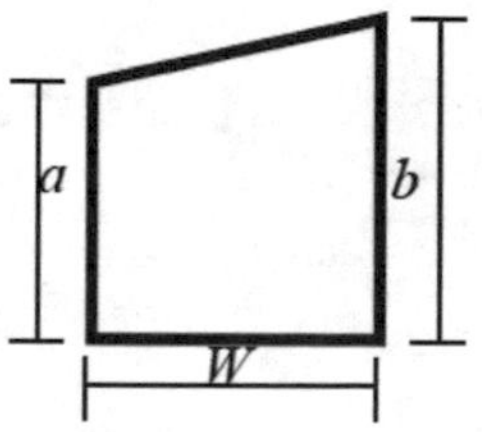

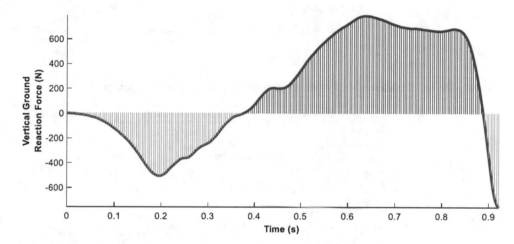

Fig. 7.11 Dimensions of a trapezoid and the equation used to compute the area of the trapezoid

Fig. 7.12 The vertical ground reaction force profile for a vertical jump, with the negative (green) and positive (red) impulse components highlighted

motion, it is then feasible to predict the motion of the center of mass during the flight phase of the jump, including the computation of jump height.

The area under the force-time curve, the impulse, can be computed using numerical integration. To compute the area under a curve it is most common to divide the curve into small sections and then model each area as a trapezoid (see Fig. 7.11).

Figure 7.12 shows the vertical ground reaction force for a vertical jump. The red indicates positive impulse and the green negative impulse; the sum of these components gives the net impulse. Note that in these curves, the force plate was set to zero when the subject stood on it; therefore during the flight phase, the force plate registers a force equal to negative the subjects body weight. If the plate had not been zeroed with the subject standing on the plate, then once the impulse curve had been calculated, the impulse due to the product of the subjects' body weight and duration of the contact phase of the jump would have to be subtracted to give the impulse responsible for the generation of the jump.

Of course it is feasible to analyze the impulse over smaller time steps. For example, if the data were sampled at 100 Hz, then for the first 0.01 s, the impulse can be computed which if the initial velocity is zero the body center of mass velocity can be computed at the end of the first one hundredth of a second. This value then gives the initial velocity for the next 0.01 s interval, so after computing the area under the curve between 0.01 and 0.02 seconds the velocity at 0.02 s can be computed and so on throughout the jump. Once the velocity profile is obtained following the same procedure, center of mass displacement can be calculated. With the vertical ground reaction force this signal can be converted to the vertical acceleration of center of mass by first subtracting the force due to body weight from the signal, and then dividing by the subject mass. Pseudo-code 7.1 illustrates the analysis of the vertical ground reaction forces for a standing start jump, and Fig. 7.13 illustrates the results from such an analysis.

Fig. 7.13 Vertical ground reaction force analysis for a vertical jump, (**a**) the vertical ground reaction force, (**b**) the center of mass velocity, and (**c**) the displacement of the center of mass

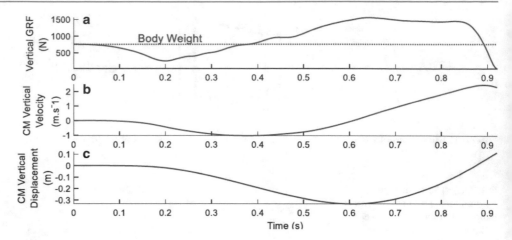

Pseudo-Code 7.1: Analysis of vertical ground reaction force of a jump

Purpose – computation of center of mass kinematics from vertical ground reaction force	
Inputs t – time base (s) Fz – vertical ground reaction force (N) weight – subjects weight (N) g – acceleration due to gravity (m/s^2)	
Outputs DispCom – displacement of center of mass VelCom – velocity of center of mass during the jump	
Fz ← Fz – weight AccCom ← Fz/(weight/g)	Remove weight from Fz Acceleration of center of mass
VelCom ← cumtrapz(t, AccCom) DispCom ← cumtrapz(t, VelCom)	Cumulative integration
VelTakeOff ← VelCom(end) JumpHeight ← VelTakeOff^2 / (g + g)	Velocity at takeoff Jump height

7.4   Center of Pressure Analysis

The center of pressure is the point where the vertical ground reaction force intercepts with the force plate (support surface). The center of pressure is a normal output from the collection force plate data; its path can be informative, for example, during quiet standing it is used as a measure of stability. The analysis of the center of pressure path in gait analysis will be described, and more extensively for the analysis of upright stance.

7.4.1   During Gait

The location of the center of pressure is important in gait analysis as it reflects the distal point of force application and can be used in the computation of the resultant joint moment (see Chap. 10). It has also been quantified during running to track the path of the center of pressure relative to the shoe footprint (Cavanagh 1980, see Fig. 7.14).

7.4.2   During Quiet Standing

It is quite common to analyze the path of the center of pressure (COP) during quiet upright standing. The COP signal is analyzed in a variety of ways and has been shown to discriminate between various populations, for example, young and old adults (e.g., Prieto et al. 1996), adolescents with and without idiopathic scoliosis (e.g., Nault et al. 2002), elderly fallers and non-fallers (e.g., Melzer et al. 2004), and athletes with and without concussions (Fino et al. 2016). Often times these measures are referred to as measures of stability but stability in most of these studies is not strictly defined, but the measures reflect metrics related to aspects of stability.

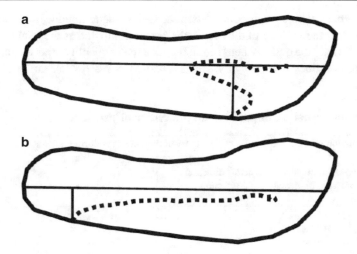

Fig. 7.14   Path of the center of pressure for two runners, (**a**) a rear foot striker and (**b**) and mid-foot striker

Fig. 7.15   Stabilograms of two subjects performing 30 seconds of quiet standing

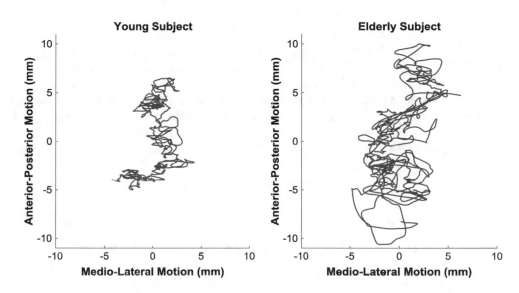

The center of pressure signal consists of two components in the two horizontal directions (x, y). Depending on how the subjects is oriented on the force plate relative to these axes, the center of pressure signal is typically referenced relative to the body axes anterior-posterior and medial-lateral. As the subject is not typically constrained to precisely place their feet on the force plate, the center of pressure values (COP_x, COP_y) is centered, giving $\overline{COP}_x(t), \overline{COP}_y(t)$, relative to the mean center of pressure position, therefore:

$$\overline{COP}_x(t) = COP_x(t) - \text{mean}(COP_x) \tag{7.39}$$

$$\overline{COP}_y(t) = COP_y(t) - \text{mean}(COP_y) \tag{7.40}$$

If the center of pressure-time series is plotted, there are two types of plots:
Statokinesiogram – plot of each component against time.
Stabilogram – plot of one component against the other component.

Figure 7.15 illustrates a stabilogram for a young (21 years old) and an older subject (70 years old)

From the center of pressure data, various metrics are determined including amplitude, root mean square distance, and path length. Another common metric is the velocity of the center of pressure computed as the total path length divided by the total time (e.g., Prieto et al. 1996). Strictly the quantity being computed is a scalar quantity; thus the metric is the speed of the center of pressure but the literature describes it as velocity. See Pseudo-code 7.2 for various measures and how they can be computed.

Pseudo-Code 7.2: Metrics which can be computed from the center of pressure

Purpose – to illustrate the computation of the basic metrics of the center of pressure path during quiet standing
Inputs
CoPap - time series of center of pressure in anterior-posterior direction
CoPml - time series of center of pressure in medial-lateral direction
t – time base, with n elements
Outputs
AmCoP – amplitude of COP displacement
RmsCoP – root mean square of COP signal
LenCoP – length of path the COP signal
SpCoP – speed of the COP signal
AmCoPap ← max(CoPap) – min(CoPap)
AmCoPml ← max(CoPml) – min(CoPml)
RmsCoPap ← (1/n) * sqrt(sum(COPap.^2))
RmsCoPml ← (1/n) * sqrt(sum(COPml.^2))
LenCoP ← sum(sqrt(CoPap.^2 + CoPml.^2))
SpCOP ← LenCoP / (t(end) – t(1))

Another popular measure is to compute an ellipse that fits the center of pressure stabilogram, and then the area of the ellipse can be used to quantify the center of pressure motion. The basic anatomy of an ellipse depends on the lengths of two axes (see Fig. 7.16).

The lengths of the axes can be used to define the coordinates of the edges of the ellipse:

$$\frac{x^2}{a^2} + \frac{y^2}{b^2} = 1 \tag{7.41}$$

If the center of the ellipse is not at the origin of the axis system, then the equation becomes:

$$\frac{(x - x_0)^2}{a^2} + \frac{(y - y_0)^2}{b^2} = 1 \tag{7.42}$$

The eccentricity of the ellipse can be computed from:

$$\varepsilon = \frac{\sqrt{a^2 - b^2}}{a} \tag{7.43}$$

An eccentricity of *0* occurs for a circle, and a ratio of nearly *1* occurs for a long and narrow ellipse. The area of the ellipse (*A*) is computed from:

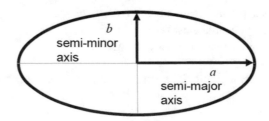

Fig. 7.16 Basic anatomy of an ellipse

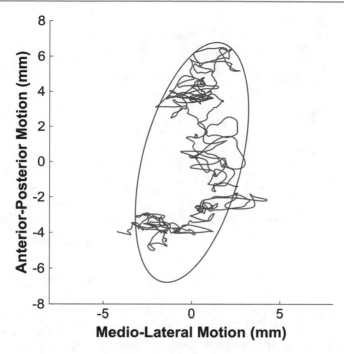

Fig. 7.17 A confidence ellipse fitted to center of pressure data so that the area of the ellipse can be used to quantify center of pressure motion. (Note the fit does not encompass all of the center of pressure values)

$$A = \pi\, a\, b \tag{7.44}$$

There have been various algorithms proposed to fit the ellipse to a series of data points (e.g., Bookstein 1979; Batschelet 1981; Gander and Hřebíček 1997), where the fit is a least-squares fit. Using any of these least-squares algorithms means not all center of pressure values are necessarily within the fitted ellipse. In biomechanics the most popular approach has been to fit a confidence ellipse, so the fitted ellipse has (1- α) probability that the fitted ellipse contains the center of the points; typically α is set to 0.95 (Press et al. 2007). The most popular confidence ellipse fitting algorithm in biomechanics is that presented by Oliveira et al. (1996), although this algorithm produces an error in the estimated area if the number of data points is small. For example, if there are 10 data points, the error is 26% and 100 data points 2.5%, but given 3000 samples, the error is 0.1% (Schubert and Kirchner 2013). Most analyses of quiet standing have subjects stand for 30 seconds, so given a sample rate of at least 100 Hz, the error would be small for most analyses of upright stance. Pseudo-code 7.3 computes the area of an ellipse using a confidence ellipse, and Fig. 7.17 illustrates an ellipse fitted to center of pressure data. See Appendix I for more details about ellipses and their modeling.

Pseudo-Code 7.3: Computation of the area of an ellipse to encompass the center of pressure data

Purpose – to illustrate the computation of the area of an ellipse encompassing the center of pressure path during quiet standing by modeling the data by a confidence ellipse.

Inputs
CoPap – time series of center of pressure in anterior-posterior direction
CoPml – time series of center of pressure in medial-lateral direction

Outputs
AreaCoP – area of ellipse fitted to COP data

CM ← cov(COPap, COPml)	Covariance matrix
[evec, eval] ← eig(CM)	Compute eigenvectors/values
Axes ← sqrt(1.96 * svd(eval))	Compute length of axes using singular value decomposition (95%)
AreaCOP ← pi * prod(axes)	Compute ellipse area

Another class of metrics of center of pressure paths is to use an analysis of the power spectrum. The power spectrum is computed, and then from this spectrum the frequencies at which the following events occur computed peak power, mean

Fig. 7.18 The power spectrum of the center of pressure data from a subject during quiet standing. The locations of the following frequencies are noted: peak power, mean power, point below which 50% of the power lies, and point below which 95% of the power lies

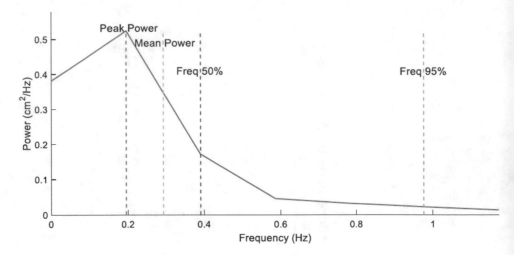

frequency, and the frequency beneath which 50% or 95% of the power exists. This analysis is illustrated in the following Pseudo-code 7.4 and representative Fig. 7.18.

As is typical of the frequency profile of quiet standing center of pressure motion, the signal exists predominantly in the low-frequency domain, mostly below 1 Hz in Fig. 7.18. The duration of the sampling dictates the lowest frequency component. The lowest frequency is often called the fundamental frequency. For example, if a signal is sampled for 20 seconds the lowest reconstructable frequency from the signal is 0.05 Hz (i.e., 1/20). Therefore the sample duration needs to be great enough to permit resolution at the low end of the spectrum (see Chap. 3 for more details).

There are other metrics extracted from the center of pressure path, but here the most typically used have been presented.

Pseudo-Code 7.4: Metrics which can be computed from the power spectrum of the center of pressure

Purpose – to illustrate the computation of the metrics obtained from the power spectrum of the center of pressure path during quiet standing
Inputs Freq – frequencies from the Fourier analysis of a time series of center of pressure data Power – power at each frequency for Fourier analysis of time series of center of pressure data t – time base, with n elements
Outputs PeakPwr – frequency at which the peak power occurs 50Pwr – frequency below which 50% of power lies 95Pwr – frequency below which 95% of power lies meanPwr – frequency at which mean power value occurs
[i] ← max(Power) PeakPwr ← Freq(i)
cumPwr ← cumtrapz(Freq, Power) totPwr ← cumPwr(end) All50Pwr ← find(cumPwr > = 0.5 *totPwr) 50Pwr ← Freq(All50Pwr(1))
All95Pwr ← find(cumPwr > = 0.95 *totPwr) 95Pwr ← Freq(All95Pwr(1))
meanPwr ← trapz(Freq, Freq.*Power)/trapz(Freq, Power)

7.5   Pressure Measurement

Pressure (p) is a derived mechanical property; this scalar quantity is force per unit area (A):

$$p = \frac{F}{A}$$

(7.45)

The units of pressure are pascals (Pa or N.m^{-2}). A force plate measures forces well but gives little indication of the pressure distributions. To understand about loadings, useful information can be obtained if the pressure distributions across the loading surface are known.

Abramson (1927) in an effort to examine the pressure beneath the feet during upright stance spread steel shot over a hard surface then placed a lead plate on top of the shot. The subjects then stood on the plate which caused the shot to penetrate the lead plate. The depth of penetration indicated the force applied to the different parts of the foot; this method is probably more a measure of foot shape than it is of the pressures applied to the foot. Elftman (1934) sought to measure the distribution of pressure beneath the foot during walking. His method for measuring pressures utilized a rubber mat with a flat upper surface and a lower studded surface. The studs were pyramidal projections. As a subject walked across the rubber mat, the pyramids deformed in proportion to the applied force; these deformations were filmed through a glass plate on which the mat rested. The time course of these deformations gave an approximation to the pressure distributions beneath the foot during walking gait.

The development of sensitive load sensors and microelectronics ushered more automatic and higher resolution methods of measuring foot pressures (e.g., Hennig et al. 1982). The technology relies on piezoceramic squares arranged in a grid pattern; if the piezoelectric sensors emit a charge the presumption is that the sensor has force applied to its surface so the local pressure can be computed. If sufficiently small piezoceramic squares are arranged in a dense grid pattern, they can provide a high-resolution time series of pressure distributions on the sole of the foot (see Fig. 7.19). While there are mats which will measure pressures, the most common application is for pressure in-soles.

The resolution of the pressure sensor, pedobarograph, is a direct function of the number of pressure-sensing elements (see Fig. 7.20). But with an increasing number of sensors comes the potential for reduced sample rate of the pressure distributions.

As each sensor is measuring the force applied to it, when the foot flat is on the floor, the sum of these forces should approximate the vertical ground reaction force. There have been a number of studies which have simultaneously measured the vertical ground reaction force from a force plate and from the sum of the forces measured by the sensors comprising the

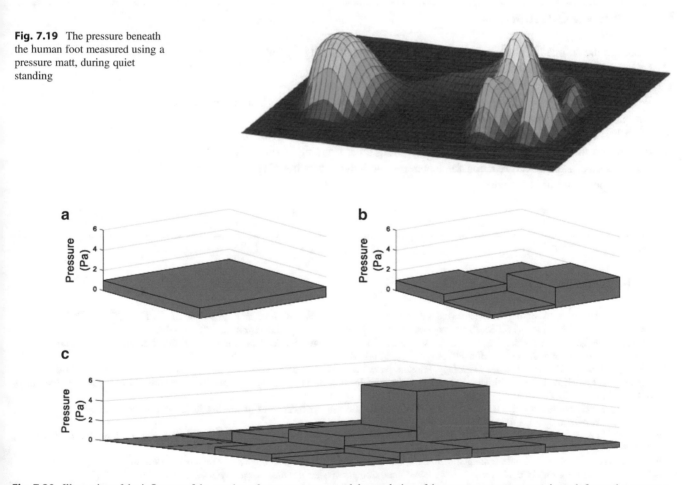

Fig. 7.19 The pressure beneath the human foot measured using a pressure matt, during quiet standing

Fig. 7.20 Illustration of the influence of the number of pressure sensors and the resolution of the pressure measurement; in each figure the same net force and pressure have been applied, but in (**a**) there is 1 sensor, (**b**) there are 4 sensors, and (**c**) there are 16 sensors

pedobarograph. Gross and Bunch (1988) performed such a study, and while the in-sole determined vertical ground reaction force captured the general force profile during running the magnitudes were low relative to the force plate measured forces. This would be expected as the foot is not always flat on the ground during running; therefore the in-soles are not always orientated, so they are measuring the vertical forces.

Pressures at the foot-shoe interface can provide useful information, for example, in footwear design to control plantar pressures in patients with diabetes (Cavanagh and Ulbrecht 1994). Pressure measures can be useful for areas other than the sole of the foot, for example, for measuring seat pressure distributions (Cardoso et al. 2018), and residual-limb-prosthesis interface pressures (Laing et al. 2019) to identify potential sites of damage and injury.

The signal from a pedobarograph is the time series of the pressures during the contact phase of locomotion. Metrics extracted from these times series include maximum pressure, often with reference to specific areas of the foot, the pressure-time integral representing the total loading on the foot, and time to the maximum pressure. Analysis of the peak pressures has been shown to correlate well with subjective feelings of comfort during walking (Jordan et al. 1997). Hills et al. (2001) measured the mean pressures under certain areas of the feet of obese and nonobese subjects, with the pressures greater for the obese. Pataky et al. (2012) showed that simple measures from a pedobarograph measured during foot-ground contacts could distinguish between the gaits of 104 subjects with a classification rate of over 99%. As with force plate data, there is a tendency to select one or two events in a time series to quantify the signal, Pataky and Goulermas (2008) demonstrated that statistical parametric mapping can provide a higher-resolution basis for the comparison of pressure distributions.

Pressure in-soles do not have the same issues as force plates, for example, fixed location and risk of force plate targeting, but there is evidence that certain experimental strategies are required to collect representative data and ensure data quality. Melvin et al. (2014) in 20 healthy subjects reported that at least 166 steps are required for subjects to acclimate to pressure in-soles and that 30 steps should be collected to ensure data quality.

7.6   Review Questions

1. What are the sources of the vertical ground reaction forces during "quiet" standing?
2. List and describe the key design criteria for a force plate.
3. Why do force plates require a high natural frequency?
4. Why might you use an analog low-pass filter for pre-processing a signal from a force plate?
5. Explain why integration of a force-time record might be of use.
6. What is the center of pressure? Describe an application where it may be of use.
7. How might the path of the center of pressure during quiet standing be quantified?
8. Sketch the vertical ground reaction force patterns for walking and jumping.
9. How do pedobarographs work?

References

Cited References

Abramson, E. (1927). Zur kenntnis der mechanik des mittelfußes. *Skandinavisches Archiv Für Physiologie, 51*(2), 175–234.
Antonsson, E. K., & Mann, R. W. (1985). The frequency content of gait. *Journal of Biomechanics, 18*(1), 39–47.
Arndt, A., Ekenman, I., Westblad, P., & Lundberg, A. (2002). Effects of fatigue and load variation on metatarsal deformation measured in vivo during barefoot walking. *Journal of Biomechanics, 35*(5), 621–628.
Batschelet, E. (1981). *Circular statistics in biology*. London; New York: Academic Press.
Bergmann, G., Graichen, F., & Rohlmann, A. (1993). Hip joint loading during walking and running, measured in two patients. *Journal of Biomechanics, 26*(8), 969–990.
Biknevicius, A. R., Mullineaux, D. R., & Clayton, H. M. (2006). Locomotor mechanics of the tölt in Icelandic horses. *American Journal of Veterinary Research, 67*(9), 1505–1510.
Bobbert, M. F., & Schamhardt, H. C. (1990). Accuracy of determining the point of force application with piezoelectric force plates. *Journal of Biomechanics, 23*(7), 705–710.
Bonde-Petersen, F. (1975). A simple force platform. *European Journal of Applied Physiology and Occupational Physiology, 34*(1), 51–54.
Bookstein, F. L. (1979). Fitting conic sections to scattered data. *Computer Graphics and Image Processing, 9*(1), 56–71.
Borelli, G. A. (1685). *De Motu Animalium*. Lugduni Batavorum,: apud Petrum Vander Aa.

Bundle, M. W., Powell, M. O., & Ryan, L. J. (2015a). Design and testing of a high-speed treadmill to measure ground reaction forces at the limit of human gait. *Medical Engineering & Physics, 37*(9), 892–897.

Cardoso, M., McKinnon, C., Viggiani, D., Johnson, M. J., Callaghan, J. P., & Albert, W. J. (2018). Biomechanical investigation of prolonged driving in an ergonomically designed truck seat prototype. *Ergonomics, 61*(3), 367–380.

Cavanagh, P. R. (1980). *The running shoe book.* Mountain View, CA: Anderson World.

Cavanagh, P. R., & Ulbrecht, J. S. (1994). Clinical plantar pressure measurement in diabetes: Rationale and methodology. *The Foot, 4*(3), 123–135.

Challis, J. H. (2001). The variability in running gait caused by force plate targeting. *Journal of Applied Biomechanics, 17*(1), 77–83.

Collins, S. H., Adamczyk, P. G., Ferris, D. P., & Kuo, A. D. (2009). A simple method for calibrating force plates and force treadmills using an instrumented pole. *Gait & Posture, 29*(1), 59–64.

Dierick, F., Penta, M., Renaut, D., & Detrembleur, C. (2004). A force measuring treadmill in clinical gait analysis. *Gait & Posture, 20*(3), 299–303.

Elftman, H. (1934). A cinematic study of the distribution of pressure in the human foot. *The Anatomical Record, 59*(4), 481–491.

Elftman, H. (1938). The measurement of the external force in walking. *Science, 88*(2276), 152–153.

Fairburn, P. S., Palmer, R., Whybrow, J., Fielden, S., & Jones, S. (2000). A prototype system for testing force platform dynamic performance. *Gait & Posture, 12*(1), 25–33.

Fino, P. C., Nussbaum, M. A., & Brolinson, P. G. (2016). Decreased high-frequency center-of-pressure complexity in recently concussed asymptomatic athletes. *Gait & Posture, 50*, 69–74.

Gander, W., & Hřebíček, J. (1997). *Solving Problems in Scientific Computing using Maple and MATLAB (4th, expanded and rev. ed).* New York: Springer.

Gola, M. M. (1980). Mechanical design, constructional details and calibration of a new force plate. *Journal of Biomechanics, 13*(2), 113–128.

Grabiner, M. D., Feuerbach, J. W., Lundin, T. M., & Davis, B. L. (1995). Visual guidance to force plates does not influence ground reaction force variability. *Journal of Biomechanics, 28*(9), 1115–1117.

Gregor, R. J., Komi, P. V., Browning, R. C., & Jarvinen, M. (1991). A comparison of the triceps surae and residual muscle moments at the ankle during cycling. *Journal of Biomechanics, 24*(5), 287–297.

Gross, T. S., & Bunch, R. P. (1988). Measurement of discrete vertical in-shoe stress with piezoelectric transducers. *Journal of Biomedical Engineering, 10*(3), 261–265.

Gruber, A. H., Edwards, W. B., Hamill, J., Derrick, T. R., & Boyer, K. A. (2017). A comparison of the ground reaction force frequency content during rear foot and non-rear foot running patterns. *Gait & Posture, 56*(Supplement C), 54–59.

Hall, M. G., Fleming, H. E., Dolan, M. J., Millbank, S. F. D., & Paul, J. P. (1996). Static in situ calibration of force plates. *Journal of Biomechanics, 29*(5), 659–665.

Heglund, N. C. (1981). A simple design for a force-plate to measure ground reaction forces. *Journal of Experimental Biology, 93*(1), 333–338.

Hennig, E. M., Cavanagh, P. R., Albert, H. T., & Macmillan, N. H. (1982). A piezoelectric method of measuring the vertical contact stress beneath the human foot. *Journal of Biomedical Engineering, 4*(3), 11568213 11568222.

Hills, A. P., Hennig, E. M., McDonald, M., & Bar-Or, O. (2001). Plantar pressure differences between obese and non-obese adults: A biomechanical analysis. *International Journal of Obesity, 25*(11), 1674 1679.

Holden, J. P., & Cavanagh, P. R. (1991). The free moment of ground reaction in distance running and its changes with pronation. *Journal of Biomechanics, 24*(10), 887–889, 891-897.

Hsieh, S. T. (2006). Three-axis optical force plate for studies in small animal locomotor mechanics. *Review of Scientific Instruments, 77*(054303), 054307.

Jordan, C., Payton, C. J., & Bartlett, R. M. (1997). Perceived comfort and pressure distribution in casual footwear. *Clinical biomechanics, 12*(3), S5.

Laing, S., Lythgo, N., Lavranos, J., & Lee, P. V. (2019). An investigation of pressure profiles and wearer comfort during walking with a transtibial hydrocast socket. *American Journal of Physical Medicine & Rehabilitation, 98*(3), 199–206.

Marey, É. J., & Demenÿ, G. (1885). *Locomotion Humaine, Mécanisme du Saut:* Gauthier-Villars, imprimeur-libraire des comptes rendus des séances de l'Académie des sciences.

Melvin, J. M. A., Preece, S., Nester, C. J., & Howard, D. (2014). An investigation into plantar pressure measurement protocols for footwear research. *Gait & Posture, 40*(4), 682–687.

Melzer, I., Benjuya, N., & Kaplanski, J. (2004). Postural stability in the elderly: A comparison between fallers and non-fallers. *Age and Ageing, 33*(6), 602–607.

Milburn, P. D. (1990). The kinetics of rugby union scrummaging. *Journal of Sports Sciences, 8*(1), 47–60.

Milner, C. E., Davis, I. S., & Hamill, J. (2006). Free moment as a predictor of tibial stress fracture in distance runners. *Journal of Biomechanics, 39*(15), 2819–2825.

More, J. J. (1977). The Levenberg-Marquardt algorithm: Implementation and theory. In G. A. Watson (Ed.), *Numerical analysis* (pp. 105–116). New York, NY: Springer.

Nault, M. L., Allard, P., Hinse, S., Le Blanc, R., Caron, O., Labelle, H., & Sadeghi, H. (2002). Relations between standing stability and body posture parameters in adolescent idiopathic scoliosis. *Spine, 27*(17), 1911–1917.

Nauwelaerts, S., & Aerts, P. (2006). Take-off and landing forces in jumping frogs. *Journal of Experimental Biology, 209*(1), 66–77.

Newton, I. (1687). *Philosophiae Naturalis Principia Mathematica.* Londini: Jussu Societatis Regiæ ac Typis Josephi Streater. Prostat apud plures Bibliopolas.

Nigg, B. M. (2001). The role of impact forces and foot pronation: A new paradigm. *Clinical Journal of Sport Medicine, 11*(1), 2–9.

Oliveira, L. F., Simpson, D. M., & Nadal, J. (1996). Calculation of area of stabilometric signals using principal component analysis. *Physiological Measures, 17*, 305–312.

Pataky, T. C., & Goulermas, J. Y. (2008). Pedobarographic statistical parametric mapping (pSPM): A pixel-level approach to foot pressure image analysis. *Journal of Biomechanics, 41*(10), 2136–2143.

Pataky, T. C., Mu, T., Bosch, K., Rosenbaum, D., & Goulermas, J. Y. (2012). Gait recognition: Highly unique dynamic plantar pressure patterns among 104 individuals. *Journal of the Royal Society Interface, 9*(69), 790–800.

Payne, A. H. (1974). A force platform system for biomechanics research in sport. In R. C. Nelson & C. A. Morehouse (Eds.), *Biomechanics IV* (pp. 502–509). Baltimore: University Park Press.

Press, W. H., Teukolsky, S. A., Vetterling, W. T., & Flannery, B. P. (2007). *Numerical recipes: The art of scientific computing*. Cambridge: Cambridge University Press.

Prieto, T. E., Myklebust, J. B., Hoffmann, R. G., Lovett, E. G., & Myklebust, B. M. (1996). Measures of postural steadiness: Differences between healthy young and elderly adults. *IEEE Transactions on Biomedical Engineering, 43*(9), 956–966.

Schubert, P., & Kirchner, M. (2013). Ellipse area calculations and their applicability in posturography. *Gait & Posture, 39*(1), 518–522.

Shorten, M., & Mientjes, M. I. V. (2011). The 'heel impact' force peak during running is neither 'heel' nor 'impact' and does not quantify shoe cushioning effects. *Footwear Science, 3*(1), 41–58.

Stergiou, N., Giakas, G., Byrne, J., & Pomeroy, V. (2002). Frequency domain characteristics of ground reaction forces during walking of young and elderly females. *Clinical biomechanics, 17*(8), 615–617.

Useful References

Bundle, M. W., Powell, M. O., & Ryan, L. J. (2015b). Design and testing of a high-speed treadmill to measure ground reaction forces at the limit of human gait. *Medical Engineering & Physics, 37*(9), 892–897.

Santos, D. A., & Duarte, M. (2016). A public data set of human balance evaluations. *PeerJ, 4*, e2648.

Anthropometry

8

Overview

In the broadest sense anthropometry is the measurement of the human body. Studies in anthropometry can focus on the dimensions of the human body (e.g., Tanner et al. 1964; Gordon et al. 1989; Claessens et al. 1991), but for mechanical analyses the anthropometric data typically must include the mass, center of mass location, and moments of inertia of the body segments.

Vitruvian Man drawn by Leonardo da Vinci (1452–1519) is in part an exploration of human anthropometry. In this picture (see Fig. 8.1), the square demonstrates that standing height is equal to wingspan, but the writing also makes reference to other body proportions, for example, that the length of the foot is one-seventh of standing height. These body proportions come from the work Marcus Vitruvius Pollio a Roman architect who in the first century BC published De architectura (Vitruvius 1914).

One of the earliest researchers to extend anthropometry to include the important mechanical properties of the human body segments was Emil Harless (1820–1862) a German physiologist. Harless made detailed measurements on two cadavers measuring segmental masses and center of locations (Harless 1860). His sample was two decapitated prisoners which suggests there may have been significant fluid loss prior to the measurement of the bodies; they were not frozen which is a typical step with fresh cadavers to minimize fluid loss. To estimate the inertial properties of the upper trunk, he modeled it as a truncated cone a method of geometric modeling of body segments that would prove popular over century after Harless's death.

The content of this chapter requires a working knowledge of matrix algebra (see Appendix A), and calculus (see Appendix J).

8.1 Body and Segment Inertial Properties

The mass, center of mass, and moments of inertia can collectively be referred to as the inertial properties; these can be for the whole body or of the segments which comprise that body. In many biomechanical studies the segmental inertial properties are required to determine whole body inertial properties.

8.1.1 Body Segments

For biomechanical analysis it is common to divide the body into different segments. For example, the upper limb can be divided into the hand, forearm, and upper arm. The decision in this case is where one segment begins and the other ends. A common approach is to adopt segment boundaries based on previous definitions (see Table 8.1). Another solution is to compute the location of an axis of rotation for the joint between segments and use this as a reference point for the start/end of a segment. Unfortunately the joints in the body do not have fixed axes of rotation (e.g., Blankevoort et al. 1990). Sometimes segments are further subdivided into separate segments for the purpose of biomechanical analyses, for example, the foot can be considered to be three separate segments rather than just one segment (e.g., Pothrat et al. 2015).

J. H. Challis, *Experimental Methods in Biomechanics*, https://doi.org/10.1007/978-3-030-52256-8_8

Fig. 8.1 Vitruvian Man drawn
by Leonardo da Vinci
around 1490

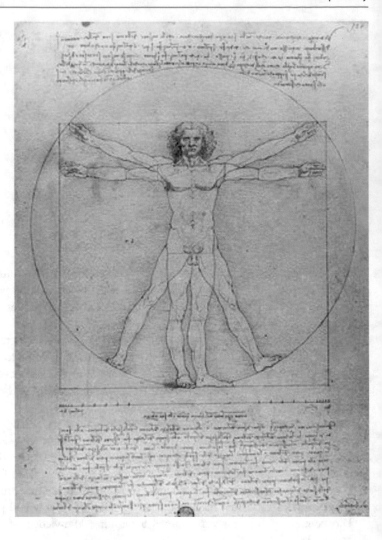

Table 8.1 The definitions of
segments by Dempster (1955)

Segment	Proximal end	Distal end
Hand	Wrist axis	Middle finger knuckle (II)
Forearm	Elbow axis	Ulnar styloid
Upper arm	Glenohumeral joint	Elbow axis
Foot	Lateral malleolus	Head metatarsal (II)
Shank	Femoral condyles	Medial malleolus
Thigh	Greater trochanter	Femoral condyles
Trunk	Greater trochanter	Glenohumeral joint
Head + neck	Cervical rib 7-thoracic rib 1	Ear canal

Generally for determining the properties of the body segments these segments are assumed to be rigid, which generally is not the case, and indeed for certain analyses this assumption of rigidity can influence the results of the analyses (e.g., Challis and Pain 2008). For the purpose of tractability of analysis, the segments are typically assumed to be rigid but the potential influence of this assumption should be considered on a study by study basis.

8.1.2 Segment Mass

The resistance of a body to linear acceleration is a function of its mass; therefore the masses of the segments are important parameters to know for mechanical analyses. The mass, which is simply the amount of matter comprising an object, should be reported in kilograms (kg). As with the other inertial properties for most analyses it is assumed to remain constant for the duration of the analysis.

Defining the mass of a segment is straightforward, but it helps when calculating the other inertial properties to consider the segment to be made up of a set of np particles each with a mass of m_i. Then the mass of the segment is computed from:

$$\text{mass} = \sum_{i=1}^{np} m_i$$

8.1.3 Segment Center of Mass

The sum of all of the forces acting on a segment can be considered to cause the acceleration of the segment center of mass. The center of mass is the point for the segment about which the mass of the segment is evenly distributed. It is the point for which the first-order mass moment is zero. On occasions, confusingly, the center of mass is also referred to as the centroid, but the instantaneous center of rotation is also sometimes caused the centroid.

For two-dimensional analysis, the center of mass is typically specified as the distance, reported in meters (m), from one end of the segment to the center of mass location along a line from one end of the segment to the other end of the segment. In three dimensions a reference frame is defined within the segment of interest, and the location of the center of mass is specified in relation to these segment axes; it therefore has three components (X, Y, Z). The center of mass can be computed from:

$$x_{CM} = \frac{1}{\text{mass}} \sum\nolimits_{i=1}^{np} m_i \, x_i \tag{8.1}$$

$$y_{CM} = \frac{1}{\text{mass}} \sum\nolimits_{i=1}^{np} m_i \, y_i \tag{8.2}$$

$$z_{CM} = \frac{1}{\text{mass}} \sum\nolimits_{i=1}^{np} m_i \, z_i \tag{8.3}$$

where x_{CM} is the location of the bodies center of mass in the X direction, m_i mass of the ith point mass, and x_i is the location of the ith point mass measured in the axis system in the X direction. The nomenclature has the same pattern for the Y and Z axes.

8.1.4 Segment Moment(s) of Inertia

The moment of inertia is a measure of an objects resistance to a change in angular velocity. For a rigid body, the moments of inertia are a function of the mass of the object and the distribution of that mass about specified axes. The moment of inertia of an object is specified with respect to a specified axis, with units of kilogram meters squared (kg.m^2).

In two-dimensional analysis the moment of inertia (I), for a body which is considered to be made up of np participles each of mass m_i, is:

$$I = \sum\nolimits_{i=1}^{np} m_i \, r_i^2 \tag{8.4}$$

Where r_i is the distance from the axis, with respect to which the moment of inertia is defined, to the ith mass. For two-dimensional analyses, the moment of inertia of a segment is typically defined relative to an axis through the segment center of mass, or with reference to a proximal joint axis, or a distal joint axis.

In three dimensions a reference frame is defined within the segment of interest, and the moments of inertia are typically specified with respect to the axes giving the inertia matrix (J), also called inertia tensor. The format of the inertia tensor is:

$$J = \begin{bmatrix} I_{XX} & -I_{XY} & -I_{XZ} \\ -I_{YX} & I_{YY} & -I_{YZ} \\ -I_{ZX} & -I_{ZY} & I_{ZZ} \end{bmatrix} \tag{8.5}$$

where I_{XX}, I_{YY}, and I_{ZZ} are the moments of inertia with the respect to the X, Y, and Z axes, respectively, and I_{XY}, I_{YX}, I_{YZ}, I_{ZY}, I_{ZX}, and I_{XZ} are the products of inertia and $I_{XY}, = I_{YX}$, $I_{YZ} = I_{ZY}$, and $I_{ZX} = I_{XZ}$.

Extending on how the moment of inertia is defined in two dimensions, the inertia tensor in three-dimensions can be computed from:

$$J = -\sum_{i=1}^{np} m_i\, D_i D_i \tag{8.6}$$

where D_i is a skew symmetrical matrix reflecting the three-dimensional location of each point mass; it is defined by:

$$D_i = \begin{bmatrix} 0 & -z_i & y_i \\ z_i & 0 & -x_i \\ -y_i & x_i & 0 \end{bmatrix} \tag{8.7}$$

The product of $D_i D_i$ gives:

$$-D_i D_i = \begin{bmatrix} \left(y_i^2 + z_i^2\right) & -x_i y_i & -x_i z_i \\ -y_i x_i & \left(x_i^2 + z_i^2\right) & -y_i z_i \\ -z_i x_i & -z_i y_i & \left(x_i^2 + y_i^2\right) \end{bmatrix} \tag{8.8}$$

From which the moments of inertia can be computed from:

$$I_{XX} = \sum \left(y_i^2 + z_i^2\right) m_i \tag{8.9}$$

$$I_{YY} = \sum \left(x_i^2 + z_i^2\right) m_i \tag{8.10}$$

$$I_{ZZ} = \sum \left(x_i^2 + y_i^2\right) m_i \tag{8.11}$$

The products of inertia are computed from:

$$I_{XY} = I_{YX} = -\sum x_i y_i m_i \tag{8.12}$$

$$I_{XZ} = I_{ZX} = -\sum x_i z_i m_i \tag{8.13}$$

$$I_{YZ} = I_{ZY} = -\sum y_i z_i m_i \tag{8.14}$$

The products of inertia are measures of symmetry of mass distribution with respect to the axes relative to which the moments of inertia are defined. If a plane is a plane of symmetry, the products of inertia associated with an axis perpendicular to that plane are zero. These products of inertia can have either sign.

8.1.4.1 Principal Axes

The moments of inertia must always be positive and their sum invariant under orientation changes of the body reference frame (but not origin changes). It is possible to define a set of axes so the products of inertia are zero; in this case the axes are called the principal axes, and the moments of inertia about these axes (diagonal terms in inertia tensor) are the principal moments of inertia. For a given inertia tensor, its eigenvalues are the principal moments of inertia, and the eigenvectors are the principal axes. See Appendix A for information on eigenvalues and eigenvectors for a matrix.

Here is an example, based on data presented by Greenwood (1965). The analysis will be of the following inertia tensor where the moments of inertia and products of inertia are with respect to a body-fixed axes A:

$$J_A = \begin{bmatrix} 150 & 0 & -100 \\ 0 & 250 & 0 \\ -100 & 0 & 300 \end{bmatrix} \text{kg.m}^2 \tag{8.15}$$

The eigenvalues of this inertia tensor are:

$$\lambda = (100, 250, 350) \tag{8.16}$$

which are therefore the values of the principal moments of inertia. The eigenvectors (v_1, v_2, v_3) are:

$$v_1 = \begin{bmatrix} -0.8944 \\ 0 \\ -9.4472 \end{bmatrix}, v_2 = \begin{bmatrix} 0 \\ 1 \\ 0 \end{bmatrix}, v_3 = \begin{bmatrix} -0.4472 \\ 0 \\ 0.8944 \end{bmatrix} \tag{8.17}$$

$$R = [v_1, v_2, v_3] = \begin{bmatrix} -0.8944 & 0 & -0.4472 \\ 0 & 1 & 0 \\ -9.4472 & 0 & 0.8944 \end{bmatrix} \tag{8.18}$$

where R is an orthogonal matrix containing the eigenvectors which represent the directions of the principal axes. Therefore the inertia matrix about the principal axes (J_P) can be computed from:

$$J_P = R J_A R^T - \begin{bmatrix} 100 & 0 & 0 \\ 0 & 250 & 0 \\ 0 & 0 & 350 \end{bmatrix} \text{kg.m}^2 \tag{8.19}$$

8.1.4.2 Parallel Axis Theorem

In some circumstances the moment of inertia is given for one axis, or set of axes, but for subsequent analysis is required for another axis, or set of axes. In these cases the parallel axis theorem can be used. The parallel axis theorem is also called the Huygens-Steiner theorem after the pioneering work of Christiaan Huygens (1629–1695) and Jakob Steiner (1796–1863). In two dimensions this theorem can be expressed as:

$$I_A = I_{CM} + m\,d^2 \tag{8.20}$$

where I_A is the moment of inertia about the parallel axis A, I_{CM} is the moment of inertia about the bodies center of mass, m is the mass of the body, and d is distance between the two axes. In three dimensions the formula is slightly more complicated but follows the same principles:

$$J_A = J_P + m[p^T p\,I - pp^T] \tag{8.21}$$

where J_A is the inertia tensor in reference frame A, J_P is the inertia tensor with reference to the principal axes (therefore the center of mass is its origin), p is the location of the objects center of mass relative to reference frame A, and I is the identity matrix.

Use of the parallel axes theorem can be illustrated using a cuboid illustrated in Fig. 8.2.

For the cuboid the inertia tensor about the inertial reference frame is:

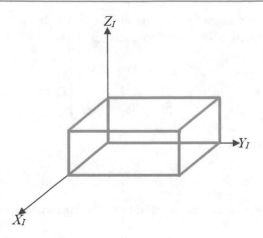

Fig. 8.2 A cuboid with dimensions of width 0.2 m (in x direction), length 0.4 m (in y direction), and height 0.1 m (in z direction) used to illustrate the parallel axis theorem. The mass of the cuboid is *8 kg*

$$J_I = \begin{bmatrix} 0.4533 & -0.1600 & -0.0400 \\ -0.1600 & 0.1333 & -0.0800 \\ -0.0400 & -0.0800 & 0.5333 \end{bmatrix} \text{kg.m}^2 \tag{8.22}$$

The inertia tensor for the principal axes would be the inertia tensor about axes through the cuboid center of mass, so $p = (0.10, 0.20, 0.05)^T$. Therefore the inertia tensor (J_P) of about the principal axes can be computed from:

$$J_P = J_I - m\left[p^T p\, I - pp^T\right] \tag{8.23}$$

$$J_P = \begin{bmatrix} 0.4533 & -0.1600 & -0.0400 \\ -0.1600 & 0.1333 & -0.0800 \\ -0.0400 & -0.0800 & 0.5333 \end{bmatrix} - 8\left[p^T p\, I - pp^T\right] \tag{8.24}$$

$$J_P = \begin{bmatrix} 0.1133 & 0 & 0 \\ 0 & 0.0333 & 0 \\ 0 & 0 & 0.1333 \end{bmatrix} \text{kg.m}^2 \tag{8.25}$$

8.1.4.3 Radius of Gyration

On occasions the moment of inertia is reported as a scaled value – the radius of gyration. The radius of gyration (*p*) is a scalar quantity with units of length obtained by dividing the moment of inertia of an object by the mass of the object (*m*) and taking the square root. Therefore the appropriate equation is:

$$p = \sqrt{\frac{I}{m}} \tag{8.26}$$

The radius of gyration and moment of inertia must be about some specified axis. In this form it represents the distance from that axis to a point at which the mass of the body can be assumed to be concentrated. In biomechanics the radius of gyration is often expressed somewhat differently, normally by making the radius of gyration dimensionless by dividing by segment length (*L*), so the scaled radius of gyration ($\widehat{p}$) is:

$$\widehat{p} = \frac{1}{L}\sqrt{\frac{I}{m}} \tag{8.27}$$

8.1.5 Whole Body Properties

The whole body has a mass, center of mass location, and moments of inertia. The whole body is modeled as a system of articulated (assumed) rigid bodies; therefore as the body segments move relative to one another, the whole body center of mass and inertia tensor can change over the duration of an analyzed task.

The total mass of the system is simply computed from the sum of the masses of the individual segments. The whole body center of mass location can be computed using the same principles which are used to determine the segmental centers of mass. To determine the whole body inertia tensor, a common reference frame must be defined for all body segments; then each segmental inertia tensor should be determined relative to this reference frame. Once this is done, the inertia tensors can be added to compute the whole body inertia tensor.

8.2 Cadaver Studies

As previously highlighted the utility of using cadavers to estimate segmental inertias has been exploited for over 150 years. The advantage with cadavers is that via dissection it is relatively easy to determine segmental inertial properties. Once these inertial properties are determined, the problem becomes how are these used for experimental subjects? In this section some of the available cadaver data will be reviewed, with methods for scaling from cadaver to an experimental subject reviewed in a subsequent section.

The early cadaver studies include the work of Harless (1860) who analyzed two specimens, Braune and Fischer (1889) who analyzed four specimens and Demspter (1955) who analyzed eight specimens. Dempster's study was one of three important studies from the Wright-Patterson Air Force Base, these studies will be reviewed as a basis for considering all cadaver studies.

8.2.1 Wright-Patterson Studies

There are three major cadaver studies from Wright-Patterson, these are Dempster (1955) which examined eight cadavers, Clauser et al. (1969) 13 cadavers, and Chandler et al. (1975) six cadavers. The general characteristics of the cadavers in these studies are presented in Table 8.2.

The general procedure with cadaver studies is to first measure the height and mass of the cadavers. Then various other anthropometric measures of limb dimensions can be made from the cadavers, but these vary between studies, for example, Dempster (1955) only reported segment lengths, while Chandler et al. (1975) also measured and reported the perimeters around joints and typically the mid-segment perimeter. After these measures the segments are dissected and then their masses measured directly. The segment volumes can be determined by water immersion and measuring the volume of displaced volume, which given their mass allows computation of segment density. The center of mass is determined by finding the segment balance point. Finally the moments of inertia were determined by exploiting the relationship between the moment of inertia and the period of swing of a pendulum. This method requires the suspension of the cadaver segment so that it can oscillate through a range of range of less than 0.17 radians under the influence of gravity (e.g., Lephart 1984). Given a known mass of the segment, center of mass location, and period of swing, the moment of inertia can be determined from:

Table 8.2 Details of major cadaver studies from Wright-Patterson, with cadaver age, mass, and height ranges

Authors	N/sex	Age (years)	Mass (kg)	Height (m)
Dempster (1955)	8 males	52–83	49–72	1.59–1.86
Clauser et al. (1969)	13 males	28–74	54–88	1.62–1.85
Chandler et al. (1975)	6 males	45–65	51–89	1.64–1.81

$$I_0 = mgl\left(\frac{T}{2\pi}\right)^2 \tag{8.28}$$

where I_0 is the moment of inertia about the axis of rotation, m is the mass of the segment, g is the acceleration due to gravity, l is the distance from the axis of rotation to the segment center of mass, and T is the time period for a complete oscillation (out and back). It has been demonstrated that this method of determining segmental inertias is accurate (e.g., Dowling et al. 2006). In Dempster's study the moments of inertia were determined for a segmental sagittal axis only; Clauser et al. (1969) did not measure the moments of inertia, while Chandler et al. (1975) measured them about three axes for each segment.

The overall sample size for these three cadavers studies is small, just 27. Unfortunately, the data from the studies cannot easily be combined as the definitions of segment boundaries varied between the studies, and as already noted not all mechanically important properties were measured in all of the studies. For example, during dissection, the cut at the knee joint was lower in Dempster than in the other two studies; there were similar issues for the head and neck.

There are general considerations for cadaver studies, when planning to extrapolate from cadaver data to a live subject. Cadavers are stored before dissection, so the method of storage is important. One of Dempster's cadavers was embalmed, while in the Clauser et al. (1969) and Chandler et al. (1975) studies, all cadavers were preserved, which raises questions about the difference between the density of the original body fluids and those which replaced them. In these studies where available the cause of death was provided, in many cases the cause of death indicates the sources of some of the cadavers were likely not very active toward the ends of their lifes. Many of the cadavers used in these studies were old at the time of death and relatively light in mass. Jensen (1993) showed that compared with the adult Canadian population, the height of the cadavers in these three studies were around the 55th percentile, but their masses were all at or below the 23rd percentile. These are some of the major issues with cadaver studies, which are counterbalanced against the relative ease with which their inertial properties can be determined compared with making these measurements on a live subject.

There has not been a cadaver study examining the anthropometry of women, which is a huge gap in the literature. This means that without data to justify the approach male cadaver data are often used for female subjects.

8.3   Imaging Studies

Imaging studies are used in two ways in relation to determining body segment inertial properties. One is to determine the surface contours to enable the external dimensions of the body to be measured. Given these body dimensions, the inertial properties of the body segments can be determined by modeling the segments, for example, as series of geometric solids. The other group of body imaging techniques exploits either X-rays, gamma rays, or magnetic resonance imaging. This group of methods provides information not only on the shape of the segments but the tissue distributions within the segment. From these tissue distributions, along with knowledge of their densities, the inertial properties of the segments can be determined.

8.3.1   Measurement of Body Contours

Weinbach (1938) took photographs from a front view and side view of subjects and after appropriate scaling measured the contours along the edges of the body from each view. These two pairs of contours were assumed to provide the diameter of the body along its length about two axes, which once the body was divided into sections permitted estimation of segmental inertial properties. Fifty years later this process became automated when subjects were rotated around their longitudinal axis as shadows were cast on their body to provide contrast, and two cameras recorded images of the subject (Jones et al. 1989). Computer algorithms then made the measurements on the recorded images. This analysis provided more detailed information than Weinbach (1938) and has been demonstrated to be as least as accurate as the more time-consuming manual option (Brooke-Wavell et al. 1994). Since then there have been a number of other systems presented for automatically measuring body shape (e.g., Sarfaty and Ladin 1993; Baca 1996; Wang et al. 2006).

One particular approach in this domain which warrants further description is that of Jensen (1978). His work is noteworthy in part because he used it to examine many populations including infants, pregnant women, and the elderly. Jensen's approach was not dissimilar to that of Weinbach (1938) as he took photographs from a front view and side view and made measurements from those images. Those measurements were used to divide the body into stacks of elliptical disks each 2 cm high, with the two radii of the ellipse taken from the photographs. Given the density of the segments, from Dempster

Table 8.3 Properties of X-rays and gamma rays

Source	Frequency (Hz)	Wavelength (nm)	Energy (eV)
X-rays	3×10^{16} to 3×10^{19}	0.01–10	10^2–10^5
Gamma rays	>10^{19}	<0.01	>10^2

N.B. eV is the unit electrovolts which is ~1.6×10^{-19} J

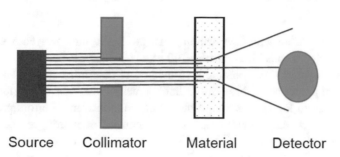

Fig. 8.3 Measurement of electromagnetic wave

(1955), the inertial properties of the ellipses were combined to compute the inertial properties of the segments. Wicke and Lopers (2003) demonstrated that this approach was accurate by comparing segment volumes with those measured by water immersion, and Sanders et al. (2015) have shown this method is reliable.

8.3.2   Body Imaging

There are four medical imaging techniques which have been used to measure body segment inertial properties; these are gamma mass scanning (GM), dual-energy X-ray absorptiometry scanning (DXA), computed tomography (CT), and magnetic resonance imaging (MRI). The first three methods rely on electromagnetic waves, which have different frequencies, wavelengths, and energies but which behave in similar ways (see Table 8.3).

The basic principles for GM and DXA are the same, relying on the generation of electromagnetic waves. When an electromagnetic wave passes through a material, it becomes weaker; if we measure the intensity of the beam before and after it passes through the material, it is possible to compute the surface density (see Fig. 8.3). The basic relationship of the strength of the beam before and after passing through a material is:

$$N_A = N_0 e^{\mu d} \tag{8.29}$$

where N_A is the intensity of beam after passing through the material, N_0 is the intensity of the beam before passing through the material, μ is the mass absorption coefficient ($\text{cm}^2.\text{g}^{-1}$), and d is the surface density (g.cm^2).

Some of the photons in the beam will pass through the material; others are scattered. The other photons in the beam do not pass through the material because of photoelectric effect or due to pair-production interactions. Typically the photons that remain in the un-scattered beam are measured, and the loss reflects attenuation of the beam. Different materials have different specific mass absorption coefficients. These mass-specific absorption coefficients also depend on the energy of the input signal. The typical GM signal has energy between 0.5 and 1.3 MeV, while with DXA there are two signals with energies of 70 and 170 keV. Different chemical elements have different absorption coefficients; composite materials reflect the combined effect of the chemicals comprising the material. Table 8.4 presents the absorption coefficients for materials typical encountered when scanning humans.

Given the different mass absorption coefficients for air and the tissues of interest it is possible with the GM and DXA scans that while recording the surface density a contour of the body is also obtained. Therefore from these images the segments can be divided into segments, and segment dimensions measured in the plane of the scan.

Both GM and DXA operate on the same basic principles while using different energy electromagnetic waves. One concern is that the systems expose subjects to radiation; while this is a legitimate concern, the radiation exposure is quite small. For example, a DXA scan has radiation with an effective dose equivalents of 5 µSv (here the Sievert, Sv, is a unit of ionizing

Table 8.4 The mass absorption coefficients (cm^2.g^{-1}) for materials common in scanning of humans

	Energy (MeV)	Air	Adipose tissue	Cortical bone	Skeletal muscle
X-ray	0.07	0.02724	0.02463	0.10448	0.02937
	0.175	0.02584	5.32800	6.18600	5.68700
γ-ray	0.5	0.02966	0.03303	0.03073	0.03269
	1.25	0.02666	0.02970	0.02745	0.02938

Data from Seltzer (1993)

radiation dose), and similar values occur with GM scanning. To put 5 μSv into perspective, a one-way transatlantic flight exposes a passenger to ~80 μSv and a spinal radiograph ~700 μSv.

Brooks and Jacobs (1975) proposed GM scanning as a method to determine segmental inertias. They evaluated their method by measuring the inertial properties of three legs of lamb directly (weighing, water immersion, finding balance point, and mounting on a pendulum). Each leg was measured three times; the maximum errors produced by GM scanning were 1.1%, 2.1%, and 4.6% for mass, center of mass location, and moment of inertia, respectively. In various publications Zatsiorsky and colleagues reported data collected on 100 male college age subjects using GM scanning (e.g., Zatsiorsky and Seluyanov 1983, 1985). In a subsequent publication Zatsiorsky et al. (1990a) also presented data on 19 female students, also of college age. Zatsiorsky et al. (1990b) reported that errors were less than 3% for all inertial properties other than the moment of inertia about the longitudinal axes where errors were up to 10%.

DXA scanning was initially used for determining bone mineral density and body composition, but this has been expanded to assess segmental inertias (e.g., Visser et al. 1999). Durkin et al. (2002) used DXA scans for determine the segmental masses, center of mass locations, and a transverse moment of inertia of 11 male subjects. As DXA scans project the three-dimensional mass distribution onto a plane, it was not feasible to estimate a moment of inertia about both anterior-posterior, and medial-lateral directions. Their validation consisted of scanning a book, which given the different mass absorption coefficients of the book compared with human tissues is not perhaps too informative, and the scanning of a cadaver limb. For the cadaver limb, the errors were 3.2%, 1.3%, 2.7%, and 8.2% for the cadaver limb mass, length, center of mass position, and moment of inertia. Clarys et al. (2010) analyzed pig cadavers via DXA scans, and via dissection and direct measurement of tissue masses. They advised some caution with DXA scans because of the assumption that tissues from different samples are assumed to have the same mass absorption coefficients. For example, muscle from different populations can have different levels of muscle fat infiltration, thus changing the mass absorption coefficients (Marcus et al. 2010).

A CT scan takes X-ray images from multiple angles, and then algorithms combine these to produce cross-sectional images of the body (a tomograph). The scan provides information about tissue distributions from which segmental inertial properties can be determined. Huang and Wu (1976) evaluated that CT scans can measure body densities to within 5%. Rodrigue and Gagnon (1983) evaluated, with 20 cadaver forearms, the accuracy of CT-derived densities. The segmental densities were overestimated by an average of 2.1%. In a CT scan of the chest there is an effective radiation dose of 6 μSv (Furlow 2010), for a whole body scan the radiation exposure would be greater. This is perhaps the reason why this method has not received broader adoption for the determination of body segment inertial properties.

MRI like CT produces cross-sectional images of the body but without radiation exposure. MRI exploits the response of atomic nuclei to a magnetic field. The magnetic field is produced by a powerful magnet, typically a solenoidal (cylindrical) magnet. The magnets used are typically 1–3 T. The tesla is a measure of magnetic flux density (1 T = 10,000 gauss); for perspective the Earth's magnetic field measures 0.5 gauss. An oscillating magnetic field is applied to the subject (the production of this oscillations is what causes the characteristic noise of MRI). The magnetic field causes excitation of the hydrogen atoms which emit a radio frequency signal; this signal is measured by a receiving coil. The signal varies for different body tissues, and signal processing allows production of the tomograph.

Normally in MRI the subject lies on a moving platform that moves the subject through the magnet, so that the body can be analyzed in slices. The pulses of the magnet signal can be varied; this is the repetition time (TR) which is the time between successive pulse sequences applied to the same slice. The time between the delivery of the pulse and the receipt of the echo signal is called time to echo (TE). When describing the use of MRI, reference is made to two relaxation times T1 (longitudinal relaxation time) and T2 (transverse relaxation time). A T1-weighted image has both short TR and TE, while a T2-weighted image has long TR and TE. By adjusting TR and TE, the nature of the image can be manipulated; for example, in a T1-weighted image, fat appears bright and fluids are dark, while in a T2-weighted image, fat has an intermediate brightness, and fluids are bright. The primary drawback of MRI is that the magnet attracts ferrous metals, meaning that exposure to MRI could be potentially injurious to a subject with, for example, implants containing ferrous metals. Common objects containing ferrous metals include cardiac pacemakers and cochlear implants.

Martin et al. (1989) examined the accuracy of MRI to estimate segmental inertial properties, with images recorded every 1.5 cm longitudinally; they did not provide other MRI operation details. They used eight baboon cadaver arms on which they measured mass (weighing), volume (water immersion), center of mass location (reaction board), and moment of inertia (mounting on pendulum). Mean errors were 6.7%, -2.4%, and 4.4% for mass, center of mass location, and transverse moment of inertia, respectively. Pearsall et al. (1994) used MRI to determine the inertial properties of the trunk of 26 male subjects. The MRI was a 1.5 T magnet, with T1 $= 500$ ms and T2 $= 20$ ms. They compared trunk circumference from MRI images with those measured directly, with the differences less than 2%. They presented linear multi-variate regression equations for estimating trunk segmental inertial properties. Bauer et al. (2007) used MRI to determine the lower limb segmental inertial properties of ten females with a mean age of 9.6 years. They used a 1.5 T MRI, with T1 $= 850$ ms. They compared their inertial properties with those derived from the regression equations of Jensen (1986). There were statistically significant differences between the two sets of inertial properties, likely in part because Jensen's data were collected on male children.

This class of methods for determining segmental inertias is attractive, particularly MRI as in addition to the inertial properties it is possible to determine aspects of, for example, muscle architecture (e.g., Blemker et al. 2007). Their drawbacks include the general availability of equipment and the possible exposure to radiation. Other methods for determining segmental inertias rely on little more than a tape measure, so are a cheaper and easier alternative. That stated these imaging techniques have the potential to provide data on which statistical models can be based, as is the case in the work of Zatsiorsky et al. (1990a, 1990b).

8.4 Geometric Solid Models

To estimate the inertial properties of the trunks of two cadavers, Harless (1860) modeled each of their upper trunks as a truncated cone. For a truncated cone given its dimensions it is feasible to derive the equations which provide its inertial properties. This approach can be generalized to other segments. Therefore given suitable geometric shapes, the segments can be modeled and their inertial properties determined. There are three prerequisite pieces of information:

1. The appropriate geometric solid and the equations to compute its inertial properties.
2. The dimensions of the segments which match the geometry of the geometric solid.
3. The density of the body segments, data typically derived from cadaver data.

8.4.1 General Principles

Within this class of techniques, all segments are modeled as a geometric solid or as a series of geometric solids. The dimensions of these solids are obtained by taking measurements on the subjects, for example:

Cylinder	$\Rightarrow$	Length and characteristic perimeter
Truncated cone	$\Rightarrow$	Length and proximal and distal perimeters
Elliptical disc	$\Rightarrow$	Height, a radius, and perimeter

For any geometric solid the volume can be computed from:

$$v = \iiint dx\, dy\, dz \tag{8.30}$$

Therefore given the density (ρ) of the solid the mass can be computed from:

$$m = \rho\, v \tag{8.31}$$

Assuming uniform density throughout a geometric solid the center of mass location (e.g., x_{CM}) can be computed from:

$$x_{CM} = \frac{1}{v} \iiint x \, dx \, dy \, dz \tag{8.32}$$

The center of mass for the other two axes is obtained after appropriate substitutions.

For any geometric solid, the components of the main diagonal of the inertia tensor (moments of inertia) can be computed from:

$$I_{XX} = \iiint \rho \left(y_i^2 + z_i^2 \right) dx \, dy \, dz \tag{8.33}$$

$$I_{YY} = \iiint \rho \left(x_i^2 + z_i^2 \right) dx \, dy \, dz \tag{8.34}$$

$$I_{ZZ} = \iiint \rho \left(x_i^2 + y_i^2 \right) dx \, dy \, dz \tag{8.35}$$

The products of inertia are computed from:

$$I_{XY} = I_{YX} = \iiint \rho \, x_i y_i \, dx \, dy \, dz \tag{8.36}$$

$$I_{XZ} = I_{ZX} = \iiint \rho \, x_i z_i \, dx \, dy \, dz \tag{8.37}$$

$$I_{YZ} = I_{ZY} = \iiint \rho \, y_i z_i \, dx \, dy \, dz \tag{8.38}$$

Appendix K demonstrates the process for estimating the inertial properties of a geometric solid and presents the formulae for common geometric shapes. As an example Table 8.5 presents the appropriate equations for an elliptical disk.

So what shapes are used? See Fig. 8.4 for sample shapes. If the forearm is used as an example, Hanavan (1964) used a single truncated cone, Jensen (1978) used stacked elliptical disks each 2 cm high, Hatze (1980) used ten elliptical disks, and Yeadon (1990) used two stacked truncated cones. For the trunk, Harless (1860) used a truncated cone stacked on an elliptical disk, Hanavan (1964) used two stacked elliptical disks, and Yeadon (1990) used two stacked stadium solids. As can be seen there are a variety of approaches in terms of the shape used, and the number of shapes.

Table 8.5 Inertial properties of an elliptical disk; here ρ refers to the density of the solid, a and b the radii of the ellipse, and h the height

Geometric solid	Mass	Center of mass	Principal moments of inertia
Elliptical disk	$\rho . \pi . a . b . h$	$\frac{h}{2}$	$I_{xx} = \frac{m.\left(3.b^2+h^2\right)}{12}$ $I_{yy} = \frac{m.\left(3.b^2+h^2\right)}{12}$ $I_{zz} = \frac{m.\left(a^2+b^2\right)}{4}$

See Fig. 8.4 for definitions of axes

Fig. 8.4 Three geometric shapes used for modeling body segments, elliptical disk (left), truncated cone (middle), and stadium solid (right)

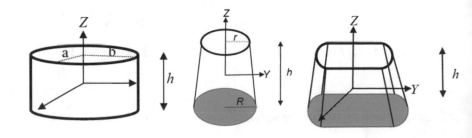

Of the shapes that are used if a truncated cone or cylinder is used, this assumes that the cross section of the body segment is a reasonable approximation to a circle. Under these conditions the geometric solid will have two of its moments of inertia equal. For the human thorax, Cornelis et al. (1978) demonstrated its unusual cross section prompting Yeadon (1990) to model the trunk as stacked stadium solids.

Irrespective of what shapes are used to model the body segments, the dimensions of the shapes must be measured on the subjects. For a truncated cone, a proximal and distal circumference and cone height are required. For an elliptical disk, one radius, circumference, and height are required. Some body segments taper along their length in which case for an elliptical disk, the choice is between making the measurements in the middle of the height of the disk, or and taking measurements distally and proximally and averaging. For a stadium solid the depth of the segment and the circumference are required both proximally and distally for the segment. Strictly the length around the perimeter of the body is being measured, but the term circumference is used here, and commonly in the literature, for convenience while acknowledging segments are both not circular or necessarily assumed to be circular by the model used.

In many models geometric solids are stacked on top of one another to allow for the subtle changes in segment dimensions which may occur along the length of a segment. In these cases the mass of each component part can simply be summed to compute whole segment mass:

$$\text{mass} = \sum_{i=1}^{ns} m_i \tag{8.39}$$

where ns is the number of geometric solids in the stack and m_i is the mass of the ith geometric solid.

The whole segment center of mass can be computed from:

$$x_{CM} = \frac{1}{\text{mass}} \sum_{i=1}^{ns} m_i \, x_i \tag{8.40}$$

where x_{CM} is the location of the whole system center of mass in the x direction and x_i is the location of the center of mass of the ith geometric solid in the x direction. This computation can similarly be performed for the other two coordinate directions.

For every geometric solid, modeling a segment, their inertia tensor will initially be specified relative to reference fixed in each solid. The first step is to make sure each reference frame has the same orientation which will change the inertia tensor (see Eq. 8.19). Then each reference frame must have the same common origin, typically the center of mass of the modeled segment. Therefore the parallel axis theorem is used so that the inertia tensor for each solid is expressed relative to the same reference system, and the segment inertia tensor can be computed from:

$$J_S = \sum_{i=1}^{ns} J_i + m_i \left[p_i^T p_i \, I - p_i p_i^T \right] \tag{8.41}$$

where J_S is the inertia tensor for the geometric solid model of the segment about a segment-based axis system, J_i is the inertia tensor for the ith geometric solid with the origin of its axis system at the center of mass of the solid, m_i is the mass of the ith geometric solid, p_i is the location of the ith geometric solids center of mass relative to the segment-based reference frame, and I is the identity matrix.

The geometric solid models require segmental density values; these are typically taken from cadaver data. Table 8.6 gives segmental density values from three major cadaver studies.

In Hatze's model he used data from collected from one cadaver by Dempster (1955) to allow for how segment density varies along the length of a segment (Hatze 1980). Ackland et al. (1988) examined the influence of assuming uniform density longitudinally in a segment on the estimation of leg inertial properties. They measured the density variation longitudinally for the leg segment, using computed tomography. For the leg they demonstrated that variable density was not necessary for accurate determination of inertial properties. Wicke et al. (2008) used DXA scanning to estimate the variation of density of the trunk and subsequently presented regression equations for estimating density along the length of the trunk (Wicke and Dumas 2008). The data were then used in a geometric solid model of the trunk, and it was shown that the estimates of segmental inertia were more sensitive to estimates of the volumes of the sub-sections of the trunk than they were to the density values (Wicke and Dumas 2010). With the trunk both mass distribution and volume can vary as a subject breathes, which means perhaps more than any other segment the inertial properties of the segment vary during activity.

For example, the following calculation illustrates how the inertial properties can be calculated by modeling the thigh as a truncated cone (see Appendix K, for all equations). The data are for a subject of mass 70 kg male with a thigh length of 0.35 m

Table 8.6 Segment densities from three major cadaver studies for human body segments (units – kg.m^{-3} × 10^3)

Segment	Study		
	Dempster (1955)	Clauser et al. (1969)	Chandler et al. (1975)
Hand	1.16	1.109	1.080
Forearm	1.13	1.098	1.052
Upper arm	1.07	1.056	1.080
Foot	1.10	1.084	1.071
Shank	1.09	1.085	1.065
Thigh	1.05	1.044	1.0195
Trunk	1.03	1.019	0.853
Head	1.11	1.070	1.056

(h), an upper thigh circumference of 0.55 m (R), knee circumference of 0.40 m (r), and a segment density of 1.05 kg. m^{-3} × 10^{-3} (Dempster 1955):

$$L_{CM} = \frac{h}{2} \frac{\left(r^2 + 2rR + 3R^2\right)}{\left(r^2 + rR + R^2\right)} \tag{8.42}$$

$$= \frac{0.35}{2} \frac{\left(0.40^2 + (2)(0.40)(0.55) + 3\left(0.55^2\right)\right)}{\left(0.40^2 + (0.40)(0.55) + 0.55^2\right)}$$

$$= 0.1568 \, m \qquad\qquad \text{(from proximal joint)}$$

$$m = \rho \frac{h}{3} \pi \left(r^2 + rR + R^2\right) \tag{8.43}$$

$$= \left(1.05 \times 10^{-3}\right) \left(\frac{0.35}{3}\right) (\pi) \left(0.40^2 + (0.4)(0.55) + 0.55\right)$$

$$= 6.65 \, kg$$

$$I_{XX} = I_{YY} = 0.0760 \text{ kg.m}^2 \tag{8.44}$$

(due to complexity calculation not shown, see Appendix K)

$$I_{ZZ} = \frac{3m}{10} \left(\frac{R^5 - r^5}{R^3 - r^3}\right) \tag{8.45}$$

$$= \frac{3(6.65)}{10} \left(\frac{0.55^5 - 0.40^5}{0.55^3 - 0.40^3}\right)$$

$$= I_{ZZ} = 0.0198 \text{ kg.m}^2$$

8.4.2 Models

In the following three whole body models will be reviewed, the models of Hanavan (1964), Hatze (1980), and Yeadon (1990), (see Tables 8.7, 8.8, and 8.9), respectively. The models illustrate the range of complexity of models used for determining body segment inertial properties.

Hanavan's (1964) model is the simplest of the three whole body models to be reviewed. The shapes used are not too complex and the number of required measurements not too onerous. Indeed if left/right limb symmetry is assumed, there are

Table 8.7 Summary of shapes used for modeling body segments by the geometric solid model of Hanavan (1964)

Body Part	Hanavan (1964)	
Foot	1 truncated cone	
Shank	1 truncated cone	
Thigh	1 truncated cone	
Hand	1 sphere	
Forearm	1 truncated cone	
Upper Arm	1 truncated cone	
Trunk	2 elliptical discs	
Head	1 ellipsoid	
#Segments	15	
#Measures	25	

Table 8.8 Summary of shapes used for modeling body segments by the geometric solid model of Hatze (1980)

Body part	Hatze (1980)	
Foot	103 unequal trapezoidal plates	**Head Model**
Shank	10 elliptical cylinders	
Thigh	1 ellipto-parabolic sect. 10 elliptical cylinders	
Hand	1 prism 1 hollow half cylinder 1 arched rectangular cuboid	
Forearm	10 elliptical cylinders	
Upper arm	1 hemisphere 10 elliptical cylinders	**Thigh Model**
Trunk	*Lower* 10 complex plates *Buttocks* 2 elliptical parabolas *Upper* unequal semiellipses, with parabolas removed	
Head	1 elliptical cylinder 1 general body of revolution	
#Segments	17	
#Measures	242	

only 25 measurements required. Rather than compute segment mass by computing the volume of the geometric solid and taking its product with the segmental density, Hanavan (1964) used the first-order regression equations of Barter (1957). These regressions equations were based on the cadaver data from Braune and Fischer (1889, $n = 3$) and Dempster (1955, $n = 8$). Once these equations were used to estimate the segment masses, any deviation between the model total mass and the subject's actual mass was evenly distributed between the segments.

Probably the most complex geometric solid model is that of Hatze (1980). There are many unique elements to this model including a shoulder segment (between the trunk segment and upper arm segment). The model divided most of the segments into multiple geometric solids, with many of the geometric solid shapes novel. This complexity of the modeling of each segment necessitates many measurements to parameterize the model (242). As previously mentioned this model assumes that the density of a segment varies along its length. For the density data, Hatze relied on the detailed analysis of one cadaver by Dempster (1955).

The model of Yeadon (1990) sits between the relative simplicity of the model of Hanavan (1964) and the complex model of Hatze (1980). The model was originally formulated for the analysis of aerial acrobatic activities; because of this each hand and forearm were considered one segment, and each foot and shank another segment. Thus the model only consisted of

11 segments in its original formulation, but since then the model has been adapted to consider the feet and hands as separate segments. The unique element of this model is the introduction of a new geometric solid, the stadium solid. This shape seems particularly appropriate for the modeling of the trunk.

8.4.3 Geometric Model Assessment

The reason geometric solid models are used is because it is hard, if not impossible, to directly measure all segmental inertial properties in vivo. Therefore it is also difficult to thoroughly assess these models as a full set of criterion data are hard to obtain. A few studies which have addressed the appropriateness of geometric solid models will be reviewed in this sub-section.

Hanavan (1964) attempted to validate his model by comparing whole body center of mass location and principal moments of inertia determined by his model with values determined experimentally. He used data from Santschi et al. (1963) who used an oscillating pendulum to measure for 66 male subjects their center of mass location and principal moments of inertia in eight different body positions. The center of mass was estimated to within 1.8 cm and the moments of inertia within 10%.

Katch et al. (1974) compared segment volumes computed by modeling the body segments as geometric solids with volumes determined by water immersion. The arms were modeled as four truncated cones, the legs as six truncated cones, the hands and feet as wedges, the trunk as four truncated cones, the neck as a truncated cone, and the head as a sphere. Errors in estimating volumes were generally less than 10%. For biomechanical analyses, volumes are typically not important but are a precursor to segment mass computation.

Jensen (1978) presented a whole body geometric model where the whole body was modeled as a series of elliptical disks each 20 mm high. Wicke and Lopers (2003) examined the ability of the elliptical zone model of Jensen (1978) to accurately measure the volume of the forearm and hand, the shank and foot, and the whole body. They examined ten male and ten female subjects and used volumes from water immersion as their criterion. Their mean percentage errors were $-0.43\% \pm 2.49$, $-0.81\% \pm 3.01$, and 2.01 ± 2.17 for the forearm and hand, the shank and foot, and the whole body, respectively.

Miller and Morrison (1975) evaluated the model of Hanavan (1964) by how well it estimated whole body mass. In the original model segment, masses were estimated using the regression equations of Barter (1957); here the equations of Clauser et al. (1969) were also used. In the original model, any deviation between the model total mass and the subjects actual mass was evenly distributed between the segments, here that adjustment was not made. Errors were 2.03% and 4.59% for equations of Barter (1957) and Clauser et al. (1969), respectively. For all whole body geometric solid models, comparison of whole body model mass and actual body mass is typically available as a partial validation.

Hatze (1980) to validate his model measured three subjects. The model was able to estimate total body mass with an error less than 0.3%. One of the test subjects was similar in stature to one of the cadavers of Dempster (1955) and their inertial properties for the forearm and trunk compared favorably. On these subjects Hatze (1980) also estimated some of the inertial properties experimentally, including water immersion method for segment volumes, use of a reaction board to estimate segmental masses and center of mass locations (Pataky et al. 2003; Miller and Nelson 1973), and a limb oscillation method for

Table 8.9 Summary of shapes used for modeling body segments by the geometric solid model of Yeadon (1990)

Body Part	Yeadon (1990)	
Foot	4 stadium solids	
Shank	2 truncated cones	
Thigh	2 truncated cones	
Hand	3 stadium solids	
Forearm	2 truncated cones	
Upper arm	2 truncated cones	
Trunk	5 stadium solids	
Head	1 semi-ellipsoid 2 truncated cones	
#Segments	11	
#Measures	95	

determining moments of inertia (Hatze 1975). Errors were 5.17% for the volumes, 10.9% for center of mass locations, and 5.03% for the moment of inertia values.

8.5   Statistical Models to Estimate Segmental Inertias

There are databases of anthropometric data which include basic anthropometry data (e.g., standing height, body mass, and limb lengths) and the segmental inertial data. One option for using these data is to find a cadaver whose height and mass, and potentially other body measures, match an experimental subject, but this scenario would not occur for many studies. Therefore the preferred approach is to scale these data to the experimental subject in some way. The following sub-sections present some of the approaches to scaling these data.

8.5.1   Simple Statistical Models

The study of Dempster (1955) can be used as an example of simple statistical modeling for the estimation of segmental inertial properties. For each of Dempster's eight male cadavers, mass and segment lengths were measured, and these were used to estimate segmental inertias. The masses of the segments are expressed as a percentage of total body mass, the location of center of mass expressed as a percentage of segment length, and segment transverse moment of inertia represented by the radius of gyration. All of these data model coefficients are based on the mean data of the cadavers, where for the limbs data from both sides were averaged. Table 8.10 presents these data.

For example, the following calculation illustrates these data for a subject of mass 70 kg male with a thigh length of 0.35 m:

$$L_{CM} = 0.35 \frac{43.3}{100} = 0.152 \, m \text{ (from proximal joint)} \tag{8.46}$$

$$\text{mass} = 70 \frac{10.0}{100} = 7 \text{ kg} \tag{8.47}$$

$$I_{CM} = \text{mass}(pL)^2 = 7 \, (0.323 \, 0.35)^2 = 0.0895 \text{ kg.m}^2 \tag{8.48}$$

The ease of use of these data can make one forget that these values are not "tuned" to the individual other than taking into account body mass and segment length. Others have taken a similar approach with cadaver data including Clauser et al. (1969). Zatsiorsky and Seluyanov (1983) used the same simple statistical technique to characterize the inertial properties of 100 young male subjects (age 23.8 ± 6.2 years), determined in live subjects using gamma mass scanning. Table 8.11 summarizes the studies which have presented simple statistical models.

Table 8.10   The coefficients for the simple statistical modeling of the data of Dempster (1955) to determine segment mass, center of mass location, and the radius of gyration

Segment	Proximal end	Distal end	Mass %BM)	Center of mass (%SL)	Radius of gyration (−)
Hand	Wrist axis	Middle finger knuckle (II)	0.6	50.6	0.297
Forearm	Elbow axis	Ulnar styloid	1.6	43.0	0.303
Upper arm	Glenohumeral joint	Elbow axis	2.8	43.6	0.322
Foot	Lateral malleolus	Head metatarsal (II)	1.45	50.0	0.475
Shank	Femoral condyles	Medial malleolus	4.65	43.3	0.302
Thigh	Greater trochanter	Femoral condyles	10.0	43.3	0.323
Trunk	Greater trochanter	Glenohumeral joint	49.7	100.0	–
Head + neck	C7-T1	Ear canal	8.1	50.0	0.495

N.B. segment mass is expressed as percentage of whole body mass (BM), the center of mass location is a distance from the proximal end of the joint expressed as a percentage of segment length (SL), and the radius gyration is about the segment center of mass.

Table 8.11 Summary of the studies which have presented simple statistical models

Study	Sample	Notes
Harless (1860)	3 males, cadavers	Mass, center of mass only
Braune and Fischer (1889)	3 male, cadavers	
Dempster (1955)	8 males, cadavers	Only sagittal plane moments of inertia
Mori and Yamamoto (1959)	6 male, cadavers	Japanese cadavers, only segment masses
Clauser et al. (1969)	13 male, cadavers	No moments of inertia computed
Chandler et al. (1978)	22 males, 17 females	Data obtained from imaging. Subjects 3–7 years old
McConville et al. (1980)	31 male subjects	Data obtained from imaging
Plagenhoef et al. (1983)	35 men, 100 females	Data obtained from water immersion
Zatsiorsky and Seluyanov (1983)	100 male, live subjects	Data obtained from gamma mass scanning.
Li and Dangerfield (1993)	140 females, 140 males	Data from water immersion, subjects were between 8 and 16 years old
Jensen and Fletcher (1994)	7 males, and 12 females	Data collected by imaging, subjects between 63 and 75 years old
Cheng et al. (2000)	8 males, live	Chinese subjects, data obtained from MRI
Durkin and Dowling (2003)	50 male, 50 females, live	Data obtained from DXA scans
Ho et al. (2013)	50 males, live	Chinese subjects, data obtained from MRI

8.5.2 Linear Multi-variate Statistical Models

In the previous sub-section, segmental inertial properties were determined using simple ratios (e.g., segment mass relative to whole body mass); with more complex statistical models the aim is to "customize the fit" to the subject. One approach is to use multi-variate linear regression equations of the format:

$$q = C_1 A_1 + C_2 A_2 + C_3 A_3 + C_4 \tag{8.49}$$

where q is some body segment inertial property, C_1, C_2, C_3, and C_4 are constants, and A_1, A_2, and A_3 are anthropometric measures made on the experimental subject.

A simple example of this type of data comes from the work of Chandler et al. (1975) who presented equations for estimating segmental inertial properties. Here are representative equations for the thigh (BM, body mass); note input units are grams (g) for mass, and the output units are g and g.m^2 for the mass and moments of inertia, respectively:

$$\text{Mass} = 0.126\,BM - 1688 \tag{8.50}$$

$$L_{CM} = 0.388\,L \text{ (from proximal joint)} \tag{8.51}$$

$$I_{AP} = 24.102\,BM - 433522 \text{ (anterior} - \text{posterior axis)} \tag{8.52}$$

$$I_{ML} = 21.186\,BM - 222796 \text{ (medial} - \text{lateral axis)} \tag{8.53}$$

$$I_L = 75.608\,BM - 260.549 \text{ (longitudinal axis)} \tag{8.54}$$

For example, the following calculation illustrates these data for a subject of mass 70 kg male with a thigh length of 0.35 m:

$$\text{mass} = 0.126\,(70000) - 1688 = 7132\,\text{g} = 7.132\,\text{kg} \tag{8.55}$$

$$L_{CM} = 0.388\,(0.35) = 0.136\,m \text{ (from proximal joint)} \tag{8.56}$$

$$I_{AP} = 24.102\,(70000) - 433522 \tag{8.57}$$

$$= 1253618\,\text{g.cm}^2 = 0.1254\,\text{kg.m}^2$$

Table 8.12 Summary of the studies which have presented multi-variate linear statistical models

Study	Sample	Notes
Chandler et al. (1975)	6 males, cadavers	Includes moments of inertia for three axes per segment
Young et al. (1983)	46 females	Data obtained by imaging
Hinrichs (1985)	6 males, cadavers	Limb segments only
Clarys and Marfell-Jones (1986)	3 males, 3 female cadavers	Only segment masses reported
Jensen (1989)	15 male subjects	Data collected by imaging, subjects measured as aged (4–20 years old)
Schneider and Zernicke (1992)	18 infants	Model data derived from geometric solid model
Zatsiorsky and Seluyanov (1983)	100 male, subjects	Data derived from gamma mass scanning
Jensen et al. (1997)	27 children	Data collected by imaging, mean age 36 weeks

$$I_{ML} = 21.186\,(70000) - 222796 \tag{8.58}$$

$$= 1260224 \text{ g.cm}^2 = 0.1260 \text{ kg.m}^2$$

$$I_L = 75.608\,(70000) - 260.549 \tag{8.59}$$

$$= 269602 \text{ g.cm}^2 = 0.0269 \text{ kg.m}^2$$

Zatsiorsky and Seluyanov (1983) used this approach with their gamma mass-scanned data from 100 young male subjects. The segmental inertias for their data could be computed using multi-variate linear regression equations, where the anthropometric measures were body mass and body height. Hinrichs (1985) used this approach for estimating segmental moments of inertia but used limb specific measures. He based his equations on the data from six male cadavers dissected and measured by Chandler et al. (1975). The following is an example equation, in this case estimating the longitudinal moment of inertia for the thigh:

$$I_{L,\text{Thigh}} = 89.242A_1 + 7.8926A_2 + 1108.7 \tag{8.60}$$

where $I_{L,\text{Thigh}}$ is the moment of inertia about the longitudinal axis for the thigh (kg.cm^2), A_1 is the knee breadth, and A_2 is the upper thigh circumference.

There is a general dearth of data on infants, but one noticeable exception is the data of Schneider and Zernicke (1992). They analyzed 18 infants up to the age of 18 months, determining their segmental inertias using the geometric solid model of Hatze (1980). The segmental inertial properties were then modeled using multi-variate regression equations for the estimation of segment masses and transverse moments of inertia. The resulting equations included various combinations of segment length, circumference, length, and child age. Table 8.12 summarizes studies using linear complex equations to estimate segmental inertias.

8.5.3   Non-linear Multi-variate Statistical Models

If estimating for a human body segment the moment of inertia, the linear equations, see previous sub-section, do not reflect the non-linear relationship between the dimensions of a segment and its moment of inertia. For example, consider a cylinder, a simple geometric shape, which is a crude approximation to the shape of some of the limb segments. The cylinder mass (m) can be computed from:

$$m = \rho\,\pi\,r^2\,L \tag{8.61}$$

where ρ is segment density, r is a characteristic segment radius, and L is segment length.

The cylinder is symmetric about its two transverse axes, so there are two moments of inertia pertinent to this shape. These can be computed from:

Fig. 8.5 Variations in (**a**) mass and (**b**) longitudinal moment of inertia for a cylinder as its dimensions vary (height and circumference)

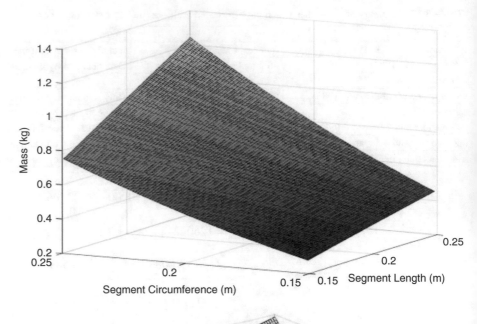

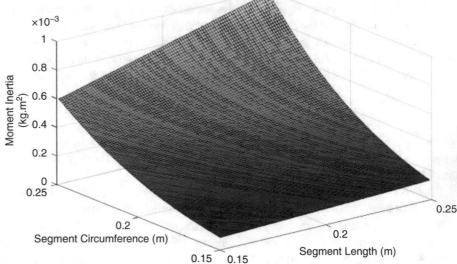

$$I_L = \frac{m\,r^2}{2} \tag{8.62}$$

$$I_T = \frac{m\left(L^2 + 3r^2\right)}{12} \tag{8.63}$$

where I_L and I_T are the longitudinal and transverse moments of inertia, respectively. The graphs (Fig. 8.5) illustrate how cylinder mass and cylinder longitudinal moment of inertia vary with segment dimensions in a non-linear fashion.

Therefore a superior way to estimate segmental inertial properties from cadaver data would be to formulate non-linear equations which take into account how segment inertias vary with segment dimensions. The equations for estimating the inertial properties of a cylinder can be restated by exploiting the relationship between the circumference of a circle and its radius. The resulting equations are:

$$m = K_1\,L\,C^2 \tag{8.64}$$

$$I_T = K_2\, C^2\, L^3 \tag{8.65}$$

$$I_L = K_3\, L\, C^4 \tag{8.66}$$

where K_1, K_2, and K_3 are constants and C is a characteristic segment circumference. Both Yeadon and Morlock (1989) and Challis and Kerwin (1992) showed that equations of this format were superior to estimating segmental limb moments of inertia compared with their linear alternatives.

Based on the gamma scanning of 100 male subjects, Zatsiorsky et al. (1990a, 1990b) presented non-linear equations based on these non-linear relationships. The equations had the following format:

$$m = K_m\, L\, C^2 \tag{8.67}$$

$$I_s = K_s\, m\, L^2 \tag{8.68}$$

$$I_F = K_F\, m\, L^2 \tag{8.69}$$

$$I_L = K_L\, m\, C^2 \tag{8.70}$$

Table 8.13 provides the appropriate coefficients, and Table 8.14 defines the segment boundaries required to determine segment lengths and where segment perimeter should be measured.

For example, the following calculation illustrate these data for a subject of mass 70 kg male with a thigh length of 0.35 m, and circumference of 0.55 m:

$$L_{CM} = 0.35\,\frac{45.49}{100} = 0.159\, m \text{ (from proximal joint)} \tag{8.71}$$

$$L = K_{\text{Len}}\, L \tag{8.72}$$

$$L = (1.083)(35) = 37.9\text{ cm}$$

$$m = K_m\, L\, C^2 \tag{8.73}$$

$$m = \left(6.664 \times 10^{-5}\right)(37.9)\left(55^2\right)$$

$$m = 7.64\text{ kg}$$

Table 8.13 The coefficients for the complex statistical modeling of the data of Zatsiorsky et al. (1990a, 1990b) to determine segment mass, center of mass location, and the moments of inertia

Segment	Center of mass (%SL)	K_m ($\times 10^{-5}$)	K_S ($\times 10^{-2}$)	K_F ($\times 10^{-2}$)	K_L ($\times 10^{-2}$)	K_{Len}
Hand	63.09	5.54	6.65	4.86	2.29	1.000
Forearm	57.26	6.26	7.55	7.03	1.51	1.000
Upper arm	55.02	9.67	10.81	9.71	2.06	0.730
Foot	55.85	6.14	7.86	7.14	1.60	1.000
Shank	40.47	5.85	8.77	8.44	1.44	1.000
Thigh	45.49	6.664	7.18	7.18	1.33	1.083
Trunk	43.70	5.64	6.23	5.27	1.18	1.465
Head	63.09	6.37	8.68	9.38	1.25	0.760

N.B. Lengths and circumferences are both measured in centimeters. These equations provide moments of inertia with units kg.cm². The coefficients are used to determine K_m segment mass, K_S sagittal axis moment of inertia, K_F frontal axis moment of inertia, and K_L longitudinal axis moment of inertia. The coefficient K_{Len} is used to scale segment lengths before use in the equations to estimate masses and moments of inertia.

Table 8.14 The body landmarks used to define segment endpoints and the perimeter measurements required for the data of Zatsiorsky et al. (1990a, 1990b)

Segment	Body landmarks	Site of perimeter measurement
Hand	Stylion, third dactylion	Metacarpus
Forearm	Radiale, stylion	Maximum perimeter
Upper arm	Acromion, radiale	Maximum perimeter
Foot	Heel, toe tip	Level of metatarsus
Shank	Tibiale, sphyrion	Maximum perimeter
Thigh	Iliospinale, tibiale	At gluteal fold
Trunk	Cervicale, hsp intersection	Mean of chest, waist, pelvis
Head	Vertex, cervicale	Max horizontal perimeter

N.B. "hsp" refers to the hip segmentation planes which are the boundaries between thighs and trunk

Table 8.15 Summary of the studies which have presented non-linear multi-variate statistical models

Study	Sample	Notes
Yeadon and Morlock (1989)	6 males, cadavers	Limb segments only
Challis and Kerwin (1992)	6 male, cadavers	Limb segments only
Zatsiorsky et al. (1990a, 1990b)	100 male, live subjects	Re-analysis of data from 1983
Ma et al. (2011)	40 male, 40 female.	Korean subjects, with data collected from a geometric solid model
Jensen et al. (1996)	15 females subjects	Data collected by imaging, subjects were pregnant
Jensen et al. (1997)	27 children	Data collected by imaging, mean age 36 weeks

$$I_s = K_s \, m \, L^2 \tag{8.74}$$

$$I_s = \left(7.18 \times 10^{-2}\right)(7.64)\left(37.9^2\right)$$

$$I_s = 787.9 \text{ kg.cm}^2 = 0.07879 \text{ kg.m}^2$$

$$I_F = K_F \, m \, L^2 \tag{8.75}$$

$$I_F = \left(7.18 \times 10^{-2}\right)(7.64)\left(37.9^2\right)$$

$$I_F = 787.9 \text{ kg.cm}^2 = 0.07879 \text{ kg.m}^2$$

$$I_L = K_L \, m \, C^2 \tag{8.76}$$

$$I_L = \left(1.33 \times 10^{-2}\right)(7.64)\left(55^2\right)$$

$$I_F = 307.4 \text{ kg.cm}^2 = 0.03074 \text{ kg.m}^2$$

There have been a number of studies of different populations which have used non-linear complex statistical models to determine segmental inertias, a summary of these are presented in Table 8.15.

8.6   Making Measurements

With many of the methods described in this chapter, it is necessary to make measurements directly on the human body; in this section some of the issues associated with these measurements will be addressed.

It is feasible that the inertial properties of a segment change during the task, and this is typically not accounted for in biomechanical studies. For example, Pain and Challis (2001) showed that the inertial properties of the shank changed as the triceps surae muscles shortened. Comparing the dorsi-flexed to the plantar-flexed position, the location of the center of mass moved closer to the proximal end of the lower leg; the moment of inertia increased by 5% about the longitudinal axis and decreased by 8% about the two transverse axes. If an experiment continues over a few days, then it is feasible the inertial properties can change in particular in the young. Lampl et al. (1992) showed the growth changes during the adolescent growth spurt occur in jumps. For example, standing height increased within one 24-h period by 0.5–1.63 cm followed by 2–28 days of stasis.

8.6.1   Operator Issues

Marks et al. (1989a, 1989b) examined the reliability with which eight basic anthropometric measures could be made. They reported reliability coefficients in excess of 0.97, for triceps and subscapular skinfolds, bitrochanteric breadth, and elbow breadth. Such reliabilities are useful, but how do they translate into the calculation of segmental inertial properties? Challis (1999) examined the variability of segmental inertias due to repeat measurements by 2 measurers on 50 subjects. Three ways to estimate the inertia properties were examined: the first used a single geometric solid to model each segment, the second used modeling segments as series of two geometric solids, and the third used multi-variate regression of equations of Zatsiorsky et al. (1990b). The precision of measurements were high for all three methods, despite requiring different measures. To evaluate the influence of the imprecision in the measures data for a subject performing a maximum vertical jump were analyzed. The data were processed using an original data set and one with the inertial properties perturbated by amounts reflecting the maximum errors due to measurement imprecision. The sagittal plane resultant joint moments were changed by 0.3%, 2.0%, and 5.2% for the ankle, knee, and hip, respectively. Larger changes in these moments could be produced by varying the filter cut-off by 0.5 Hz.

For Hanvan's model a total of 37 anthropometric measures are required (Hanavan 1964). The number of measures can be reduced to 25 by measuring 1 side of the body only. For more complex models, such reductions in the number of measures are attractive, but is this justified? Laubach and McConville (1967) examined the difference between body sides for 21 measures made on the limbs, for between 42 and 117 subjects. They found statistically significant differences between the two sides of the body for eight of the anthropometric measures. These results correspond with the results from a cadaver study which showed asymmetry in lower limb muscle masses (Chhibber and Singh 1970).

8.6.2   Inter-Operator Issues

Bennet and Osborne (1986) had 8 trained measurers make 63 anthropometric measures on 8 male and 8 female subjects. They reported poor consistency between measurers with errors which were worst for their female subjects. Kouchi et al. (1999) performed a similar study examining 32 anthropometric measures on 37 subjects. There were large inter-measurer errors for five of the measurements, but these were reduced when the measurers made measurements on a second day, suggesting that practice can reduce inter-measurer differences. Challis (1999) also examined the inter-measurer variability of segmental inertias determined for 50 subjects for 3 methods for determining segmental inertias. The inter-measurer precisions of measurements were high for all three methods, comparable to intra-measurer precisions. Collectively these results suggest that with appropriate training, multiple measurers (operators) can be tolerated for determining segmental inertias.

8.7   Animal Studies

The focus in the preceding has been studies on humans, but the techniques described can also be applied to other animals. In equine biomechanics Buchner et al. (1997) measured the three-dimensional inertial properties of six Dutch Warmblood horses and then developed linear equations for the estimation of these inertial properties. In a similar study, Nauwelaerts and Clayton (2018) measured the inertial properties of the limbs of a variety of breeds of 30 horses, and presented a scaling method to estimate these properties.

Cheng and Scott (2000) determined the inertial properties of the body segments of six *Macaca mulatta* and three *Macaca fascicularis* (Old World monkeys). The inertial properties were determined using measurement techniques similar to Dempster (1955). They also examined simple statistical models for the estimation of inertial properties. Schoonaert et al. (2007) determined the inertial properties of the body segments of 53 *Pan troglodytes* (chimpanzees). They determined the inertial properties by modeling the segments as geometric solids.

8.8   Review Questions

1. Define the following: mass, moment of inertia, center of mass, principal moments of inertia, and principal axes.
2. Describe the limitations of the major cadaver studies used for determining human body segment inertial parameters.
3. Outline how Dempster (1955) used his cadaver data to estimate body segment inertial parameters.
4. Contrast two candidate statistical models for estimating body segment inertial parameters.
5. How can medical imaging techniques be used to determine human body segment inertial parameters?
6. What geometric shapes can be used for the modeling of human body segments? What are the relative merits of the different geometric solid models of the human body?

References

Cited References

Ackland, T. R., Henson, P. W., & Bailey, D. A. (1988). The uniform density assumption: Its effect upon estimation of body segment inertial parameters. *International Journal of Sport Biomechanics, 4*(2), 146–155.

Baca, A. (1996). Precise determination of anthropometric dimensions by means of image processing methods for estimating human body segment parameter values. *Journal of Biomechanics, 29*(4), 563–567.

Barter, J. T. (1957). *Estimation of the mass of body segments (Vol. WADC technical report 57–260).* Ohio: Wright-Patterson Air-Force Base.

Bauer, J. J., Pavol, M. J., Snow, C. M., & Hayes, W. C. (2007). MRI-derived body segment parameters of children differ from age-based estimates derived using photogrammetry. *Journal of Biomechanics, 40*(13), 2904–2910.

Bennett, K. A., & Osborne, R. H. (1986). Interobserver measurement reliability in anthropometry. *Human Biology, 58*(5), 751–759.

Blankevoort, L., Huiskes, R., & de Lange, A. (1990). Helical axes of passive knee joint motions. *Journal of Biomechanics, 23*(12), 1219–1229.

Blemker, S. S., Asakawa, D. S., Gold, G. E., & Delp, S. L. (2007). Image-based musculoskeletal modeling: Applications, advances, and future opportunities. *Journal of Magnetic Resonance Imaging, 25*(2), 441–451.

Braune, W., & Fischer. (1889). Uber den schwerpunkt des menschlichen körpers, mit rucksicht auf die aiisrustung des deutschen Infanteristen. (The Center of Gravity of the Human Body as Related to the Equipment of the German Infantry). *Abhandlungen der Mathematisch-Physischen Klasse der Königlich-Sächsischen Gesellschaft der Wissenschaften, 26,* 561–672.

Brooke-Wavell, K., Jones, P. R., & West, G. M. (1994). Reliability and repeatability of 3-D body scanner (LASS) measurements compared to anthropometry. *Annals of Human Biology, 21*(6), 571–577.

Brooks, C. B., & Jacobs, A. M. (1975). The gamma mass scanning technique for inertial anthropometric measurement. *Medicine and Science in Sports, 7*(4), 290–294.

Buchner, H. H. F., Savelberg, H. H. C. M., Schamhardt, H. C., & Barneveld, A. (1997). Inertial properties of Dutch warmblood horses. *Journal of Biomechanics, 30*(6), 653–658.

Challis, J. H. (1999). Precision of the estimation of human limb inertial parameters. *Journal of Applied Biomechanics, 15*(4), 418–428.

Challis, J. H., & Kerwin, D. G. (1992). Calculating upper limb inertial parameters. *Journal of Sports Sciences, 10,* 275–284.

Challis, J. H., & Pain, M. T. G. (2008). Soft tissue motion influences skeletal loads during impacts. *Exercise and Sport Sciences Reviews, 36*(2), 71–75.

Chandler, R. F., Clauser, C. E., McConville, J. T., Reynolds, H. M., & Young, J. W. (1975). *Investigation of the inertial properties of the human body (AMRL technical report 74–137).* Ohio: Wright Patterson Air Force Base.

Chandler, R. F., Snow, C. C., & Young, J. W. (1978). Computation of mass distribution characteristics of children. In *Applications of human biostereometrics (NATO)* (Vol. 166, pp. 159–166). Washington, D.C.: SPIE.

Cheng, C. K., Chen, H. H., Chen, C. S., Lee, C. L., & Chen, C. Y. (2000). Segment inertial properties of Chinese adults determined from magnetic resonance imaging. *Clinical Biomechanics, 15,* 559–566.

Cheng, E. J., & Scott, S. H. (2000). Morphometry of Macaca mulatta forelimb. I. Shoulder and elbow muscles and segment inertial parameters. *Journal of Morphology, 245*(3), 206–224.

Chhibber, S. R., & Singh, I. (1970). Asymmetry in muscle weight and one-sided dominance in the human lower limbs. *Journal of Anatomy, 106*(3), 553–556.

Claessens, A. L., Veer, F. M., Stijnen, V., Lefevre, J., Maes, H., Steens, G., & Beunen, G. (1991). Anthropometric characteristics of outstanding male and female gymnasts. *Journal of Sports Sciences, 9*(1), 53–74.

Clarys, J. P., & Marfell-Jones, M. J. (1986). Anatomical segmentation in humans and the prediction of segmental masses from intra-segmental anthropometry. *Human Biology, 58*(5), 771–782.

Clarys, J. P., Scafoglieri, A., Provyn, S., Louis, O., Wallace, J. A., & De Mey, J. (2010). A macro-quality evaluation of DXA variables using whole dissection, ashing, and computer tomography in pigs. *Obesity, 18*(8), 1477–1485.

Clauser, C. E., McConville, J. T., & Young, J. W. (1969). *Weight, Volume and Center of Mass of Segments of the Human Body (AMRL technical report 69–70).* Ohio: Wright Patterson Air Force Base.

Cornelis, J., Van Gheluwe, B., Nyssen, M., & van den Berghe, F. (1978). A photographic method for the tridimensional reconstruction of the human thorax. In *Applications of Human Biostereometrics (NATO)* (Vol. 166, pp. 294–300). Washington: SPIE.

Dempster, W. T. (1955). *Space Requirements of the Seated Operator (WADC technical report 55–159)*. Ohio: Aerospace Medical Research Laboratory, Wright-Patterson Air Force Base.

Dowling, J. J., Durkin, J. L., & Andrews, D. M. (2006). The uncertainty of the pendulum method for the determination of the moment of inertia. *Medical Engineering & Physics, 28*(8), 837–841.

Durkin, J. L., & Dowling, J. J. (2003). Analysis of body segment parameter differences between four human populations and the estimation errors of four popular mathematical models. *Journal of Biomechanical Engineering, 125*(4), 515–522.

Durkin, J. L., Dowling, J. J., & Andrews, D. M. (2002). The measurement of body segment inertial parameters using dual energy X-ray absorptiometry. *Journal of Biomechanics, 35*(12), 1575–1580.

Furlow, B. (2010). Radiation dose in computed tomography. *Radiologic Technology, 81*(5), 437–450.

Gordon, C. C., Churchill, T., Clauser, C. E., Bradtmiller, B., McConville, J. T., Tebbetts, I., & Walker, R. A. (1989). *1988 anthropometric survey of U.S. Army personnel: Correlation coefficients and regression equations part 1 statistical techniques, landmarks, and measurement definitions*. Natwick: United State Army.

Greenwood, D. T. (1965). *Principles of Dynamics*. New Jersey: Pretice-Hall, Inc., Englewood Cliffs.

Hanavan, E. P. (1964). *A mathematical model of the human body (AMRL technical report 64–102)*. Ohio: Wright-Patterson Air Force Base.

Harless, E. (1860). Die statischen momente der menschlichen gliedmassen [the static moments of the component masses of the human body]. *Abhandlungen der Mathemat - Physikalischen Classe Der Koeniglichen Bayerischen Akademie der Wissenschaften, 8*(71–97), 259–294.

Hatze, H. (1975). A new method for the simultaneous measurement of the moment of inertia, the damping coefficient and the location of the Centre of mass of a body segment in situ. *European Journal of Applied Physiology, 34*, 217–226.

Hatze, H. (1980). A mathematical model for the computational determination of parameter values of anthropomorphic segments. *Journal of Biomechanics, 13*(10), 833–843.

Hinrichs, R. N. (1985). Regression equations to predict segmental moments of inertia from anthropometric measurements. *Journal of Biomechanics, 18*(8), 621–624.

Ho, W. H., Shiang, T. Y., Lee, C. C., & Cheng, S. Y. (2013). Body segment parameters of young Chinese men determined with magnetic resonance imaging. *Medicine and Science in Sports Exercise, 45*(9), 1759–1766.

Huang, H. K., & Wu, S. C. (1976). The evaluation of mass densities of the human body in vivo from CT scans. *Computers in Biology and Medicine, 6*, 337–343.

Jensen, R. K. (1978). Estimation of the biomechanical properties of three body types using a photogrammetric method. *Journal of Biomechanics, 11*(8–9), 349–358.

Jensen, R. K. (1986). Body segment mass, radius and radius of gyration proportions of children. *Journal of Biomechanics, 19*(5), 359–368.

Jensen, R. K. (1989). Changes in segment inertia proportions between 4 and 20 years. *Journal of Biomechanics, 22*(6–7), 529 536.

Jensen, R. K. (1993). Human Morphology: Its role in the mechanics of movement. *Journal of Biomechanics, 26*(Supplement 1), 81–94.

Jensen, R. K., & Fletcher, P. (1994). Distribution of mass to the segments of elderly males and females. *Journal of Biomechanics, 27*(1), 89–96.

Jensen, R. K., Treitz, T., & Doucet, S. (1996). Prediction of human segment inertias during pregnancy. *Journal of Applied Biomechanics, 12*(1), 15–30.

Jensen, R. K., Treitz, T., & Sun, H. (1997). Prediction of infant segment inertias. *Journal of Applied Biomechanics, 13*(3), 287–299.

Jones, P. R. M., West, G. M., Harris, D. H., & Read, J. B. (1989). The Loughborough anthropometric shadow scanner (LASS). *Endeavour, 13*(4), 162–168.

Katch, V., Weltman, A., & Gold, E. (1974). Validity of anthropometric measurements and the segment-zone method for estimating segmental and total body volume. *Medicine and Science in Sports, 6*(4), 271–276.

Kouchi, M., Mochimaru, M., Tsuzuki, K., & Yokoi, T. (1999). Interobserver errors in anthropometry. *Journal of Human Ergology, 28*(1/2), 15–24.

Lampl, M., Veldhuis, J. D., & Johnson, M. L. (1992). Saltation and stasis: A model of human growth. *Science, 258*(5083), 801–803.

Laubach, L. L., & McConville, J. T. (1967). Notes on anthropometric technique: Anthropometric measurements — Right and left sides. *American Journal of Physical Anthropology, 26*(3), 367–369.

Lephart, S. A. (1984). Measuring the inertial properties of cadaver segments. *Journal of Biomechanics, 17*(7), 537–543.

Li, Y., & Dangerfield, P. H. (1993). Inertial characteristics of children and their application to growth study. *Annals of Human Biology, 20*(5), 433–454.

Ma, Y., Lee, K., Li, L., & Kwon, J. (2011). Nonlinear regression equations for segmental mass-inertial characteristics of Korean adults estimated using three-dimensional range scan data. *Applied Ergonomics, 42*(2), 297–308.

Marcus, R. L., Addison, O., Kidde, J. P., Dibble, L. E., & Lastayo, P. C. (2010). Skeletal muscle fat infiltration: Impact of age, inactivity, and exercise. *The Journal of Nutrition, Health & Aging, 14*(5), 362–366.

Marks, G. C., Habicht, J. P., & Mueller, W. H. (1989a). Reliability, dependability and precision of anthropometric measurements. *Amercian Journal of Epidemiology, 130*(3), 578–587.

Marks, G. C., Habicht, J. P., & Mueller, W. H. (1989b). Reliability, dependability and precision of anthropometric measurements. *Amercian Journal of Epidemiology, 130*(3), 578–587.

Martin, P. E., Mungiole, M., Marzke, M. W., & Longhill, J. M. (1989). The use of magnetic resonance imaging for measuring segment inertial properties. *Journal of Biomechanics, 22*(4), 367–376.

McConville, J. T., Clauser, C. E., Churchill, T. D., Cuzzi, J., & Kaleps, I. (1980). *Anthropometric Relationships of Body and Body Segment Moments of Inertia*: Wright-Patterson Air Force Base, Ohio 45433.

Miller, D. I., & Morrison, W. E. (1975). Prediction of segmental parameters using the Hanavan human body model. *Medicine and Science in Sports, 7*(3), 207–212.

Miller, D. I., & Nelson, R. C. (1973). *Biomechanics of sport*. London: Henry Kimpton.

Mori, M., & Yamamoto, T. (1959). Die massenanteile der einzelnen körperabschnitte der Japaner. *Cells, Tissues, Organs, 37*(4), 385–388.

Pain, M. T. G., & Challis, J. H. (2001). A high resolution technique for determining body segment inertial parameters and their variation due to soft tissue motion. *Journal of Applied Biomechanics, 17*(4), 326–334.

Pataky, T. C., Zatsiorsky, V. M., & Challis, J. H. (2003). A simple method to determine body segment masses in vivo: Reliability, accuracy and sensitivity analysis. *Clinical Biomechanics, 18*(4), 364–368.

Pearsall, D. J., Reid, J. G., & Ross, R. (1994). Inertial properties of the human trunk of males determined from magnetic resonance imaging. *Annals of Biomedical Engineering, 22*(6), 692–706.

Plagenhoef, S. C., Evans, F. G., & Abdelnour, T. (1983). Anatomical data for analysing human motion. *Research Quarterly, 54*(2), 169–178.

Pothrat, C., Authier, G., Viehweger, E., Berton, E., & Rao, G. (2015). One- and multi-segment foot models lead to opposite results on ankle joint kinematics during gait: Implications for clinical assessment. *Clinical biomechanics, 30*(5), 493–499.

Rodrigue, D., & Gagnon, M. (1983). The evaluation of forearm density with axial tomography. *Journal of Biomechanics, 16*(11), 907–913.

Sanders, R. H., Chiu, C. Y., Gonjo, T., Thow, J., Oliveira, N., Psycharakis, S. G., Payton, C. J., & McCabe, C. B. (2015). Reliability of the elliptical zone method of estimating body segment parameters of swimmers. *Journal of Sports Science & Medicine, 14*(1), 215–224.

Nauwelaerts, S., & Clayton, H. M. (2018). Evaluation of a pictorial method to obtain subject-specific inertial properties in equine limb segments. *Journal of Morphology, 279*(7), 997–1007.

Santschi, W. R., DuBois, J., & Omoto, C. (1963). *Moments of Inertia and Centers of Gravity of the Living Human Body*: Wright-Patterson Air Force Base, Ohio.

Sarfaty, O., & Ladin, Z. (1993). A video-based system for the estimation of the inertial properties of body segments. *Journal of Biomechanics, 26*(8), 1011–1016.

Schneider, K., & Zernicke, R. F. (1992). Mass, center of mass, and moment of inertia estimates for infant limb segments. *Journal of Biomechanics, 25*(2), 145–148.

Schoonaert, K., D'Aout, K., & Aerts, P. (2007). Morphometrics and inertial properties in the body segments of chimpanzees (Pan troglodytes). *Journal of Anatomy, 210*(5), 518–531.

Seltzer, S. M. (1993). Calculation of photon mass energy-transfer and mass energy-absorption coefficients. *Radiation Research, 136*(2), 147–170.

Tanner, J. M., Whitehouse, R. H., & Jarman, S. (1964). *The physique of the Olympic athlete; a study of 137 track and field athletes at the XVIIth Olympic games, Rome 1960, and a comparison with weight-lifters and wrestlers*. London: G. Allen and Unwin.

Visser, M., Fuerst, T., Lang, T., Salamone, L., & Harris, T. B. (1999). Validity of fan-beam dual-energy X-ray absorptiometry for measuring fat-free mass and leg muscle mass. Health, Aging, and Body Composition Study–Dual-Energy X-ray Absorptiometry and Body Composition Working Group. *Journal of Applied Physiology, 87*(4), 1513–1520.

Vitruvius, P. (1914). *Vitruvius, the ten books on architecture* (M. H. Morgan & H. L. Warren, trans.). Cambridge: Harvard university press; etc.

Wang, J., Gallagher, D., Thornton, J. C., Yu, W., Horlick, M., & Pi-Sunyer, F. X. (2006). Validation of a 3-dimensional photonic scanner for the measurement of body volumes, dimensions, and percentage body fat. *The American Journal of Clinical Nutrition, 83*(4), 809–816.

Weinbach, A. P. (1938). Contour maps, center of gravity, moment of inertia and surface area of the human body. *Human Biology, 10*(3), 356–371.

Wicke, J., & Dumas, G. A. (2010). Influence of the volume and density functions within geometric models for estimating trunk inertial parameters. *Journal of Applied Biomechanics, 26*(1), 26–31.

Wicke, J., Dumas, G. A., & Costigan, P. A. (2008). Trunk density profile estimates from dual X-ray absorptiometry. *Journal of Biomechanics, 41*(4), 861–867.

Wicke, J., & Lopers, B. (2003). Validation of the volume function within Jensen's (1978) elliptical cylinder model. *Journal of Applied Biomechanics, 19*(1), 3–12.

Wicke, J., & Dumas, G. A. (2008). Estimating segment inertial parameters using fan-beam DXA. *Journal of Applied Biomechanics, 24*(2), 180–184.

Yeadon, M. R. (1990). The simulation of aerial movement – II. A mathematical inertia model of the human body. *Journal of Biomechanics, 23*(1), 67–74.

Yeadon, M. R., & Morlock, M. (1989). The appropriate use of regression equations for the estimation of segmental inertia parameters. *Journal of Biomechanics, 22*(6/7), 683–689.

Young, J. W., Chandler, R. F., Snow, C. C., Robinette, K. M., Zehner, G. F., & Lofberg, M. S. (1983). *Anthropometric and mass distribution characteristics of the adult female (report FAA-AM-83-16)*. Oklahoma City: Civil Aeromedical Institute, Federal Aviation Administration.

Zatsiorsky, V. M., & Seluyanov, V. (1983). The mass and inertial characteristics of the main segments of the human body. In H. Matsui & K. Kobayashi (Eds.), *Biomechanics VIII-B* (pp. 1152–1159). Baltimore: University Park Press.

Zatsiorsky, V. M., & Seluyanov, V. (1985). Estimation of the mass and inertia characteristics of the human body by means of the best predictive regression equations. In D. A. Winter, R. W. Norman, R. P. Wells, K. C. Hayes, & A. E. Patla (Eds.), *Biomechanics IX-B* (pp. 233–239). Champaign: Human Kinetics Publishers.

Zatsiorsky, V. M., Seluyanov, V., & Chugunova, L. (1990a). Methods of determining mass-inertial characteristics of human body segments. In G. G. Chernyi & S. A. Regirer (Eds.), *Contemporary problems of biomechanics* (pp. 272–291). Moscow: Mir Publishers.

Zatsiorsky, V. M., Seluyanov, V., & Chugunova, L. (1990b). In vivo body segment inertial parameters determination using a gamma-scanner method. In N. Berme & A. Cappozzo (Eds.), *Biomechanics of human movement: Applications in rehabilitation, sports and ergonomics* (pp. 186–202). Worthington: Bertec Corporation.

Useful References

Contini, R. (1972). Body segment parameters, part II. *Artificial Limbs, 16*(1), 1–19.

Drillis, R., Contini, R., & Bluestein, M. (1964). Body segment parameters. *Artificial Limbs, 8*(1), 44–66.

Hay, J. G. (1973). The center of gravity of the human body. *Kinesiology, III*, 120–144.

Hay, J. G. (1974). Moment of inertia of the human body. *Kinesiology, IV*, 43–52.

Reid, J. G., & Jensen, R. K. (1990). Human body segment inertia parameters: A survey and status report. *Exercise and Sport Sciences Reviews, 18*, 225–241.

Kinematics

Overview

Kinematics is the study of the motion of a point, points, bodies, or system of bodies without considering their masses or the forces which cause the motion. When the Weber brothers examined human walking in the early nineteenth century, their experimental apparatus consisted of a stopwatch, measuring tape, and telescope (Weber and Weber 1992). This apparatus allowed them to measure step length, step cadence, and walking speed. The increased sophistication of experimental apparatus since then means that more detailed descriptions of the kinematics of the human body are now feasible.

There are three major issues with describing the kinematics of the human body:

- The computation of some of the kinematic properties is sensitive to the noise which corrupts the experimental data.
- That aspects of the kinematics depend on choices made by the experimenter, for example, how reference frames are defined for the human body segments.
- That there is not a single option for how to quantify the kinematics, for example, to describe the orientation of the shank, one could use Euler angles, Cardan angles, or quaternions (to give just three options).

The content of this chapter requires a working knowledge of matrix algebra (see Appendix A) and trigonometry (see Appendix D).

9.1 Two-Dimensional Analysis

For a single body, for example the forearm, in two-dimensional space it has three degrees of freedom. Therefore, we need to define two coordinates to define its position and one angle to define its orientation, and then we have defined its pose (position and orientation). If we have two segments moving relative to one another, for example the forearm and upper arm articulating at the elbow joint, their relative orientation gives joint orientation.

It is common for mechanical analysis of human movement to be performed in two dimensions, making an assumption that the movement is actually constrained to a single place. This may be the case, for example, for reaching along the surface of a tabletop but this is not always the case.

9.1.1 Segment Position

To define the position of a rigid body we need to define a reference frame. The primary reference to define is an inertial reference frame, often referred to as a laboratory reference frame. The presumption is that this reference frame does not move, which is a prerequisite for applying Newtonian mechanics. This reference system is typically an X, Y, Z system, but as we are only examining two-dimensional motion then we only require two of these axes (see Fig. 9.1). With a reference frame for two-dimensional analysis, the X axis does need to be considered missing; it is just coming out of the page and provides a reference for defining an axis of rotation for angles in two dimensions.

If a rigid body translates from one position to another, then the motion can be described with the following equation:

© The Editor(s) (if applicable) and The Author(s), under exclusive license to Springer Nature Switzerland AG 2021
J. H. Challis, *Experimental Methods in Biomechanics*, https://doi.org/10.1007/978-3-030-52256-8_9

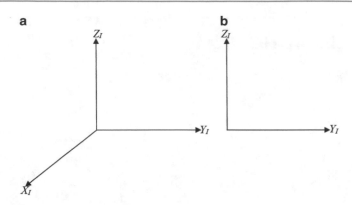

Fig. 9.1 Three- and two-dimensional reference frames. (**a**) A three-dimensional inertial reference frame and (**b**) a two-dimensional inertial reference frame

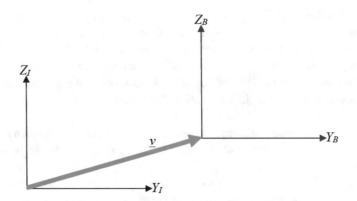

Fig. 9.2 The translation vector, $\underline{v}$, from the origin of the two-dimensional inertial reference frame (Y_I, Z_I) to the origin of the two-dimensional body-fixed reference frame (Y_B, Z_B)

$$\begin{bmatrix} y_2 \\ z_2 \end{bmatrix} = \begin{bmatrix} y_1 \\ z_1 \end{bmatrix} + \underline{v} \tag{9.1}$$

where y_2, z_2 are the coordinates of a point at time 2, y_1, z_1 are the coordinates of a point at time 1, and $\underline{v}$ is the vector describing the translation of the body which occurs from time 1 to time 2.

Another way to visualize this is to consider the inertial reference frame and a reference frame attached to the body of interest. In this case (see Fig. 9.2), the vector $\underline{v}$ describes the position of the origin of the body-fixed reference frame relative to the inertial reference frame.

Defining the motion using an inertial reference frame and body-fixed reference frame may seem to overcomplicate the analysis, but this is how motion in three dimensions is defined so this representation for a two-dimensional analysis is a useful stepping-stone to three-dimensional analyses.

9.1.2   Segment Orientation

The orientation of a segment in two dimensions can be easily defined by drawing a reference line from one end of the segment to the other and then quantifying the angle the line makes relative to the horizontal. Confusingly, sometimes, orientation is defined relative to the vertical. In motion analysis the line on the segment is typically between the distal and the proximal joint centers. Such an approach works but here the problem of defining segment orientation is cast in terms of matrix algebra. This has two advantages: the first is it presents a natural segue to considering orientations in three dimensions, and the second is that the matrix approach opens up some computational options which can improve the robustness of the computations required to determine segment orientation.

Fig. 9.3 The same point measured in two different reference frames. (**a**) I(nertial) and (**b**) B(ody fixed)

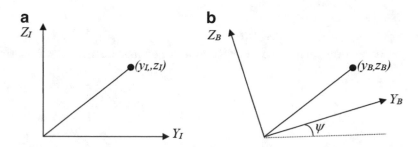

To illustrate the matrices, required for defining rigid body orientation in two dimensions, the derivation of one of them will be described. If the rotation is about the x axis, then this can be visualized by examining the Y and Z axes only (the X axis would be perpendicular to the plane of the page) (see Fig. 9.3).

The same point can be measured in both reference frames, $p_I = (y_I, z_I)$, $p_B = (y_B, z_B)$. Using basic trigonometry a point measured in one reference frame can be described by the measurements made in the other reference frame:

$$y_B = y_I \cos(\psi) + z_I \cos\left(\frac{\pi}{2} + \psi\right) \tag{9.2}$$

$$z_B = y_I \cos\left(\frac{\pi}{2} - \psi\right) + z_I \cos(\psi) \tag{9.3}$$

Using trigonometric identities, these two equations can be simplified:

$$\cos\left(\frac{\pi}{2} + \psi\right) = -\sin(\psi) \quad \rightarrow \quad y_B = y_I \cos(\psi) - z_I \sin(\psi) \tag{9.4}$$

$$\cos\left(\frac{\pi}{2} - \psi\right) = \sin(\psi) \quad \rightarrow \quad z_B = y_I \sin(\psi) + z_I \sin(\psi) \tag{9.5}$$

These equations can now be put in matrix form:

$$\begin{bmatrix} y_B \\ z_B \end{bmatrix} = \begin{bmatrix} \cos(\psi) & -\sin(\psi) \\ \sin(\psi) & \cos(\psi) \end{bmatrix} \begin{bmatrix} y_I \\ z_I \end{bmatrix} \tag{9.6}$$

which can be expressed as:

$$p_B = R_{IB} p_I \tag{9.7}$$

where R_{IB} is the attitude matrix which transforms points from the inertial to the body-fixed reference frame. It is from this attitude matrix that the orientation of the rigid body can be extracted. So:

$$R_{IB} = \begin{bmatrix} \cos(\psi) & -\sin(\psi) \\ \sin(\psi) & \cos(\psi) \end{bmatrix} \tag{9.8}$$

There are a number of important properties of the attitude matrix; these are:

1. The matrix is proper orthogonal, which means the rows are all orthogonal to one another and similarly the columns.
2. As the matrix is proper orthogonal, the determinant of the attitude matrix is *1* (if the determinant was -1, then the matrix would represent a reflection). A determinant of *1* indicates the length of any vector multiplied by this matrix is unchanged.
3. The attitude matrix belongs to the special-orthogonal group of order two ($R \in SO(2)$).
4. As the attitude matrix belongs to the special-orthogonal group, the inverse of this matrix also belongs to this group, as does the result of the product of any matrices in this group.
5. The transpose of the attitude matrix is equal to its inverse; therefore:

$$R^T = R^{-1} \tag{9.9}$$

To confirm point 2, the determinant of the attitude matrix, a *2 x 2* matrix, is computed from:

$$\det(R) = r_{1,1}r_{2,2} - r_{1,2}r_{2,1}$$

$$= \cos(\psi)\cos(\psi) + \sin(\psi)\sin(\psi)$$

$$= \cos(\psi)^2 + \sin(\psi)^2 = 1 \tag{9.10}$$

Cayley (1843) identified how a proper orthogonal matrix is a function of a skew-symmetric matrix, P:

$$R = (I - P)(I + P)^{-1} \tag{9.11}$$

where I is the identity matrix and P is a skew-symmetric matrix which has the following format:

$$P\{p\} = \begin{bmatrix} 0 & p \\ -p & 0 \end{bmatrix} \tag{9.12}$$

Therefore, P will only have one unique element, which is the angle that can be extracted from R.

9.1.3   Segment Pose

If a rigid body in two dimensions undergoes translation and rotation, then the new pose (position and orientation) of a point on that body can be described by:

$$p_B = Rp_I + \underline{v} \tag{9.13}$$

On occasions, the attitude matrix and position vector can be combined into one matrix; this is referred to as a homogenous transformation matrix:

$$\begin{bmatrix} y_B \\ z_B \\ 1 \end{bmatrix} = \begin{bmatrix} \cos(\psi) & -\sin(\psi) & v_y \\ \sin(\psi) & \cos(\psi) & v_z \\ 0 & 0 & 1 \end{bmatrix} \begin{bmatrix} y_I \\ z_I \\ 1 \end{bmatrix} \tag{9.14}$$

9.1.4   Joint Orientation

The orientation of a joint is typically defined by the motion of the distal segment with respect to the proximal joint. For example, if a reference frame is embedded in the upper arm, and one in the forearm, the mapping from forearm-based reference frame to the upper arm reference frame provides the elbow joint angle:

$$x_I = R_{PI}x_P + \underline{v}_{PI}(\text{mapping from proximal to inertial reference frame}) \tag{9.15}$$

$$x_I = R_{DI}x_D + \underline{v}_{DI}(\text{mapping from distal to inertial reference frame}) \tag{9.16}$$

where x_I is points measured in the inertial reference frame, R_{PI} attitude matrix mapping from proximal to inertial reference frame, x_P points measured in proximal reference frame, $\underline{v}_{PI}$ translation vector from origin of proximal reference frame to origin of inertial reference frame, R_{DI} attitude matrix mapping from distal to inertial reference frame, x_D points measured in proximal reference frame, and $\underline{v}_{DI}$ translation vector from distal of proximal reference frame to origin of inertial reference frame. Note that the output from a motion analysis system will measure the points in an inertial reference frame, and the positions of points in the body-fixed reference frames are defined prior to the collection of data of the activity of interest.

The matrix describing the orientation of the joint, R_J, can be computed from R_{PI} and R_{DI}:

$$R_J = R_{DP} = R_{IP}R_{DI} \tag{9.17}$$

where $R_{IP} = R_{PI}{}^T$

9.1.5   Calculating Segment Pose

Given an inertial reference frame and a reference frame defined for a rigid body, the transformation of coordinates measured in rigid body reference frame to the inertial reference frame to the rigid body reference can be represented by:

$$y_i = \begin{bmatrix} \cos(\psi) & -\sin(\psi) \\ \sin(\psi) & \cos(\psi) \end{bmatrix} x_i + \underline{v} = Rx_i + \underline{v} \tag{9.18}$$

where y_i is the position of point i on rigid body measured in inertial reference frame, x_i is the position of point i body-fixed reference frame, $\underline{v}$ is the position of the origin of rigid body reference frame in the inertial reference frame, and R is the attitude matrix.

In a least-squares sense, the task of determining the attitude matrix and translation vector is equivalent to minimizing:

$$\frac{1}{n}\sum_{i=1}^{n}(Rx_i + \underline{v} - y_i)^T(Rx_i + \underline{v} - y_i) \tag{9.19}$$

where n is the number of common landmarks measured in both reference frames ($n \geq 2$). With noiseless data, this equation should be equal to zero for the appropriate values of R and $\underline{v}$. In reality there will be errors corrupting the data so this least-squares estimate finds the R and $\underline{v}$ which minimize the equation.

The mean vectors ($\bar{x}$ and $\bar{y}$) are computed:

$$\bar{x} = \frac{1}{n}\sum_{i=1}^{n}x_i \tag{9.20}$$

$$\bar{y} = \frac{1}{n}\sum_{i=1}^{n}y_i \tag{9.21}$$

which can be utilized to compute $\underline{v}$ once R is computed:

$$\underline{v} = \bar{y} - R\bar{x} \tag{9.23}$$

Substitution of these means into Eq. 9.19 can eliminate $\underline{v}$ from the equation:

$$\frac{1}{n}\sum_{i=1}^{n}(Rx_i - y_i + R\bar{x} + \bar{y})^T(Rx_i - y_i + R\bar{x} + \bar{y}) \tag{9.24}$$

This expression can be further simplified by substituting in the following:

$$\dot{x} = x_i - \bar{x} \tag{9.25}$$

$$\acute{y} = y_i - \bar{y} \tag{9.26}$$

which gives after appropriate simplification:

$$\frac{1}{n}\sum\nolimits_{i=1}^{n}\left(\acute{y}_i^T\acute{y}_i + \acute{x}_i^T\acute{x}_i - 2\acute{y}_i^T R\acute{x}\right) \tag{9.27}$$

The sum of squares is minimized by setting the derivative of the equation with respect to the attitude angle (ψ) to zero, and then the following equation is obtained:

$$0 = \frac{2}{n}\sum\nolimits_{i=1}^{n}\cos{(\psi)}\left(\acute{y}_{xi}\acute{x}_{yi} - \acute{y}_{yi}\acute{x}_{xi}\right) + \sin{(\psi)}\left(\acute{y}_{xi}\acute{x}_{yi} - \acute{y}_{yi}\acute{x}_{xi}\right) \tag{9.28}$$

where the subscripts $_x$, $_y$ refer to the horizontal and vertical components of the vectors, respectively. Therefore, if P and Q are defined as:

$$P = \sum\nolimits_{i=1}^{n}\left(\acute{y}_{xi}\acute{x}_{yi} - \acute{y}_{yi}\acute{x}_{xi}\right) \tag{9.29}$$

$$Q = \sum\nolimits_{i=1}^{n}\left(\acute{y}_{xi}\acute{x}_{yi} - \acute{y}_{yi}\acute{x}_{xi}\right) \tag{9.30}$$

then the angle ψ can be computed:

$$\psi = \tan^{-1}\left(\frac{P}{Q}\right) \tag{9.31}$$

In all programming languages, the inverse tangent is an option (e.g., *atan(y/x)*), but this returns values over the interval $-\pi/2$ to $\pi/2$ (the returned value is limited to two quadrants). The majority of programming languages also have a function *atan2*, which uses all four quadrants and returns values over the interval $-\pi$ to π.

The angle can be computed by simply placing two markers along the axis of the segment of interest; the advantage of the approach presented here is twofold. One is that the estimate of the orientation angle is optimal in a least-squares sense, and the other is it allows the easy incorporation of more than two markers placed on the segment.

9.1.6   Center of Rotation

For a body moving in a plane, the center of rotation is the fixed point about which the body can be considered to rotate. The center of rotation for human joints in two dimensions is a kinematic variable which is popular for the assessment of joint function. The center of rotation is normally represented by the finite center of rotation (FCR) which relates to a measure taken from a single finite displacement. For example, Frankel et al. (1971) used the FCR as a way of quantifying damage to human knee joint surfaces or ligaments. Selbie et al. (1993) used the FCR to quantify the mobility of the cervical spine of the cat. Alexander (1981) assessed the FCR of the dog knee joint so that the moment arms of the muscles crossing the joint could be computed. Of course during the move from initial to final position, there is a continuous (instantaneous) center of rotation (ICR). In certain instances, a series of FCR are evaluated for small movement steps as an approximation to the ICR.

If the attitude matrix and position vector are known at two instants (e.g., t_1, t_2), then it is possible to compute the FCR using the following relationship:

$$\mathrm{FCR} = \underline{p} + \left(2\tan\left(\frac{\Delta\psi}{2}\right)\right)^{-1}\begin{bmatrix} 0 & -1 \\ 1 & 0 \end{bmatrix}\Delta\underline{v} \tag{9.32}$$

where $\underline{p} = \frac{1}{2}(\underline{v}(t_1) - \underline{v}(t_2))$, $\Delta\psi = \psi(t_1) - \psi(t_2)$, and $\Delta\underline{v} = \underline{v}(t_1) - \underline{v}(t_2)$.

When only a pair of markers are used, it is possible to visualize this approach geometrically (see Figure 9.4). The point P lies in the middle of a line between the midpoint of the pair of landmarks in the initial and the midpoint of the pair in the final position. This midpoint line represents the vector $\Delta\underline{v}$. The FCR lies along a line passing through P and perpendicular to this

Fig. 9.4 Illustrates for a rigid body moving from one position to another the estimation of the FCR

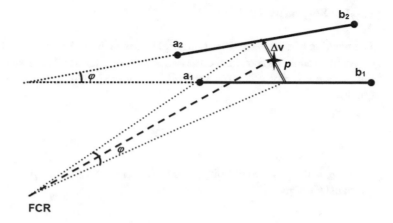

midpoint line. The angle φ represents the relative attitude of the body in the two positions and also represents the angle between the two lines from the midpoints intersecting at the FCR.

The classic method for the estimation of the FCR is the technique of Reuleaux (1875), who graphically demonstrated that the FCR is the point of intersection of the mid-perpendiculars of two distinct landmark displacement vectors. This approach assumes that the pairs of landmark coordinates are error-free, but if there are errors in the landmark positions for small angles of rotation, the errors in the FCR estimated using the method of Reuleaux can be significant. Despite these limitations, this method is often used in biomechanics.

The instantaneous center of rotation (ICR) or centroid is the point on a rigid body which has zero velocity. The basic transformation equation can be stated using a time index:

$$y(t) = R(t)x(t) + \underline{v}(t) \tag{9.33}$$

If this equation is differentiated, and then the ICR should correspond to a point with zero velocity so setting $\dot{y}(t) = 0$ at ICR $(t) = y(t)$, then:

$$\text{ICR}(t) = \underline{v}(t) + \begin{bmatrix} 0 & -1 \\ 1 & 0 \end{bmatrix} \left(\dot{\underline{v}}(t) \big/ \dot{\psi}(t) \right) \tag{9.34}$$

As outlined in Chaps. 4 and 11, the computation of differential quantities is susceptible to the influence of noise which is why some researchers prefer to approximate the ICR by the FCR.

9.1.7 Differential Quantities

For analysis in two dimensions, the computation of derivatives is straightforward, and standard numerical differentiation techniques (Chap. 4) can be used to find the time derivatives of the linear position and angular orientation. To compute angular orientation data from position data requires some non-linear transformation; in doing this, the properties of the noise corrupting the position data is transformed (warped) and so is harder to remove from the computed angular data. This means in the calculation of angular data, the position data should be suitably filtered to reduce noise levels prior to the computation of angular data.

9.2   Three-Dimensional Analysis

For a single body, for example the forearm, in three-dimensional space it has six degrees of freedom. Therefore, we need to define three coordinates to define its position and three angles to define its orientation, and then we have defined its pose (position and orientation). If we have two segments moving relative to one another, for example, the forearm and upper arm articulating at the elbow joint, we need to define their relative pose to quantify the motion of the elbow.

9.2.1   Segment Position

The position of a body or a point on a rigid body in three-dimensional space can be specified by considering the location of the origin of a reference frame attached to the body of interest relative to the origin of an inertial reference frame. In this case (see Fig. 9.5), the vector $\underline{v}$ describes the position of the origin of the body-fixed reference frame relative to the inertial reference frame.

In equation form:

$$p_B = p_I + \underline{v} \tag{9.35}$$

where p_B is the position of point in body-fixed reference frame, p_I is the position of point in inertial reference frame, and $\underline{v}$ is the translation vector.

9.2.2   Segment Orientation: General Principles

A cube moves from one orientation to another, rotating about some fixed point on the cube, a vertex in Fig. 9.6. Therefore, common points on the cube measured in orientation 1 and then orientation 2 will have different values, except for the point of rotation.

The cube's change in orientation can be represented by:

$$p_2 = R\, p_I \tag{9.36}$$

where p_2 is the position of points on cube in position 2, R is a 3×3 attitude matrix, and p_I is the position of points on cube in position 1. The matrix R describes the rotational motion from one cube orientation to the other. Here the matrix is described as the attitude matrix but is also called the rotation matrix and transformation matrix.

If a set of axes is defined affixed to the cube (X_2, Y_2, Z_2), then in position 1, those axes align with the axes of the reference frame (X_1, Y_1, Z_1), but in position 2 they have moved. Therefore, there are two coordinate systems which have a set of basis vectors $((X_1, Y_1, Z_1), (X_2, Y_2, Z_2))$. As the basis vectors are unit vectors, their dot product is the cosine of the angle between them; therefore, these dot products are referred to as the direction cosines. The attitude matrix contains the direction cosines:

$$R = \begin{bmatrix} X_2 \cdot X_1 & Y_2 \cdot X_1 & Z_2 \cdot X_1 \\ X_2 \cdot Y_1 & Y_2 \cdot Y_1 & Z_2 \cdot Y_1 \\ X_2 \cdot Z_1 & Y_2 \cdot Z_1 & Z_2 \cdot Z_1 \end{bmatrix} \tag{9.37}$$

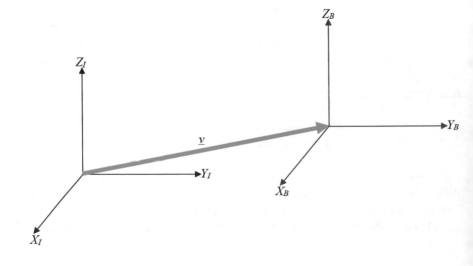

Fig. 9.5 The translation vector, $\underline{v}$, from the origin of the three-dimensional inertial reference frame (X_I, Y_I, Z_I) to the three-dimensional body-fixed reference frame (X_B, Y_B, Z_B)

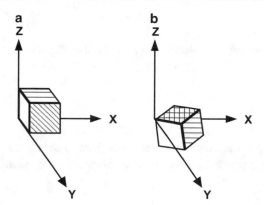

Fig. 9.6 Three-dimensional rotation of a cube, about a vertex of the cube, (**a**) original orientation, and (**b**) rotated orientation

where $X_2 \cdot X_1$ is the dot product between two unit vectors ($X_2 \cdot X_1 = |X_2||X_2| \cos (\theta) = 1\ 1 \cos (\theta) = \cos (\theta)$) which yields the direction cosines between the unit vectors defined by the X axes in reference frames 2 and 1, respectively and similarly for the other matrix components. See Appendix L for details on the dot product.

As in the two-dimensional case, there are important properties of the three-dimensional attitude matrix. These important properties can be summarized as follows:

1. The matrix is proper orthogonal, which means the rows are all orthogonal to one another and similarly the columns.
2. As the matrix is proper orthogonal, the determinant of the attitude matrix is 1 (if the determinant was -1, then the matrix would represent a reflection). A determinant of 1 indicates the length of any vector multiplied by this matrix is unchanged.
3. The attitude matrix belongs to the special-orthogonal group of order three ($R \in SO(3)$).
4. As the attitude matrix belongs to the special-orthogonal group, the inverse of this matrix also belongs to this group, as does the result of the product of any matrices in this group.
5. The transpose of the attitude matrix is equal to its inverse; therefore:

$$R^T = R^{-1} \tag{9.38}$$

The attitude matrix consists of nine direction cosines, but they are not particularly useful for describing three-dimensional rotations, for example, plotting the nine direction cosines against time does not convey the nature of the motion being analyzed. Using the relationship between a proper orthogonal matrix and the skew-symmetric matrix P, initially proposed by Cayley (1843), the following can be written:

$$R = (I - P)(I + P)^{-1} \tag{9.39}$$

where I is the identity matrix and P is a skew-symmetric matrix which has the following format:

$$P\{p\} = \begin{bmatrix} 0 & -p_3 & p_2 \\ p_3 & 0 & -p_1 \\ -p_2 & p_1 & 0 \end{bmatrix} \tag{9.40}$$

As P has only three unique elements, it should be feasible to parameterize the attitude matrix by three elements. There are popular three parameter descriptors of the direction cosine matrix, for example, Cardan angles (Sect. 9.2.5) and Euler angles (Sect. 9.2.6). There are also a number of four parameters descriptions of the direction cosine matrix including the helical axis (Sect. 9.2.4), and quaternions (Sect. 9.2.7). The options for describing rigid body orientation are much greater than those for describing rigid body position.

9.2.3 Segment Pose

If a rigid body in three dimensions undergoes translation and rotation, then the new pose (position and orientation) of a point on that body can be described by:

$$p_2 = Rp_1 + \underline{v} \tag{9.41}$$

where p_2 are points measured in reference frame 2, R is a 3×3 attitude matrix, p_1 are points measured in reference frame 1, and $\underline{v}$ is a 3×1 vector describing the translation from reference frame 1 to reference frame 2.

On occasions, the attitude matrix and position vector can be combined into one matrix; this is referred to as a homogenous transformation matrix:

$$\begin{bmatrix} x_2 \\ y_2 \\ z_2 \\ 1 \end{bmatrix} = \begin{bmatrix} R_{3\times3} & \underline{v}_{3\times1} \\ 0_{1\times3} & 1 \end{bmatrix} \begin{bmatrix} x_I \\ y_I \\ z_I \\ 1 \end{bmatrix} \tag{9.42}$$

9.2.4 Segment Orientation: Helical Axis

The French mathematician Michel Floréal Chasles (1793–1880) proposed a theorem concerning the motion of a rigid body (Chasles 1830). Chasles theorem states that the motion of a rigid body can be considered a translation along, and a rotation about, a suitable axis in space. In light of this theorem, the motion of a rigid body or motion of one rigid body relative to another rigid body can be defined by the motion along and around a helical axis.

Helical Axis Format

The finite helical axis (FHA) describes the motion of a rigid body from one position to another and can be described as translation along, and a rotation about, a directed line in space. The helical axis parameters can be determined from the attitude matrix and translation vector. The parameters describing the finite helical axis are the angle of rotation (θ) about the axis, the unit direction vector ($\underline{e}$) of the axis, the amount of translation (u) along the axis, and the location of a point (p) on the helical axis.

The term screw axis is sometimes used for the helical axis, or Mozzi axis.

Geometric Interpretation of the Helical Axis

For any rigid body undergoing movement, that motion can be visualized as a translation along and rotation about a directed line in space (see Figure 9.7).

Direction Cosine Matrix from Helical Axis Parameters

The matrix R can be determined from the helical axis parameters using Rodrigues rotation formula (Rodrigues 1840):

$$R = I \cos(\theta) + (1 - \cos(\theta))\underline{e}\,\underline{e}^T + \sin(\theta)S\{\underline{e}\} \tag{9.43}$$

where $S\{\}$ generates a skew-symmetric matrix from a vector, for example, for vector $\underline{a} = (a_1, a_2, a_3)$:

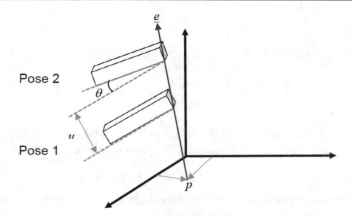

Fig. 9.7 Representation of a finite helical axis, for a rigid body in two poses. The helical axis parameters represented as the the angle of rotation (θ) about the axis, the unit direction vector ($\underline{e}$) of the axis, the amount of translation (u) along the axis, and the location of a point (p) on the helical axis

$$
S\{a\} = \begin{bmatrix} 0 & -a_3 & a_2 \\ a_3 & 0 & -a_1 \\ -a_2 & a_1 & 0 \end{bmatrix}
\tag{9.44}
$$

or in a more explicit form, from:

$$
R(e,\theta) = \begin{bmatrix} e_x e_x(1-\cos\theta)+\cos\theta & e_x e_y(1-\cos\theta)-e_z\sin\theta & e_x e_z(1-\cos\theta)+e_y\sin\theta \\ e_x e_y(1-\cos\theta)+e_z\cos\theta & e_y e_y(1-\cos\theta)+\cos\theta & e_y e_z(1-\cos\theta)-e_x\sin\theta \\ e_x e_z(1-\cos0)-e_y\cos\theta & e_y e_z(1-\cos\theta)+e_x\cos\theta & e_z e_z(1-\cos\theta)+\cos\theta \end{bmatrix}
\tag{9.45}
$$

Helical Axis Parameters from Direction Cosine Matrix

The eigenvalues of the direction cosine matrix are $(1, e^{i\theta}, e^{-i\theta})$; the trace of the matrix is the sum of the eigenvalues:

$$
tr(R) = 1 + e^{i\theta} + e^{-i\theta}
$$

$$
= 1 + 2\cos\theta
\tag{9.46}
$$

Therefore, the angle of rotation (θ) about the axis can determined from:

$$
\cos\theta = \frac{1}{2}(r_{11} + r_{22} = r_{33} - 1)
\tag{9.47}
$$

If $\cos\theta = 1$, then the attitude matrix is $R(e,\theta) = I$, and the rotation axis is undefined, but in this case, the presumption would be there is no rotation.

If $-1 < \cos\theta < 1$, then the unit direction vector ($\underline{e}$) of the axis is calculated from:

$$
\sin(\theta)\underline{e} = \frac{1}{2}\begin{bmatrix} r_{3,2} - r_{2,3} \\ r_{1,3} - r_{3,1} \\ r_{2,1} - r_{1,2} \end{bmatrix}
\tag{9.48}
$$

The unit direction vector is equivalent to the eigenvector corresponding to the unit eigenvalue.

The amount of translation (u) along the axis can be computed from:

$$u = \underline{e}^T \underline{v} \tag{9.49}$$

Finally, the location of a point (p) on the FHA is given by:

$$p = \frac{1 + \cos\theta}{2\sin^2\theta}(I - R^T)\underline{v} \tag{9.50}$$

The ratio of the translation along and rotation about the helical axis is referred to as the pitch of the axis ($u/_\theta$). If the pitch is less than zero, then the axis is right-handed, and left-handed if the pitch is greater than zero.

Helical Angles
This angle definition is based around the parameters describing the finite helical axis: helical rotation (θ) and the unit direction vector of the axis ($\underline{e}$). These two parameters have been combined to give a three-parameter representation of rigid body attitude:

$$\underline{\varphi} = \begin{pmatrix} \varphi_1 \\ \varphi_2 \\ \varphi_3 \end{pmatrix} = \theta\underline{e} = \begin{pmatrix} \theta e_x \\ \theta e_y \\ \theta e_z \end{pmatrix} \tag{9.51}$$

where φ_1, φ_2, φ_3 are the helical angles.

This type of parameterization has been called the helical angles by Woltring (1991). These angles have the advantage that they are clearly identifiable as angles compared with, for example, quaternions. They also do not suffer from singularities as do Cardanic or Eulerian angle conventions. To their detriment, they are hard to visualize. Woltring (1991) has shown that these angles are not as adversely affected by noise as either Cardanic or Eulerian angles.

9.2.5 Segment Orientation: Cardan Angles

Gerolamo Cardano (1501–1576) was an Italian polymath, whose expertise included medicine, math, physics, biology, and gambling. His success in gambling arose, in part, because of his unique knowledge of probability theory. He described a gimbal mechanism, in which a gyroscope or compass could be mounted, which consisted of three concentric rings – it is this mechanism which forms the bases of the angle system which bears his name.

Cardan Angle Format
Cardan angles can be described as an ordered sequence of rotations about three coordinate axes. In this case, the rotations use each of the three axes once. The three component matrices are:

$$R_x(\alpha) = \begin{bmatrix} 1 & 0 & 0 \\ 0 & \cos(\alpha) & -\sin(\alpha) \\ 0 & \sin(\alpha) & \cos(\alpha) \end{bmatrix} \tag{9.52}$$

$$R_y(\beta) = \begin{bmatrix} \cos(\beta) & 0 & \sin(\beta) \\ 0 & 1 & 0 \\ -\sin(\beta) & 0 & \cos(\beta) \end{bmatrix} \tag{9.53}$$

$$R_Z(\gamma) = \begin{bmatrix} \cos(\gamma) & -\sin(\gamma) & 0 \\ \sin(\gamma) & \cos(\gamma) & 0 \\ 0 & 0 & 1 \end{bmatrix} \tag{9.54}$$

where α, β, and γ are angles of rotation about the X, Y, and Z axes, respectively. If the rotations are completed in the following sequence, X, then Y, and Z, the matrix sequence can be written as:

$$R_{XYZ} = R_Z(\gamma)R_Y(\beta)R_X(\alpha) \tag{9.55}$$

The angle sequence is X, Y, and then Z axes; note that the matrices are listed in reverse order as the rotation about the X axis must be first, then about the Y, and finally about the Z. This particular angle sequence is often referred to as the yaw-pitch-roll sequence, in part because it is commonly used in aircraft navigation.

As matrix multiplication is not commutative, the order of multiplication is important, for example, generally:

$$R_{XYZ} \neq R_{ZYX} \tag{9.56}$$

Figure 9.8 shows how simply changing the sequence of the initial two rotations changes the orientation of the object, highlighting the importance of the sequence order on the rotations achieved by a Cardanic angle sequence.

For Cardanic angles, the complete set of potential sequences is X-Y-Z, X-Z-Y, Y-Z-X, Y-X-Z, Z-X-Y, and Z-Y-X. For both Cardan angles and Euler angles (see next sub-section), different sequences of rotations can be used, and the angles determined will in part depend on the sequence of rotations adopted. Each angle sequence gives a different resulting matrix in terms of the sine and cosines of the angles; as a consequence, the angles extracted from these matrices will differ between angle sequences. Table 9.1 gives examples of the different angles extracted from the same direction cosine matrix given the different angle sequences.

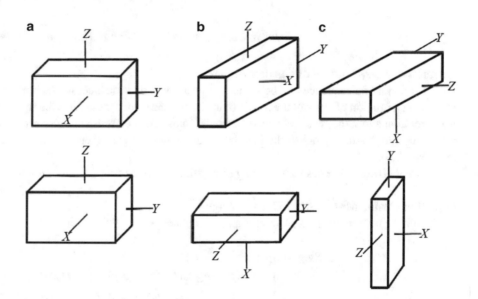

Fig. 9.8 Two rotation sequences (Z and then Y vs. Y and then Z). Upper sequence shows (**a**) initial configuration, (**b**) 90° rotation about the Z axis, and (**c**) rotation of 90° degrees about the Y axis. Lower sequence shows (**a**) initial configuration, (**b**) 90° rotation about the Y axis, and (**c**) rotation of 90° degrees about the Z axis. Note that the final configuration of the object is different when the angle sequence is different

Table 9.1 Illustration of the influence of different angle sequences on the resulting amounts of rotations about each axis

	Rotation about specified axis (degrees)		
Angle sequence	X	Y	Z
X-Y-Z	60.00	5.00	−10.00
X-Z-Y	59.12	5.08	−9.96
Y-Z-X	59.63	11.15	−0.72
Y-X-Z	59.62	9.93	−1.42
Z-X-Y	57.70	21.21	−18.89
Z-Y-X	59.49	11.15	−0.73

This sequence effect is demonstrated for the six different Cardanic angle sequences

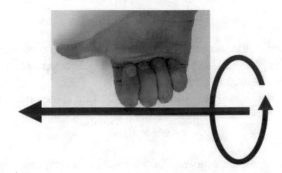

Fig. 9.9 Illustration of positive angle of rotation about an axis, where the thumb is aligned with the axis and the curl of the fingers is in the direction of a positive rotation about that axis

Fig. 9.10 The $X - Y - Z$ angle sequence, with corresponding angles of α, β, and γ. Note the blue axis is the intended axis of rotation

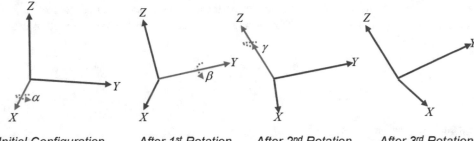

Initial Configuration   After 1st Rotation   After 2nd Rotation   After 3rd Rotation

Geometric Interpretation of Cardan Angles

Each Cardan angle sequence can be visualized as an ordered sequence of rotations about a set of axes. The X-Y-Z sequence starts with a rotation of α about the X axis of the inertial reference frame, β is the rotation about the rotated Y axis, and finally γ is the rotation about the twice relocated (rotated) Z axis. To get a sense of these rotations, make a fist with your right hand but with your thumb sticking out. If the thumb is aligned with the axis, then the curl of the fingers is a positive rotation about the axis (Fig. 9.9).

Figure 9.10 demonstrates the changing orientations of the axes as each rotation in a Cardanic angle sequence is performed.

Direction Cosine Matrix from Cardan Angles

The resulting matrix for the X, Y, and Z angle sequence is:

$$
\begin{aligned}
R_{XYZ} &= R_Z(\gamma)R_Y(\beta)R_X(\alpha) \\
&= \begin{bmatrix} c(\gamma)c(\beta) & c(\gamma)s(\beta)s(\gamma) - s(\gamma)c(\alpha) & c(\gamma)s(\beta)c(\alpha) + s(\gamma)s(\alpha) \\ s(\gamma)c(\beta) & s(\gamma)s(\beta)s(\alpha) + c(\gamma)c(\alpha) & s(\gamma)s(\beta)c(\alpha) - c(\alpha)s(\gamma) \\ -s(\beta) & c(\beta)s(\alpha) & c(\beta)c(\alpha) \end{bmatrix}
\end{aligned}
\tag{9.57}
$$

Note that $cos(\alpha)$ and $sin(\alpha)$ are represented by $c(\alpha)$ and $s(\alpha)$, similarly for the other angles.

See Appendix M for the matrices associated with all six Cardan angle sequences.

Cardan Angles from Direction Cosine Matrix

The Cardan angles (X-Y-Z sequence) can be calculated from:

$$
\sin(\beta) = -r_{3,1} \tag{9.58}
$$

$$
\sin(\alpha) = \frac{r_{3,2}}{\cos(\beta)} \tag{9.59}
$$

$$
\sin(\gamma) = \frac{r_{2,1}}{\cos(\beta)} \tag{9.60}
$$

If the angles are expected to exceed $\frac{\pi}{2}$, then a superior way to compute these angles is to use the *atan2* function:

$$\beta = atan2\left(-r_{3,1}, \sqrt{r_{1,1}^2 + r_{2,1}^2}\right) \tag{9.61}$$

$$\alpha = atan2\left(\frac{r_{3,2}}{\cos(\beta)}, \frac{r_{3,3}}{\cos(\beta)}\right) \tag{9.62}$$

$$\gamma = atan2\left(\frac{r_{2,1}}{\cos(\beta)}, \frac{r_{1,1}}{\cos(\beta)}\right) \tag{9.63}$$

Irrespective of how the angles are computed, the terminal angles (α, γ) are undefined for β angles of $\pm(2n + 1)\frac{\pi}{2}$ $(n = 0, 1, 2,...)$, because to compute these two angles division by $\cos(\beta)$ is required, which is division by zero for β values of $\pm(2n + 1)\frac{\pi}{2}$.

9.2.6 Segment Orientation: Euler Angles

Leonhard Euler (1707–1783) was a Swiss mathematician and physicist. His contributions span many branches of math and mechanics and include complex numbers, geometry, calculus, and number theory. In mechanics he developed a law related to angular momentum, and a system for defining angles in three dimensions. He was so prolific that his collected works require 75 volumes, with nearly 50% of the work produced after he went blind. Euler proposed an angle rotation sequence where the third rotation is about the first axis (e.g., Z-X-Z), but of course the second rotation will have moved the first axis to a new orientation. The term Euler angles is often used as a generic term for all three-dimensional angle sets, but historically Euler angles require a repeated rotation and Cardan angles do not. Therefore, to make it clear which rotation sequence is being used, it is easier to specifically identify as Euler angles those angle sequences with the terminal (first and last) rotations about a common axis, and Cardan angles those with three unique axes of rotations.

Euler Angle Format
Euler angles (pronounced oiler) can be described as an ordered sequence of rotations about three coordinate axes. In this case, the rotations use one axis twice, for example, Z-X-Z; the terminal rotations are about the same axes. The three component matrices would be:

$$R_Z(\alpha) = \begin{bmatrix} \cos(\alpha) & -\sin(\alpha) & 0 \\ \sin(\alpha) & \cos(\alpha) & 0 \\ 0 & 0 & 1 \end{bmatrix} \tag{9.64}$$

$$R_x(\beta) = \begin{bmatrix} 1 & 0 & 0 \\ 0 & \cos(\beta) & -\sin(\beta) \\ 0 & \sin(\alpha\beta) & \cos(\beta) \end{bmatrix} \tag{9.65}$$

$$R_Z(\gamma) = \begin{bmatrix} \cos(\gamma) & -\sin(\gamma) & 0 \\ \sin(\gamma) & \cos(\gamma) & 0 \\ 0 & 0 & 1 \end{bmatrix} \tag{9.66}$$

where α, β, and γ are angles of rotation about the Z, X, and Z axes, respectively, remembering that the middle rotation will have reorientated the Z axis, so the third rotation will be about a differently orientated Z axis. If the rotations are completed in the following sequence, Z then X, and Z, the matrix sequence can be written as:

$$R_{ZXZ} = R_Z(\gamma)R_X(\beta)R_Z(\alpha) \tag{9.67}$$

Fig. 9.11 The *Z-X-Z* angle
sequence, with corresponding
angles of α, β, and γ. Note the blue
axis is the intended axis of
rotation

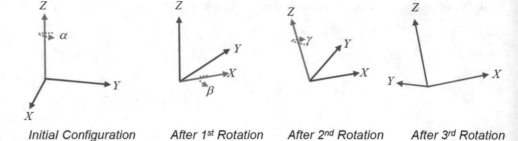

Initial Configuration   After 1st Rotation   After 2nd Rotation   After 3rd Rotation

The Eulerian angles involve repetition of rotations about one particular axis; the complete set is *X-Y-X*, *X-Z-X*, *Y-X-Y*, *Y-Z-Y*, *Z-X-Z*, and *Z-Y-Z*.

Geometric Interpretation of Euler Angles

Each Euler angle sequence can be visualized as an ordered sequence of rotations about a set of axes. The *Z-X-Z* sequence starts with a rotation of α about the *Z* axis of the reference frame, β is the rotation of the about the rotated *X* axis, and finally γ is the rotation about the relocated (rotated) *Z* axis. Figure 9.11 demonstrates the changing orientations of the axes as each rotation in a Euler angle sequence is performed.

Direction Cosine Matrix from Euler Angles

The resulting matrix for the *Z*, *X*, and *Z* angle sequence is:

$$R_{ZXZ} = R_Z(\gamma)R_X(\beta)R_Z(\alpha)$$

$$= \begin{bmatrix} c(\alpha)c(\beta)c(\gamma) - s(\alpha)s(\gamma) & -c(\alpha)c(\beta)s(\gamma) - s(\alpha)\cos(\gamma) & c(\alpha)s(\beta) \\ s(\alpha)c(\beta)c(\gamma) + c(\alpha)s(\gamma) & -s(\alpha)c(\beta)s(\gamma) + c(\alpha)c(\gamma) & s(\alpha)s(\beta) \\ -s(\beta)c(\gamma) & s(\beta)s(\gamma) & c(\beta) \end{bmatrix} \qquad (9.68)$$

Note that $\cos(\alpha)$ and $\sin(\alpha)$ are represented by $c(\alpha)$ and $s(\alpha)$, similarly for the other angles.
See Appendix M for the matrices associated with all six Euler angle sequences.

Euler Angles from Direction Cosine Matrix

The Eulerian angles (*Z-X-Z* sequence) can be calculated from:

$$\cos(\beta) = r_{3,3} \qquad (9.69)$$

$$\sin(\alpha) = \frac{r_{2,3}}{\sin(\beta)} \qquad (9.70)$$

$$\cos(\gamma) = \frac{-r_{1,3}}{\sin(\beta)} \qquad (9.71)$$

If the angles are expected to exceed $\frac{\pi}{2}$, then a superior way to compute these angles is to use the *atan2* function:

$$\beta = atan2\left(\sqrt{r_{3,1}^2 + r_{3,2}^2}, r_{3,3}\right) \qquad (9.72)$$

$$\alpha = atan2\left(\frac{r_{2,3}}{\sin(\beta)}, \frac{r_{1,3}}{\sin(\beta)}\right) \qquad (9.73)$$

$$\gamma = atan2\left(\frac{r_{3,2}}{\sin{(\beta)}}, \frac{-r_{3,1}}{\sin{(\beta)}}\right) \tag{9.74}$$

Irrespective of how the angles are computed the terminal angles (α, γ) are undefined for β angles of $\pm n.\pi$ $(n = 0, 1, 2,\ldots)$, because to compute these two angles require division by $\sin(\beta)$, which is division by zero for β values of $\pm n\pi$.

9.2.7   Segment Orientation: Quaternions

Quaternions are a way of representing three-dimensional attitude that is numerically robust, compared with Euler and Cardan angles, but is at first viewing somewhat less intuitive. Compared with other approaches (e.g., Euler angles, Cardan angles), the quaternion does not suffer from singularities and therefore avoids the gimbal lock (see later sub-section for explanation of gimbal lock). The quaternion represents the direction cosine matrix as a homogenous quadratic function of the components of the quaternion; unlike other approaches it does not require trigonometric or other transcendental function evaluations. If storage is an issue the direction cosine matrix can be represented by four parameters as opposed to all nine elements of the matrix.

The invention of quaternions is generally attributed to William Rowan Hamilton (1805–1865) in 1843, but a similar approach was suggested in 1840 by Rodrigues, and the general idea was communicated in a letter by Leonhard Euler sent to Christian Goldbach in 1748. That stated Hamilton developed the area significantly. Indeed, he had wrestled with the problem of generalizing complex numbers in a plane to three dimensions for over a decade and had a sudden insight when walking with his wife on October 16, 1843. He was so excited by this insight that he carved the key formula into the Broome Bridge, which crosses the Royal Canal in Dublin.

Quaternion Format
The set of quaternions is associated with a four-dimensional vector space, so there are four elements associated with each quaternion. Somewhat confusingly the way in which quaternions are written varies, but they all mean the same thing and it depends on application and interpretation which approach is the most appropriate. The approach used here is a vector with four components:

$$q = (q_0, \; q_1, \; q_2, \; q_3) = (q_0, \; q) \tag{9.75}$$

This representation is the four-tuple form of the quaternion. It should be noted that on occasions the quaternion in this form moves the first element to the end $(q_1, \; q_2, \; q_3, \; q_0)$. Here unit quaternions will be used, so its norm is equal to 1 ($|q| = 1$).

Geometric Interpretation of Quaternions
There are two ways in which a quaternion can be visualized. The first is that each quaternion corresponds to a point on a hypersphere (Fig. 9.12).

A quaternion, $q = (q_0, \; q_1, \; q_2, \; q_3)$, can also be considered a rotation of angle Ω, about an axis $\underline{e}$, where:

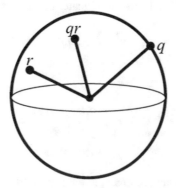

Fig. 9.12 Quaternions represented on a hypersphere, where q and r are quaternions, and qr is the quaternion resulting from their product

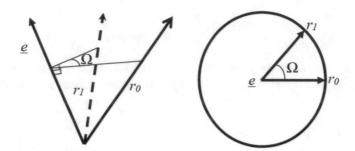

Fig. 9.13 The transformation of a vector r_0 by a rotation of Ω about a line $\underline{e}$ to vector $r1$. The left image shows the general representation of the transformation, and the right image shows a view in a plane normal to the axis of rotation

Table 9.2 Some unit quaternions, with explanation of their action

q_0	q_1	q_2,	q_3	Description
1	0	0	0	Identity quaternion, no rotation
0	1	0	0	180° rotation around X axis
0	0	1	0	180° rotation around Y axis
0	0	0	1	180° rotation around Z axis
$\sqrt{0.5}$	$\sqrt{0.5}$	0	0	90° rotation around X axis
$\sqrt{0.5}$	0	$\sqrt{0.5}$	0	90° rotation around Y axis
$\sqrt{0.5}$	0	0	$\sqrt{0.5}$	90° rotation around Z axis

N.B. – these are all unit quaternions, so $\sqrt{q_0^2 + q_1^2 + q_2^2 + q_3^2} = 1$, and that $\cos^{-1}(\sqrt{0.5}) = 45°$, $\cos^{-1}(0) = 90°$

$$q_0 = \pm \cos \frac{\Omega}{2} \tag{9.76}$$

$$\begin{bmatrix} q_1 \\ q_2 \\ q_3 \end{bmatrix} = \pm \underline{e} \, \sin \frac{\Omega}{2} \tag{9.77}$$

where $0 \leq \Omega \leq \pi$.

Therefore, each rotation has two representations, $(q_0, \; q_1, \; q_2, \; q_3)$ and $(-q_0, \; -q_1, \; -q_2, \; -q_3)$; to avoid confusion, quaternion representations of orientations can be restricted to one half of the hypersphere.

A quaternion can be directly visualized as a directed line in space about which there is a rotation. For example (see Fig. 9.13), if a vector r_0 is transformed by a rotation matrix to vector r_1, then then this transformation can be visualized as a rotation (Ω) about a line ($\underline{e}$).

Table 9.2 gives some example quaternions which help illustrate their geometric properties.

Direction Cosine Matrix from Quaternions

The direction cosine matrix (R) can be computed from a unit quaternion using:

$$R = (q_0^2 - \boldsymbol{q}^T \boldsymbol{q})I + 2\boldsymbol{q}\boldsymbol{q}^T - 2q_0 S\{\boldsymbol{q}\} \tag{9.78}$$

which when expanded gives:

$$R(q) = \begin{bmatrix} q_0^2 + q_1^2 - q_2^2 - q_3^2 & 2(q_1 q_2 - q_3 q_0) & 2(q_1 q_3 - q_2 q_0) \\ 2(q_1 q_2 + q_3 q_0) & q_0^2 - q_1^2 + q_2^2 - q_3^2 & 2(q_2 q_3 - q_1 q_0) \\ 2(q_1 q_3 - q_2 q_0) & 2(q_2 q_3 + q_1 q_0) & q_0^2 - q_1^2 - q_2^2 + q_3^2 \end{bmatrix} \tag{9.79}$$

Note this matrix can be represented using quaternions in a number of different ways, and they are all equivalent.

Quaternions from Direction Cosine Matrix

The quaternion can be computed from a direction cosine matrix in a variety of ways. The following four vectors could each represent the quaternion of a direction cosine matrix:

$$x_0 = \begin{bmatrix} 1 + r_{1,1} + r_{2,2} + r_{3,3} \\ r_{2,3} - r_{3,2} \\ r_{3,1} - r_{1,3} \\ r_{1,2} - r_{2,1} \end{bmatrix} \tag{9.80}$$

$$x_1 = \begin{bmatrix} r_{2,3} - r_{3,2} \\ 1 + r_{1,1} - r_{2,2} - r_{3,3} \\ r_{1,2} + r_{2,1} \\ r_{1,3} - r_{3,1} \end{bmatrix} \tag{9.81}$$

$$x_2 = \begin{bmatrix} r_{3,1} - r_{1,3} \\ r_{2,1} + r_{1,2} \\ 1 + r_{2,2} - r_{3,3} - r_{1,1} \\ r_{2,3} + r_{3,2} \end{bmatrix} \tag{9.82}$$

$$x_3 = \begin{bmatrix} r_{1,2} - r_{2,1} \\ r_{3,1} + r_{1,3} \\ r_{3,2} + r_{2,3} \\ 1 + r_{3,3} - r_{1,1} - r_{2,2} \end{bmatrix} \tag{9.83}$$

The unit quaternion is found by normalizing any of these vectors:

$$q = \pm x_i / \|x_i\| \tag{9.84}$$

To minimize numerical errors the vector selected for normalization should be the one with the largest norm. The norm for each vector does not need to be computed to make this selection; if the trace of the rotation matrix ($r_{1,1} + r_{2,2} + r_{3,3}$) is larger than $r_{i,i}$ (where $i = 1,3$), then the first listed vector (x_0) should be used; otherwise, if $r_{1,1}$ is the largest, then use the second listed vector (x_1), similarly for the other vectors.

Quaternions have a number of useful features but also require special rules for their math; this math and some of their features are outlined in Appendix N and in Challis (2020a).

9.2.8 Joint Orientation

Defining orientation and position (pose) in three dimensions hinges around mapping from one reference frame to another. Therefore, for a rigid body moving in three dimensions the mapping is from an inertial reference frame to a reference frame "attached" to the rigid body. When analyzing pairs of interconnected rigid bodies the relative orientation provides information about joint angles. To analyze joint orientations reference frames must be attached to both segments' either side of a joint (proximal reference frame and a distal reference frame). The orientation of a joint is typically defined by the motion of the distal segment with respect to the proximal joint. For example, if a reference frame is embedded in the thigh, and one in the shank, the mapping from the shank-based reference frame to thigh-based reference frame provides the knee joint angles:

$$x_I = R_{PI}x_P + v_{PI} \text{ (mapping from proximal to inertial reference frame)} \tag{9.85}$$

$$x_I = R_{DI}x_D + v_{DI} \text{ (mapping from distal to inertial reference frame)} \tag{9.86}$$

where x_I is points measured in the inertial reference frame, R_{PI} attitude matrix mapping from proximal to inertial reference frame, x_I points measured in proximal reference frame, v_{PI} translation vector from origin of proximal reference frame to origin of inertial reference frame, R_{DI} attitude matrix mapping from distal to inertial reference frame, x_D points measured in proximal reference frame, and v_{DI} translation vector from distal of proximal reference frame to origin of inertial reference frame.

The matrix describing the orientation of the joint, R_J, can be computed from R_{PI} and R_{DI}:

$$R_J = R_{DP} = R_{IP}R_{DI} \tag{9.87}$$

where $R_{IP} = R_{PI}{}^T$.

Once the matrix describing the orientation of the joint, R_J, has been determined, the orientation of the joint can be defined using any of the methods previously described for specifying segment orientation.

Obviously, the experimenter has a choice of descriptors of joint orientation, but this choice is further complicated as the nature of, for example, the Cardan angles depends on the sequence selected and how the axes are defined for the segments. For human joints, there are a variety of proposals as to how these reference fames should be defined (e.g., Wu and Cavanagh 1995; Wu et al. 2002, 2005).

9.2.9 Gimbal Lock, Singularities, and Codman's Paradox

The singularities which arise in the Cardan and Euler angle representations of rotations in three dimensions are typically referred to as gimbal lock. This gimbal lock can be viewed mathematically or with reference to the motion of a gimbal mechanism. A gimbal or gyroscope consists of three concentric rings, with axes through each ring representing a rotation. The two inner rings can be rotated, so they completely overlap with one another and the rotation about one axis cannot be separated from rotation about the other (see Fig. 9.14).

From a mathematical perspective, considering the Cardanic sequence X-Y-Z (angles α, β, and γ), the attitude matrix is:

$$R_{XYZ} = \begin{bmatrix} c(\gamma)c(\beta) & c(\gamma)s(\beta)s(\alpha) - s(\gamma)c(\alpha) & c(\gamma)s(\beta)c(\alpha) + s(\gamma)s(\alpha) \\ s(\gamma)c(\beta) & s(\gamma)s(\beta)s(\alpha) + c(\gamma)c(\alpha) & s(\gamma)s(\beta)c(\alpha) - c(\gamma)s(\alpha) \\ -s(\beta) & c(\beta)s(\alpha) & c(\beta)c(\alpha) \end{bmatrix} \tag{9.88}$$

If the middle rotation $\beta = \pi/2$, then the matrix becomes (remember $\cos\left(\frac{\pi}{2}\right) = 0$, $\sin\left(\frac{\pi}{2}\right) = 1$):

$$R_{XYZ} = \begin{bmatrix} 0 & \cos(\gamma)\sin(\alpha) - \sin(\gamma)\cos(\alpha) & \cos(\gamma)\cos(\alpha) + \sin(\gamma)\sin(\alpha) \\ 0 & \sin(\gamma)\sin(\alpha) + \cos(\gamma)\cos(\alpha) & \sin(\gamma)\cos(\alpha) - \cos(\gamma)\sin(\alpha) \\ -1 & 0 & 0 \end{bmatrix} \tag{9.89}$$

which can be further simplified to:

Fig. 9.14 Illustration of gimbal lock, (**a**) the three potential axes of rotation, (**b**) with two of these axes aligned producing gimbal lock

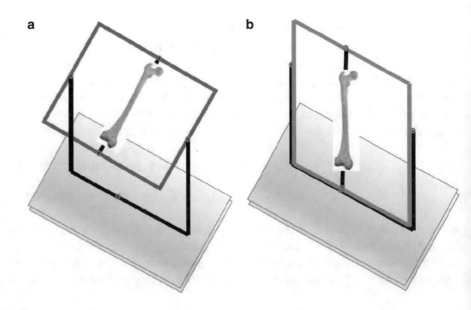

Fig. 9.15 Illustration of
Codman's paradox, (**a**) initial
position with the "thumb" toward
the front, (**b**) result of rotation of
the arm to the horizontal, (**c**) result
of rotation of the arm to the front,
and (**d**) rotation to bring the arm
back down to the side but now the
"thumb" is pointing to the side

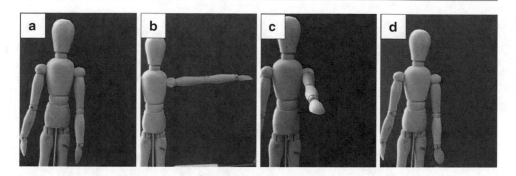

$$R_{XYZ} = \begin{bmatrix} 0 & \sin(\alpha - \gamma) & \cos(\alpha - \gamma) \\ 0 & \cos(\alpha - \gamma) & -\sin(\alpha - \gamma) \\ -1 & 0 & 0 \end{bmatrix} \tag{9.90}$$

The matrix illustrates that the rotation depends only on the difference between the two angles $(\alpha - \gamma)$ and therefore only has one degree of freedom instead of two. The rotation of $\beta = \pi/2$ means motions of angles α and γ result in rotations about the same axis.

To avoid gimbal lock, a common approach is to select the sequence of rotations for a Cardan or Euler sequence so the system never approaches, let alone reaches, a singularity. Another approach is to use a representation of a body's orientation which does not have singularities (e.g., quaternions).

With the helical axis description of motion, the angle of rotation is poorly defined if the angle is close to $\pm\pi$. Also with helical angles, $R(e, \theta) = R(-e, -\theta)$ which can also cause confusion. Quaternions can be represented by positions on the surface of hypersphere; this representation means that a rigid body orientation can be visualized using two quaternions, (q_0, q_1, q_2, q_3) and $(-q_0, -q_1, -q_2, -q_3)$; to avoid this ambiguity, quaternions can be constrained to either the top or bottom half of the hypersphere.

Codman's paradox refers to the situation where three rotations result in what might be considered a spurious rotation about the middle axis in the angle sequence. It was first identified by Codman in 1934 when examining the function of the shoulder. If the arm in its initial position is hanging by the side with the thumb toward the front and the fingers pointing down, first rotate the arm to the horizontal (wing position), then rotate the arm to the front (fingers are now pointing straight ahead), and then bring the arm back down to the side. In this final position, the arm has undergone an axial rotation; therefore, the thumb is now pointing to the side (inward) (see Fig. 9.15).

Codman's paradox is a function of how the sequencing of angles works; this can be illustrated using a Cardan angle sequence:

$$\text{1st Rotation} \quad R_y\left(-\frac{\pi}{2}\right) = \begin{bmatrix} \cos\left(-\frac{\pi}{2}\right) & 0 & -\sin\left(-\frac{\pi}{2}\right) \\ 0 & 1 & 0 \\ \sin\left(-\frac{\pi}{2}\right) & 0 & \cos\left(-\frac{\pi}{2}\right) \end{bmatrix} = \begin{bmatrix} 0 & 0 & -1 \\ 0 & 1 & 0 \\ 1 & 0 & 0 \end{bmatrix} \tag{9.91}$$

$$\text{2nd Rotation} \quad R_Z\left(\frac{\pi}{2}\right) = \begin{bmatrix} \cos\left(\frac{\pi}{2}\right) & -\sin\left(\frac{\pi}{2}\right) & 0 \\ \sin\left(\frac{\pi}{2}\right) & \cos\left(\frac{\pi}{2}\right) & 0 \\ 0 & 0 & 1 \end{bmatrix} = \begin{bmatrix} 0 & -1 & 0 \\ 1 & 0 & 0 \\ 0 & 0 & 1 \end{bmatrix} \tag{9.92}$$

$$\text{3rd Rotation}\quad R_x\left(\frac{\pi}{2}\right) = \begin{bmatrix} 1 & 0 & 0 \\ 0 & \cos\left(\frac{\pi}{2}\right) & -\sin\left(\frac{\pi}{2}\right) \\ 0 & \sin\left(\frac{\pi}{2}\right) & \cos\left(\frac{\pi}{2}\right) \end{bmatrix} = \begin{bmatrix} 1 & 0 & 0 \\ 0 & 0 & 1 \\ 0 & -1 & 0 \end{bmatrix} \tag{9.93}$$

The attitude matrix comprising all of these rotations is:

$$R = R_x\left(\frac{\pi}{2}\right)R_Z\left(\frac{\pi}{2}\right)R_y\left(-\frac{\pi}{2}\right) = \begin{bmatrix} 0 & -1 & 0 \\ 1 & 0 & 0 \\ 0 & 0 & 1 \end{bmatrix} = R_Z\left(\frac{\pi}{2}\right) \tag{9.94}$$

Therefore, these three orthogonal rotations of a rigid body about a universal joint are equivalent to a single $\pi/2$ rotation about the middle axis in the rotation sequence.

9.2.10 Axes of Rotation

The definition of rigid body attitude can be considered as the motion of a moving reference frame to align it with some other reference frame. But it is also feasible to define attitude relative to a fixed reference frame. These two ways of defining the attitude of a rigid body are compatible. Three rotations about three axes of a moving reference frame define the same orientation as a fixed reference frame if the three rotations are taken in the opposite order about the axes of the fixed frame; therefore, as an example:

$$R_{XYZ} \quad = \quad R_{ZYX} \tag{9.95}$$

$$\text{Moving Frame} \quad \rightarrow \quad \text{Fixed Frame}$$

9.2.11 Calculating Segment Pose: Direct Kinematics

There are two approaches for determining segment pose: one assumes that each body segment has the potential for six degrees of freedom of motion and uses markers on the segment to determine the motion of a reference frame attached to the segment relative to an inertial reference frame. This approach is referred to as direct kinematics and is the focus of this sub-section. The other approach is inverse kinematics, where for the system under analysis, a model is specified, giving the degrees of freedom and range of motion limits; the markers on the segments are then used to determine the segment poses (see the following sub-section). To some extent, both approaches give equivalent results (Kainz et al. 2016), and the appropriate approach likely depends on the application.

To determine the contents of the direction cosine matrix, and the translation vector requires the measurements of the position of three or more noncollinear points defined on the rigid body under consideration in two sets of reference frames, for example, an inertial reference frame and a body-fixed reference frame. Methods for computing these transformation parameters have been the focus of research in many domains including spacecraft kinematics (e.g., Markley 1993), photogrammetry (e.g., Schut 1968), and biomechanics (e.g., Veldpaus et al. 1988). There are a number of methods for performing this calculation (see Chap. 5 in Markley and Crassidis 2014); the bases of many of these methods were solutions to Problem 65-1 proposed by Grace Wahba which sought algorithms for finding the least-squares estimation of rigid body attitude (Wahba 1965). This challenge has resulted in a number of solutions (e.g., Farrell et al. 1966; Bar-Itzhack and Reiner 1984). In the following, a least-squares solution is presented which is based around using the singular value decomposition of a matrix formed from points measured in two reference frames (Challis 1995); this particular method has some advantages if the points used to define the rigid body have a poor spatial distribution.

The basic rigid body transformation equation used to transform points measured in one reference frame to another is:

$$y_i = Rx_i + \underline{v} \tag{9.96}$$

where y_i, x_i *are* points on a rigid body measured in two different reference frames, R is the attitude matrix, and $\underline{v}$ is a translation vector. Using a least-squares method, the problem of determining R and $\underline{v}$ is equivalent to minimizing:

$$\frac{1}{n}\sum_{i=1}^{n}(Rx_i + \underline{v} - y_i)^T(Rx_i + \underline{v} - y_i) \tag{9.97}$$

where n is the number of noncollinear points measured in both reference frames ($n \geq 3$). If the data were accurate (and noiseless) the result of this equation would be zero, but in reality this does not occur so R and $\underline{v}$ are selected to make the result as close to zero as possible.

The mean vectors ($\bar{x}$ and $\bar{y}$) are computed:

$$\bar{x} = \frac{1}{n}\sum_{i=1}^{n}x_i \tag{9.98}$$

$$\bar{y} = \frac{1}{n}\sum_{i=1}^{n}y_i \tag{9.99}$$

Given these mean vectors, the cross-dispersion matrix C can be computed:

$$C = \frac{1}{n}\sum_{i=1}^{n}(y_i - \bar{y})^T(x_i - \bar{x}) \tag{9.100}$$

The singular value decomposition of C is computed:

$$C = UDV^T \tag{9.101}$$

where U is a 3×3 orthogonal matrix, consisting of vectors u_1, u_2, u_3; D is a 3×3 diagonal matrix, whose elements are nonnegative real values (the singular values); and V is a 3×3 orthogonal matrix consisting of vectors v_1, v_2, v_3. See Appendix H for details of the singular value decomposition.

Given the singular value decomposition of the cross-dispersion matrix the attitude matrix, R, can be computed:

$$R = U\begin{bmatrix} 1 & 0 & 0 \\ 0 & 1 & 0 \\ 0 & 0 & \det(UV^T) \end{bmatrix}V^T \tag{9.102}$$

Then the vector $\underline{v}$ can be determined using the mean vectors:

$$\underline{v} = \bar{y} - R\bar{x} \tag{9.103}$$

A full derivation of this method for determining the attitude matrix and translation vector is presented in Challis (1995). See Pseudo-code 9.1, for an example of the implementation of this algorithm.

The number of nonzero singular values in matrix D indicates the rank of matrix C. If C has a rank of 3, then the points have a three-dimensional distribution, which would be the aim when identifying points to use to define three-dimensional kinematics. If the choice of points is poor or some markers, for example, are lost during data recording, then the usable data from the remaining points may have a planar distribution in which case the rank of C will be 2. In this case, only the first and second columns of U and V are defined, with the third column can be approximated via the following cross-products (see Appendix O):

$$u_3 = u_1 \times u_2 \tag{9.104}$$

$$v_3 = v_1 \times v_2 \tag{9.105}$$

Pseudo-Code 9.1: Determination of Attitude Matrix and Translation Vector Using Singular Value Decomposition Method

Purpose – computation of attitude matrix and translation vector of a set of points measured in two positions (or in two reference frames)	
Inputs	
x – X, Y, Z coordinates of points in position 1	
y – X, Y, Z coordinates of the same points in position 2	
n – number of points (n must be $>= 3$)	
Outputs	
R – 3 x 3 attitude matrix	
v – 3 x 1 translation vector	
Xmean $\leftarrow$ sum$(x)/n$	Compute mean values
Ymean $\leftarrow$ sum$(y)/n$	
For i = 1 to n	Compute differences
$\quad$ xd(i) $\leftarrow$ x(i) – Xmean	
$\quad$ yd(i) $\leftarrow$ y(i) – Ymean	
end for	
$C \leftarrow$ zeros(3,3)	Initialize matrix C
for i = 1 to n	
$\quad C \leftarrow C +$ xd(i) * yd(i)'	Compute cross-dispersion matrix (C)
end for	
[U, D, V] $\leftarrow$ svd(C)	Singular value decomposition
$d \leftarrow$ det (U * V)	Compute determinant
$R \leftarrow$ U * diag(1, 1, d) * V	Compute attitude matrix
$v \leftarrow$ Ymean – R * Xmean	Compute translation vector

The worst case would be if the points were all located on a line, in which case the rank of the cross-dispersion matrix would be *1*. In this case the attitude of the rigid body would be poorly defined but can be approximated by extracting the rotation angle (θ) and rotation axis ($\underline{e}$) from:

$$\cos(\theta) = u_1^T v_1 \qquad (9.106)$$

$$\sin(\theta)\underline{e} = u_1 \times v_1 \qquad (9.107)$$

Then the matrix R can be approximated using Rodrigues rotation formula (Rodrigues 1840):

$$R = I\cos(\theta) + (1 - \cos(\theta))\underline{e}\,\underline{e}^T + \sin(\theta)S\{\underline{e}\} \qquad (9.108)$$

where $S\{\}$ generates a skew-symmetric matrix from a vector, for example, for vector $\underline{a} = (a_1, a_2, a_3)$:

$$S\{a\} = \begin{bmatrix} 0 & -a_3 & a_2 \\ a_3 & 0 & -a_1 \\ -a_2 & a_1 & 0 \end{bmatrix} \qquad (9.109)$$

Once the attitude matrix has been computed, angles, or quaternions, can be extracted from it.

9.2.12 Calculating Segment Pose: Inverse Kinematics

In inverse kinematics the system being analyzed is modeled. In the model the joint axes are specified, as are the degrees of freedom for each segment. In addition, there may be constraints placed on the maximum angular excursion. The segmental poses and joint orientations are determined by minimizing the difference between the models' predicted marker positions and their measured positions (Lu and O'Connor 1999).

In the inverse kinematics approach, the model of the linkage system is defined by nf degrees of freedom, which are defined by the generalized coordinates $\boldsymbol{a} = (a_1, a_2, \ldots, a_{nf})$. This model of the linkage system can be customized to an experimental subject, for example, axes of rotation might differ between subjects. Reference frames are defined for the segments and for an inertial reference frame. The transformation from markers measured in the inertial reference frame (typically measured via a motion analysis system) has coordinates $\boldsymbol{b} = (b_1, b_2, \ldots, b_{nm})$, and body-fixed reference frames have coordinates $\boldsymbol{c} = (c_1, c_2, \ldots, c_{nm})$, where nm is the number of markers. Therefore:

$$b_i = R\, c_i + v \quad i = 1, nm \tag{9.110}$$

The attitude matrix (R) defining rotations from body-fixed reference frame to the inertial reference frame is a function of the generalized coordinates ($R(\boldsymbol{a})$), as is the translation vector between the two reference frames ($v(\boldsymbol{a})$). Therefore, to determine the generalized coordinates, the following equation is minimized:

$$\min \sum_{i=1}^{nm} \|b_i - R(\boldsymbol{a})\, c_i + v(\boldsymbol{a})\|^2 \tag{9.111}$$

Depending on how the equations are configured the generalized coordinates used may differ for markers associated with different segments, as attitude matrices and translation vectors change with segment. It is feasible to add constraints, for example, joint ranges of motion (e.g., $-\pi < a_1 < \pi/2$ radians). Optimization algorithms, for example (Gill et al. 2002), are used to minimize the equation and determine the generalized coordinates.

9.2.13 Differential Quantities

For analysis in three dimensions, the computation of derivatives of the displacement data is straightforward, and standard numerical differentiation techniques can be used (Chap. 4 and Appendix F). The computation of the derivatives of the angular data is more complex as will be outlined. In addition, it should be remembered that to compute angular orientation data from position data requires some non-linear transformations; in doing this the properties of the noise corrupting the position data are transformed (warped) and so are harder to remove from the computed angular data. This means that prior to the calculation of the angular data the position data should be suitably filtered to reduce noise levels.

Angular Velocity
The angular velocity is the rate of change of the orientation of one reference frame with respect to another; therefore, the angular velocities cannot simply be computed from the differentiation of the orientation angles. To compute angular velocity, consider the basic transformation matrix:

$$y_i = R_{BI} x_i \tag{9.112}$$

where y_i, x_i are points on a rigid body measured in inertial and body-fixed reference frames and R_{BI} is the attitude matrix mapping from the body to inertial reference frames. Differentiate Eq. 9.112 to get the velocity of the points:

$$\dot{y}_i = \dot{R}_{BI} x_i + R_{BI} \dot{x}_i \tag{9.113}$$

Since the time derivative of the points measured in the body-fixed reference frame is zero ($\dot{x}_i = 0$), the final term in the equation can be eliminated. The equation element x_i can be expressed in terms of y_i ($x_i = R_{IB} y_i = R_{BI}^T y_i$), so Eq. 9.113 can be simplified to:

$$\dot{y}_i = \dot{R}_{BI} R_{BI}^T y_i \tag{9.114}$$

The linear velocity of points is related to their position and the angular velocity (ω) of the body:

$$\dot{y}_i = \omega \times y_i \tag{9.115}$$

This equation can also be expressed using the skew-symmetric matrix: ($A\{\omega\}$),

$$\dot{y}_i = A\{\omega\} y_i \tag{9.116}$$

where $A\{\omega\} = \begin{bmatrix} 0 & -\omega_Z & \omega_Y \\ \omega_Z & 0 & -\omega_x \\ -\omega_Y & \omega_X & 0 \end{bmatrix}$ which means the angular velocities can be computed from:

$$A\{\omega\} = \dot{R}_{IB} R_{IB}^T \tag{9.117}$$

or, in a more general form, referred to as Poisson's equation (Wittenburg 1977):

$$A\{\omega\} = \dot{R} R^T \tag{9.118}$$

For example, if rotation occurs about the Z axis, only then these same equations apply although the resulting solution is relatively simple:

$$R_Z(\gamma) = \begin{bmatrix} \cos(\gamma) & -\sin(\gamma) & 0 \\ \sin(\gamma) & \cos(\gamma) & 0 \\ 0 & 0 & 1 \end{bmatrix} \tag{9.119}$$

Then:

$$\dot{R}_Z(\gamma) = \dot{\gamma} \begin{bmatrix} -\sin(\gamma) & -\cos(\gamma) & 0 \\ \cos(\gamma) & -\sin(\gamma) & 0 \\ 0 & 0 & 0 \end{bmatrix} \tag{9.120}$$

The angular velocities can be computed from:

$$A\{\omega\} = \dot{R} R^T$$

$$= \dot{\gamma} \begin{bmatrix} -\sin(\gamma) & -\cos(\gamma) & 0 \\ \cos(\gamma) & -\sin(\gamma) & 0 \\ 0 & 0 & 0 \end{bmatrix} \begin{bmatrix} \cos(\gamma) & \sin(\gamma) & 0 \\ -\sin(\gamma) & \cos(\gamma) & 0 \\ 0 & 0 & 1 \end{bmatrix}$$

$$= \dot{\gamma} \begin{bmatrix} 0 & -1 & 0 \\ 1 & 0 & 0 \\ 0 & 0 & 0 \end{bmatrix} = \begin{bmatrix} 0 & -\omega_Z & 0 \\ \omega_Z & 0 & 0 \\ 0 & 0 & 0 \end{bmatrix} \tag{9.121}$$

which shows we only have an angular velocity different from zero about the Z axis.

The transformations involved in the computation of the angular velocities are non-linear which means the signal spectrum will be warped, which if the attitude matrix is derived from noisy position data can be problematic. The position data should be low-pass filtered prior to the computation of the angular velocities to minimize the influence of this spectrum warping. As an alternative the angular velocities can be computed from:

$$A\{\omega(t)\} = \frac{R(t + \Delta t)R^T(t - \Delta t) - R(t - \Delta t)R^T(t + \Delta t)}{4\Delta t} \tag{9.122}$$

An alternative approach allows the computation of the angular velocities from the quaternions:

$$\begin{bmatrix} \omega_X \\ \omega_Y \\ \omega_Z \end{bmatrix} = \begin{bmatrix} -q_1 & -q_0 & -q_3 & q_2 \\ -q_2 & q_3 & q_0 & -q_1 \\ -q_3 & -q_2 & q_1 & q_0 \end{bmatrix} \begin{bmatrix} \dot{q}_0 \\ \dot{q}_1 \\ \dot{q}_2 \\ \dot{q}_3 \end{bmatrix} \tag{9.123}$$

Instantaneous Helical Axis

In two dimensions, the instantaneous center of rotation (ICR) or centroid is the point on a rotating rigid body which has a velocity of zero; in three dimensions a unique point with a velocity of zero cannot be found, but the rigid body movement may be viewed as having a translation velocity along, and a rotation velocity about, a directed line in space. This description is called the instantaneous helical axis (IHA), although sometimes the term screw is substituted for helical. The rotational speed ($\bar{\omega}$) about the IHA is computed from:

$$\bar{\omega} = \sqrt{\omega^T \omega} = |\omega| \tag{9.124}$$

The unit direction vector ($\underline{e}_I$) of the axis is calculated from:

$$\underline{e}_I = \frac{\omega}{\sqrt{\omega^T \omega}} = \frac{\omega}{\bar{\omega}} \tag{9.125}$$

The translation speed ($\dot{u}$) along the axis is calculated from:

$$\dot{u} = \dot{\underline{v}}^T \underline{e}_I \tag{9.126}$$

Finally, the location of a point (p_I) on the IHA is given by:

$$p_I = p + \frac{\omega \times \dot{\underline{v}}}{\bar{\omega}^2} \tag{9.127}$$

9.2.14 Averaging Angular Kinematics

In the following the computation of the mean of a set of three-dimensional orientation angles, and the mean pivot of rotation from a time series of instantaneous helical axis (IHA) will be presented.

Mean Three-Dimensional Orientation Angles

The three-dimensional angular kinematics are not defined in three-dimensional Euclidean space as are linear vectors but exist on the surface of a non-linear manifold. Therefore, the computation of the mean orientation is not simply a case of, for example, averaging a set of Cardan angles. For example, if there are rotations about each axis:

$$R = R_z\left(\frac{\pi}{2}\right) R_y\left(\frac{\pi}{2}\right) R_x\left(-\frac{\pi}{2}\right) = \begin{bmatrix} 0 & 0 & 1 \\ 0 & 1 & 0 \\ -1 & 0 & 0 \end{bmatrix} \qquad (9.128)$$

So the corresponding Cardan angles are $\left(-\frac{\pi}{2}, \frac{\pi}{2}, \frac{\pi}{2}\right)$; imagine that the other Cardan angle set to average is $\left(0, \frac{\pi}{2}, 0\right)$; if the angles were simply averaged, the averaged set of angles would be $\left(-\frac{\pi}{4}, \frac{\pi}{2}, \frac{\pi}{4}\right)$, but the matrix corresponding to this second Cardan sequences is:

$$R = R_z(0)\, R_y\left(\frac{\pi}{2}\right) R_x(0) = R_z\left(\frac{\pi}{4}\right) R_y\left(\frac{\pi}{2}\right) R_x\left(-\frac{\pi}{4}\right) = \begin{bmatrix} 0 & 0 & 1 \\ 0 & 1 & 0 \\ -1 & 0 & 0 \end{bmatrix} \qquad (9.129)$$

Clearly, there is an ambiguity in the calculation of the averaging of Cardan angles or Euler angles. As the definition of an angular orientation exists on the surface of a non-linear manifold, simple averaging cuts across the surface of the manifold rather than appropriately traversing the surface of the manifold. If, for example, there are two orientations to average and these are defined by R_1 and R_2, then the angle which maps between these two orientations can be computed from:

$$\cos\theta = \frac{1}{2}\left(tr\left(R_1^T R_2\right) - 1\right) \qquad (9.130)$$

The error in averaging the Cardan or Euler angles extracted from R_1 and R_2 is (Moakher 2002):

$$d_E = 2\sqrt{2}\left|\sin\frac{\theta}{2}\right| \qquad (9.131)$$

which indicates the error arising from the averaging of angles is negligible if the angular distance (θ) between the two orientations is small. Problems akin to this exist when computing angular range of motion (Michaud et al. 2014).

Challis (2006, 2020b) has presented a procedure for determining mean angles which does not cause these problems. If the mean of a sequence of angles is to be computed from m measures, the attitude matrices associated with each angle set should be used (R_i), with the mean matrix computed (E):

$$E = \frac{1}{m}\sum_{i=1}^{m} R_i \qquad (9.132)$$

In the example given, the resulting matrix (E) would be sufficient for computing the mean angles. In other cases, the mean matrix cannot always be proper orthogonal; therefore, the the singular value decomposition of the mean matrix is then computed:

$$E = UDV^T \qquad (9.133)$$

where U is a 3×3 orthogonal matrix, consisting of vectors u_1, u_2, u_3; D is a 3×3 diagonal matrix, whose elements are nonnegative real values (the singular values); and V is a 3×3 orthogonal matrix, consisting of vectors v_1, v_2, v_3. See Appendix H for details of the singular value decomposition. Then the mean attitude matrix ($\overline{R}$) from which the mean angles can be determined is computed from:

$$\overline{R} = UV^T \qquad (9.134)$$

Mean Pivot of Rotation

For some analyses of three-dimensional motion, it is useful to have a mean point of rotation. From a sequence of IHA, it is feasible to compute the mean pivot of rotation; similar analysis can be performed on a time series of FHA measured to approximate the IHA. To compute the mean pivot point, first define a new matrix:

$$G_i = I - \underline{e_{I,i}}\,\underline{e_{I,i}}^T \quad i = 1, n \tag{9.135}$$

where $\underline{e_{I,i}}$ is unit direction vector of the IHA for sample value i and n is the number of samples. The mean matrix is computed from:

$$\widehat{G} = \frac{1}{n}\sum_{i=1}^{n} G_i \tag{9.136}$$

The mean pivot (p_{opt}) which is overall closest to all n IHA can be computed in a least-squares sense from:

$$p_{\text{opt}} = \widehat{G}^{-1}\frac{1}{n}\sum_{i=1}^{n} G_i\, p_{I,i} \tag{9.137}$$

9.3 Review Questions

1. Write the equation which describes the rotation and translation of a rigid body from one location to another.
2. How might the attitude matrix be parameterized? What are the merits of the different systems?
3. Describe Cardanic and Eulerian angles.
4. With Cardanic and Eulerian angles, explain what is meant by a singularity. How can these be avoided?
5. Describe in equation form Codman's paradox.
6. Describe the finite and instantaneous helical axes. What are the parameters used to describe these axes?
7. What is the difference between finite and instantaneous helical axes?

References

Cited References

Alexander, R. M. (1981). Analysis of force platform data to obtain joint forces. In D. Dowson & V. Wright (Eds.), *An introduction to the bio-mechanics of joints and joint replacement* (pp. 30–35). London: Mechanical Engineering Publications Ltd..

Bar-Itzhack, I. Y., & Reiner, J. (1984). Recursive attitude determination from vector observations: Direction cosine matrix identification. *Journal of Guidance, Control, and Dynamics, 7*(1), 51–56. https://doi.org/10.2514/3.56362.

Cayley, A. (1843). On the motion of rotation of a solid body. *Cambridge Mathematics Journal, 3*(1843), 224–232.

Challis, J. H. (1995). A procedure for determining rigid body transformation parameters. *Journal of Biomechanics, 28*(5), 733–737.

Challis, J. H. (2006). Statistically appropriate computation of mean segment orientation. *Journal of Biomechanics, 39*, S558.

Challis, J. H. (2020a). Quaternions as a solution to determining the angular kinematics of human movement. *BMC Biomedical Engineering, 2*(1), 5.

Challis, J. H. (2020b). Determining the average attitude of a rigid body. *Journal of Biomechanics, 98*, 109492.

Chasles, M. (1830). Note sur les propriétés générales du système de deux corps semblables entr'eux. *Bulletin des Sciences Mathématiques, Astronomiques, Physiques et Chemiques, 14*, 321–326.

Codman, E. (1934). *The shoulder: Rupture of the supraspinatus tendon and other lesions in or about the subacromial Bursa*. Boston: T. Todd.

Farrell, J. L., Stuelpnagel, J. C., Wessner, R. H., & Velman, J. R. (1966). Problem 65-1: A least squares estimate of satellite attitude. *SIAM Review, 8*(3), 384–386.

Frankel, V. H., Burstein, A. H., & Brooks, D. B. (1971). Biomechanics of Internal Derangement of the Knee: Pathomechanics as determined by analysis of the instant centers of motion. *The Journal of Bone and Joint Surgery, A53*(5), 945–977.

Gill, P. E., Murray, W., & Saunders, M. A. (2002). SNOPT: An SQP algorithm for large-scale constrained optimization. *SIAM Journal on Optimization, 12*(4), 979–1006.

Kainz, H., Modenese, L., Lloyd, D. G., Maine, S., Walsh, H. P. J., & Carty, C. P. (2016). Joint kinematic calculation based on clinical direct kinematic versus inverse kinematic gait models. *Journal of Biomechanics, 49*(9), 1658–1669.

Lu, T. W., & O'Connor, J. J. (1999). Bone position estimation from skin marker co-ordinates using global optimisation with joint constraints. *Journal of Biomechanics, 32*(2), 129–134.

Markley, F. L. (1993). Attitude determination using vector observations – A fast optimal matrix algorithm. *Journal of the Astronautical Sciences, 41*(2), 261–280.

Markley, F. L., & Crassidis, J. L. (2014). *Fundamentals of spacecraft attitude determination and control*. New York: Springer.

Michaud, B., Jackson, M. I., Prince, F., & Begon, M. S. (2014). Can one angle be simply subtracted from another to determine range of motion in three-dimensional motion analysis? *Computer Methods in Biomechanics and Biomedical Engineering, 17*(5), 507–515.

Moakher, M. (2002). Means and averaging in the group of rotations. *SIAM Journal on Matrix Analysis and Applications, 24*(1), 1–16.

Reuleaux, F. (1875). *The kinematics of machinery: Outline of a theory of machines* (A. B. W. Kennedy, Trans.). London: Macmillan.

Rodrigues, O. (1840). Des lois géometriques qui regissent les déplacements d' un systéme solide dans l' espace, et de la variation des coordonnées provenant de ces déplacement considérées indépendant des causes qui peuvent les produire. *Journal de Mathématiques Pures et Appliquées, 5,* 380–440.

Schut, G. H. (1968). Formation of strips from independent models. *Photogrammetric Engineering, 34*(7), 690–695.

Selbie, W. S., Thomson, D. B., & Richmond, F. J. R. (1993). Sagittal-plane mobility of the cat cervical spine. *Journal of Biomechanics, 26*(8), 917–927.

Veldpaus, F. E., Woltring, H. J., & Dortmans, L. J. M. G. (1988). A least-squares algorithm for the equiform transformation from spatial marker co-ordinates. *Journal of Biomechanics, 21*(1), 45–54.

Wahba, G. (1965). Problem 65-1: A least squares estimate of spacecraft attitude. *SIAM Review, 7*(3), 409.

Weber, W. E., & Weber, E. W. (1992). *Mechanics of the human walking apparatus* (P. Maquet & R. Furlong, Trans.). Berlin: Springer.

Woltring, H. J. (1991). Representation and calculation of 3-D joint movement. *Human Movement Science, 10*(5), 603–616.

Wittenburg, J. (1977). *Dynamics of systems of rigid bodies.* Stuttgart: B.G. Teubner.

Wu, G., & Cavanagh, P. R. (1995). ISB recommendations for standardization in the reporting of kinematic data. *Journal of Biomechanics, 28*(10), 1257–1261.

Wu, G., Siegler, S., Allard, P., Kirtley, C., Leardini, A., Rosenbaum, D., Whittle, M., D'Lima, D. D., Cristofolini, L., Witte, H., Schmid, O., & Stokes, I. (2002). ISB recommendation on definitions of joint coordinate system of various joints for the reporting of human joint motion–part I: Ankle, hip, and spine. International Society of Biomechanics. *Journal of Biomechanics, 35*(4), 543–548.

Wu, G., van der Helm, F. C., Veeger, H. E., Makhsous, M., Van Roy, P., Anglin, C., Nagels, J., Karduna, A. R., McQuade, K., Wang, X., Werner, F. W., & Buchholz, B. (2005). ISB recommendation on definitions of joint coordinate systems of various joints for the reporting of human joint motion–part II: Shoulder, elbow, wrist and hand. *Journal of Biomechanics, 38*(5), 981–992.

Specific References

Anglin, C., & Wyss, U. P. (2000). Review of arm motion analyses. *Journal of Biomedical Engineering, 214*(5), 541–555.

Bell, A. L., Brand, R. A., & Pedersen, R. (1989). Prediction of hip joint centre location from external landmarks. *Human Movement Science, 8*(1), 3–16.

Cerveri, P., Lopomo, N., Pedotti, A., & Ferrigno, G. (2005). Derivation of centers and axes of rotation for wrist and fingers in a hand kinematic model: Methods and reliability results. *Annals of Biomedical Engineering, 33*(3), 402–412.

Chao, E. Y., & Morrey, B. F. (1978). Three-dimensional rotation of the elbow. *Journal of Biomechanics, 11*(1–2), 57–73.

Dixon, P. C., Böhm, H., & Döderlein, L. (2012). Ankle and midfoot kinetics during normal gait: A multi-segment approach. *Journal of Biomechanics, 45*(6), 1011–1016.

Grip, H., Sundelin, G., Gerdle, B., & Karlsson, J. S. (2007). Variations in the axis of motion during head repositioning--a comparison of subjects with whiplash-associated disorders or non-specific neck pain and healthy controls. *Clinical biomechanics, 22*(8), 865–873.

Karduna, A. R., McClure, P. W., & Michener, L. A. (2000). Scapular kinematics: Effects of altering the Euler angle sequence of rotations. *Journal of Biomechanics, 33*(9), 1063–1068.

Lafortune, M. A., Cavanagh, P. R., Sommer Iii, H. J., & Kalenak, A. (1992). Three-dimensional kinematics of the human knee during walking. *Journal of Biomechanics, 25*(4), 347–357.

Okita, N., Meyers, S. A., Challis, J. H., & Sharkey, N. A. (2013). Segmental motion of forefoot and Hindfoot as a diagnostic tool. *Journal of Biomechanics, 46*(15), 2578–2585.

Salvia, P., Woestyn, L., David, J. H., Feipel, V., Van Sint Jan, S., Klein, P., & Rooze, M. (2000). Analysis of helical axes, pivot and envelope in active wrist circumduction. *Clinical biomechanics, 15*(2), 103–111.

Sangeux, M., Pillet, H., & Skalli, W. (2014). Which method of hip joint centre localisation should be used in gait analysis? *Gait & Posture, 40*(1), 20–25.

Stokdijk, M., Biegstraaten, M., Ornek, W., de Boer, Y. A., Veeger, H. E. J., & Rozing, P. M. (2000). Determining the optimal flexion-extension axis of the elbow in vivo – A study of interobserver and intraobserver reliability. *Journal of Biomechanics, 33*(9), 1139–1145.

van den Bogert, A. J., Reinschmidt, C., & Lundberg, A. (2008). Helical axes of skeletal knee joint motion during running. *Journal of Biomechanics, 41*(8), 1632–1638.

Woltring, H. J., Long, K., Osterbauer, P. J., & Fuhr, A. W. (1994). Instantaneous helical axis estimation from 3-D video data in neck kinematics for whiplash diagnostics. *Journal of Biomechanics, 27*(12), 1415–1432.

Useful References

Altmann, S. L. (1986). *Rotations, quaternions, and double groups.* New York: Clarendon.

Brennan, A., Deluzio, K., & Li, Q. (2011). Assessment of anatomical frame variation effect on joint angles: A linear perturbation approach. *Journal of Biomechanics, 44*(16), 2838–2842.

Liegeois, A. (1985). The rigid body: Configuration and motion. In *Robot technology: Performance and computer-aided design* (Vol. 7, pp. 9–28). London: Kogan Page.

Morais, J. P., Georgiev, S., & Sprößig, W. (2014). *Real quaternionic calculus handbook.* Basel: Springer.

Sommer, H. J. (1992). Determination of first and second order instant screw parameters from landmark trajectories. *Journal of Mechanical Design, 114*(2), 274–282.

Woltring, H. J. (1994). 3-D attitude representation of human joints: A standardization proposal. *Journal of Biomechanics, 27*(12), 1399–1414.

Woltring, H. J., Huiskes, R., de Lange, A., & Veldpaus, F. E. (1985). Finite centroid and helical axis estimation from noisy landmark measurements in the study of human joint kinematics. *Journal of Biomechanics, 18*(5), 379–389.

Overview

Inverse dynamics is the procedure which allows the estimation of the resultant joint forces and moments. Winter (1980) claimed:

> One of the most valuable biomechanical variables to have for the assessment of any human movement is the time history of the moments of force at each joint.

The resultant joint moments have been used extensively as a way of obtaining insight into muscular coordination.

The computation of resultant joint forces and moments requires information on segment kinematics and inertial properties. These data can be combined with information from a force plate to permit computation of the resultant joint forces and moments. As such, these forces and moments require data from a number of sources, and a number of numerical steps. In light of this level of complexity the study of Elftman (1939), performed before the development of computers is an impressive feat of industry.

Human movement is fueled by chemical energy, so the tracking of mechanical energy is an obvious way of trying to understand the link between the metabolic costs of movement and mechanical output. It turns out that the establishment of this link has been more challenging than at first thought. Despite this challenge, the determination of the mechanical energy associated with human movement has provided some valuable insights into human movement.

The computation of resultant joint forces and moments and segmental and joint energetics requires data from motion analysis (Chap. 6), kinematic analysis (Chap. 9), the computation of derivatives (Chap. 4), body segment inertial properties (Chap. 8), and possibly ground reaction forces (Chap. 7). In addition to the material from previous chapters, this chapter requires a working knowledge of matrix algebra (see Appendix A).

10.1 Inverse Dynamics

Inverse dynamics is the procedure which allows the estimation of resultant joint forces and moments. The term moment is used here to refer to the rotational equivalent of force. Sometimes these moments are also referred to as the moment of force or torque. Paul (1978) has argued that the term torque should be reserved for the rotational force associated with the relative angular movement between the ends of an object about its long axis. In the literature, all three terms are used to refer to the same thing, but to avoid confusion, the term moment will be used to refer to the angular equivalent of a force which causes a tendency to twist about a given axis.

The resultant moments have been particularly popular as a means of interpreting human motion (e.g., Vaughan 1996). The resultant moment can be considered the result of three potential moments at a joint:

$$T = \sum_{i=1}^{nm} r_{m,i} \times F_{m,i} + \sum_{j=1}^{nl} r_{L,j} \times F_{L,j} + \sum_{k=1}^{nc} r_{c,k} \times F_{c,k} \tag{10.1}$$

where T is the vector of resultant joint moments; nm, nl, nc are the number of muscles, ligaments, and articular contact points, respectively; $r_{m,i}$, $r_{L,j}$, $r_{c,k}$ are the vectors from the joint center to the line of action of the force of the muscles, ligaments, and articular contact points, respectively; and $F_{m,\,i}$, $F_{L,\,j}$, $F_{c,\,k}$ are the vectors of forces produced by the muscles, ligaments, and articular contacts, respectively. Here $\times$ refers to the cross-product (see Appendix O for details).

Equation 10.1 indicates that a resultant joint moment is the net effect of the moments about a joint due to structures crossing that joint, the muscles and ligaments, along with the bone forces. For healthy joints within a reasonable range of motion the resultant joint moment is dominated by the moment due to the muscles. Often when the resultant joint moment is computed it is referred to as the muscle moment, but this may not always be the case, for example, if the joint under investigation has some joint degeneration thus increasing articular contact forces and potentially the moments due to these forces.

On occasions, the resultant joint force has been confused with the joint contact force. The bone-on-bone forces are the sum of all forces acting at (across) the joint. In Fig. 10.1, the joint reaction force is the same for both conditions (50 N), but the muscular forces differ; they produce zero bone-on-bone forces in scenario 1 and a bone-on-bone force of 50 N in scenario 2. If bone-on-bone forces are required, there needs to be a model of some kind to estimate these forces (e.g., Morrison 1968; Scott and Winter 1990) or appropriate instrumentation of a joint (e.g., Bergmann et al. 2001).

The computation of the resultant joint forces and moments requires a number of parameters and variables, where in the computation process a parameter is a constant factor, and a variable is a numerical value which varies. The parameters include masses of the segments, center of mass location for each segment, and moments of inertia for each segment. The required variables include the positions of the joints, accelerations of the centers of mass, and segment angular velocities and accelerations and can be augmented by the inclusion of the external forces and moments. Figure 10.2 shows how the data are combined to compute the resultant forces and moments.

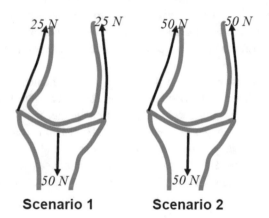

Scenario 1   **Scenario 2**

Fig. 10.1 In both scenarios, the joint reaction forces are the same (50 N), but the bone-on-bone forces differ

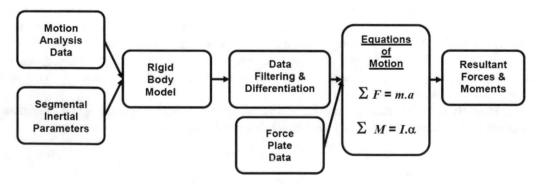

Fig. 10.2 Various inputs required for the computation of joint resultant forces and moments. Strictly, the data from a force plate are not necessarily required, but if there is an external contact, the force provided from the force plate can improve the accuracy of the computed resultant joint forces and moments

10.1.1 Common Assumptions

The typical assumptions made when computing the resultant joint forces and moments are the following:

1. The human body can be modeled as a series of rigid bodies.
2. These rigid bodies have fixed masses, center of mass locations, and moments of inertia. If the segments are rigid bodies, these fixed values are a necessary condition.
3. The centers of rotation for the joints are known.
4. The joints allow no translation.

Each of the assumptions will be discussed in the following paragraphs.

Human body segments are not rigid (see, e.g., Pain and Challis 2001), but assuming that they are, make the analysis tractable. This lack of rigidity means the inertial properties can change during an activity (Pain and Challis 2001). To date, there are no measurement techniques which permit determination of the resultant joint forces and moments assuming the segments are nonrigid, although simulation models have suggested that the soft tissue motion can make an important contribution to resultant joint forces and moments during movements with a large impact (Pain and Challis 2006).

For an inverse dynamics analysis a point for the action of the resultant joint force at each joint is required. This point is often referred to as the joint center. Approaches for estimating the location of these joint centers include anatomy-based (e.g., Weinhandl and O'Connor 2010), statistics-based (e.g., Bell et al. 1989), X-ray-based (e.g., Kirkwood et al. 1999), or recorded functional movements (e.g., Piazza et al. 2004). A mean pivot point from the helical axes describing the motion of a joint can also serve as an estimate of the joint center. Such an approach ignores the apparent motion of the helical axis with joint motion (e.g., Blankevoort et al. 1990), suggesting the joint center is not in a fixed location. Indeed, it is feasible that the location of helical axes, and therefore the joint center, might vary under different joint loadings. An additional complication is that some joints can also translate (e.g., Karduna et al. 1997).

Additional assumptions are made if the resultant joint moment is assumed to be equal to the resultant muscular moment (see previous section).

10.1.2 Formulating the Equations

The equations for computing the resultant joint forces and moments can be determined using a number of different mechanical approaches; they all give the same final result. These approaches include Newton-Euler (e.g., Luh et al. 1980), Lagrangian (e.g., Lewis et al. 1993), and Kane's method (Kane and Levinson 1985). The resulting equations may look different but they are equivalent (Silver 1982). To some extent, the eventual formulation depends on the generalized coordinates used to define the system of interest and whether the moments are determined in a sequential manner or all determined simultaneously. In three dimensions with a large number of segments the equations can become quite cumbersome, and the programming of the equations requires close attention to basic bookkeeping of variables. As a consequence software packages which automatically determine the equations of motion for a user-defined system are the preferred approaches for complex systems (e.g., Kurz et al. 2010; Lot and Da Lio 2004).

10.1.3 Newton-Euler Approach

When performing inverse dynamics using a Newton-Euler approach there are two basic equations involved in this process. The first equation is due to Newton:

$$\sum F_i = m \, a_{cm} \tag{10.2}$$

which states that the sum of all of the forces (F_i) acting on a rigid body is equal to the product of the mass (m) of the body and the acceleration of the center of mass (a_{cm}) of the body. For example, Fig. 10.3 shows a segment with forces acting at its proximal (F_1) and distal (F_2) ends and a force due to the segment weight (mg).

Fig. 10.3 The forces acting on a single segment and the resultant force. Note that here $\sum F_i = m\,a_{cm}$, with $F_3 = mg$

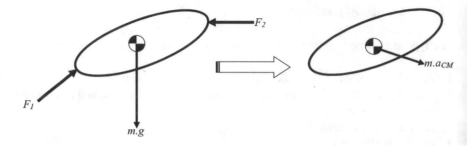

Fig. 10.4 The moments (T_1, T_2) acting on a single segment and the resultant moment they produce, for this two-dimensional case $\sum T_i = I\alpha$

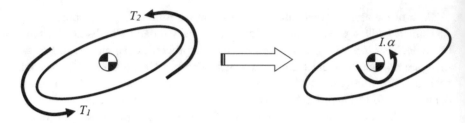

The following equations describe the system:

$$\sum F_i = m\,a_{cm}$$
$$= F_1 + F_2 + m\,g = m\,a_{cm} \tag{10.3}$$

Under certain scenarios, there may be additional forces applied to the segment, but as long as these are known, they are easily incorporated into Eq. 10.3. Generally only three forces are considered, a distal force, a vertically downward acting force due to segment weight, and a proximal force. In an inverse dynamics computation typically the distal force is known, as are the weight of the segment and the segment acceleration; thus, one force vector is unknown and can be computed. The first distal force is either a segment free in space in which case the force is zero or the segment is in contact with the ground so the distal force can be measured with a force plate. For the next segment in the sequence the previously computed proximal reaction force becomes the distal force using Newton's third law (to every action, there is an equal and opposite reaction).

The second equation is due to Euler; his rotation equation is:

$$\sum T_i = J\,\alpha + \omega \times (J\,\omega) \tag{10.4}$$

where T_i are the moments acting on the rigid body, J is the inertia tensor (see Sect. 8.1.4), α are the angular accelerations of the body, and ω are the angular velocities of the body.

If the axes in the segment represent the principal axes, then these equations can be expanded to:

$$T_X = I_{XX}\,\alpha_X + (I_{ZZ} - I_{YY})\omega_Y\omega_Z \tag{10.5}$$

$$T_Y = I_{YY}\,\alpha_Y + (I_{XX} - I_{ZZ})\omega_Z\omega_X \tag{10.6}$$

$$T_Z = I_{ZZ}\,\alpha_Z + (I_{YY} - I_{XX})\omega_X\omega_Y \tag{10.7}$$

A planar system helps to illustrate the application of the equation due to Euler (Fig. 10.4). The net effect of all moments acting on a rigid body to cause rotation about the segment center of mass is equal to the product of the moment of inertia of the segment and the angular acceleration about the center of mass for this two-dimensional system.

For a rigid body the moments (T_i) acting on it can be determined from any force applied (F_i), and the vector of that force from the point of force application to the axis of rotation (r_i), here assumed to be the segment center of mass. Therefore:

$$T_i = F_i \times r_i \tag{10.8}$$

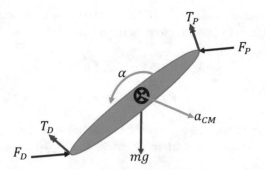

Fig. 10.5 Free body diagram showing forces and moments applied to a segment and the resulting linear and angular accelerations they produce of and around the segment center of mass. Note double-headed arrows indicate moments

For the analysis of a rigid body, the forces applied to the body, unless with a line of action through the center of mass, will produce moments about the center of mass. Therefore, from Fig. 10.3, the moments would be computed using:

$$T_1 = F_1 \times r_1 \tag{10.9}$$

$$T_2 = F_2 \times r_2 \tag{10.10}$$

where the vectors r_1, r_2 are from their respective points of force application on the segment to the segment center of mass. For a segment in the human body, the moments acting on the system give the following equation (see Fig. 10.5):

$$J\,\alpha + \omega \times (J\,\omega) = T_D + T_P + (F_D \times r_D) + (F_P \times r_P) \tag{10.11}$$

where T_D is the resultant moment applied to the segment distally, T_P is the resultant moment applied to the segment proximally, F_D is the force vector applied to the segment distally, F_P is the force vector applied to the segment proximally, r_D is the vector from the point of application of the distal force to the segment to the segment center of mass, and r_P is the vector from the point of application of the proximal force to the segment to the segment center of mass. Under certain scenarios, there may be additional forces, but their moment contributions are easily incorporated into Eq. 10.11.

10.1.4 Inverse Dynamics in Three Dimensions

With a Newton-Euler approach to inverse dynamics the resultant joint moments can be computed sequentially starting from a distal segment, and then for the next segment in the chain, and so on up the chain. In the example, consider that the most distal segment is the foot.

Step 1 – reaction forces:

$$\sum F_i = m\,a_{cm}$$
$$= F_D + F_P + m\,g = m\,a_{cm} \tag{10.12}$$

where F_D is the distal reaction force acting on the foot and F_P is the proximal reaction force at the ankle joint. The distal reaction force must be known either because the foot is free in space in which case the force is zero or because the foot is in contact with the ground in which case the distal reaction force is provided by a force plate. The mass of the foot is derived from anthropometric data, and the acceleration of the segment center of mass is determined by the double differentiation of body segment position data, typically from motion analysis data. The proximal reaction force (ankle joint reaction force) is then computed from:

$$F_D = m\,a_{cm} - F_P - m\,g \tag{10.13}$$

These computations are all performed on the data using their measurements in an inertial reference frame, but for use in the subsequent steps, they must be given in the segment reference frame; therefore:

$$F_{D,S1} = R_{I \to S1} \, F_D \tag{10.14}$$

$$F_{P,S1} = R_{I \to S1} \, F_P \tag{10.15}$$

where $F_{D,\,S1}$ is the distal reaction force referenced to segment 1 reference frame, $R_{I \to S1}$ is an attitude matrix transforming from the inertial (lab) reference frame to segment 1 reference frame, and $F_{P,\,S1}$ is the proximal reaction force referenced to segment 1 reference frame.

Step 2 – moments:

$$\sum T_i = J_{S1} \, \alpha_{S1} + \omega_{S1} \times (J_{S1} \, \omega_{S1})$$

$$= T_{D,S1} + T_{P,S1} + (F_{D,S1} \times r_D) + (F_{P,S1} \times r_P) \tag{10.16}$$

where J_{S1} is the inertia tensor of segment 1, α_{S1} is the vector of segmental angular accelerations of segment 1, ω_{S1} is the vector of segment 1 angular velocities, $T_{D,\,S1}$ is the vector of moments applied to the distal end of the segment with reference to the segment-based reference frame, $T_{P,\,S1}$ is the vector of moments applied to the proximal end of the segment with reference to the segment-based reference frame, and r_D and r_P are the vectors from the distal point and proximal points of force application to the segment center of mass in segment 1 reference frame. The inertia tensor is determined from anthropometric data. The angular velocity and acceleration of the segments are determined from the single and double differentiation of body segment angular position data, typically derived from motion analysis data. The distal resultant moments must be known either because the foot is free in space in which case the moments are zero or because the foot is in contact with the ground in which case the distal moment vector is provided by a force plate. The proximal resultant joint moments are then determined from:

$$T_{P,S1} = J_{S1} \, \alpha_{S1} + \omega_{S1} \times (J_{S1} \, \omega_{S1}) - T_{D,S1} + (F_{D,S1} \times r_D) + (F_{P,S1} \times r_P) \tag{10.17}$$

Step 3 – another segment

The proximal reaction forces and moments have been determined for segment 1, so now the next segment is analyzed (segment 2). A basic requirement for determining the proximal reaction forces and moments for segment 2 requires the information from segment 1.

Using the proximal reaction forces for segment 1, the distal reaction forces for segment 2 can be determined:

$$F_D(segment\ 2) = -F_P(segment\ 1) \tag{10.18}$$

where both sets of reaction forces are measured with respect to the inertial reference frame.

Using the proximal resultant joint moments from segment 1, the distal resultant joint moment for segment 2 can be determined. The first step is to reference these moments relative to the inertial reference frame:

$$T_{P,I} = R_{S1 \to I} \, T_{P,S1} \tag{10.19}$$

where $T_{P,\,I}$ are the proximal moments for segment 1 in the inertial reference frame and $R_{S1 \to I}$ is an attitude matrix transforming from segment 1 reference frame to the inertial (lab) reference frame. Then:

$$T_{D,S2} = -R_{I \to S2} \, T_{P,I} \tag{10.20}$$

where $T_{D,\,S2}$ are the distal moments for segment 2 in the reference frame for segment 2 and $R_{I \to S2}$ is an attitude matrix transforming from the inertial (lab) reference frame to segment 2 reference frame.

The sequence is then computed for the next joint in the system and completed up the chain of segments.

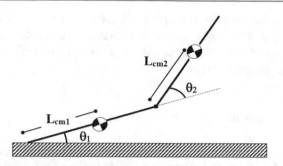

Fig. 10.6 Planar two-link system. Note θ_1, θ_2 are the angles of the first and second segments, respectively, and L_{CM1}, L_{CM2} are the distances from the joint center to the center of mass for first and second segments, respectively

10.1.5 State-Space Representation

The equations for inverse dynamics can be set up using a state-space representation, which means that in contrast to the Newton-Euler approach from the previous section, all of the moments are computed at once. A typical formulation, for a system with n rotational degrees of freedom, has the following structure:

$$T = M(\theta)\ddot{\theta} + v(\theta, \dot{\theta}) + G(\theta) \tag{10.21}$$

where T is a $n \times 1$ vector of joint moments, $M(\theta)$ is the $n \times 1$ inertia matrix, $v(\theta, \dot{\theta})$ is an $n \times 1$ vector of centrifugal/Coriolis terms, $G(\theta)$ is a $n \times 1$ vector of gravity terms, and $\theta, \dot{\theta}$, and $\ddot{\theta}$ are the generalized coordinates and their first and second derivatives.

For example, for a planar two-link system (see Fig. 10.6), the following equations can be derived:

$$M(\theta) = \begin{bmatrix} m_1 L_{CM1}^2 + I_1 + m_2\left(L_1^2 + L_{CM2}^2 + 2L_1 L_{CM2} \, \cos\left(\theta_2\right)\right) + I_2 & m_2 L_1 L_{CM2} \cos\left(\theta_2\right) + m_2 L_{CM2}^2 + I_2 \\ m_2 L_1 L_{CM2} \cos\left(\theta_2\right) + m_2 L_{CM2}^2 + I_2 & m_2 L_{CM2}^2 + I_2 \end{bmatrix} \tag{10.22}$$

$$v(\dot{\theta}) = \begin{bmatrix} -m_2 \, L_1 \, L_{CM2} \, \sin\left(\theta_2\right)\left(2\dot{\theta}_1\dot{\theta}_1 + \dot{\theta}_2^2\right) \\ m_2 \, L_1 \, L_{CM2} \, \sin\left(\theta_2\right)\left(\dot{\theta}_1^2\right) \end{bmatrix} \tag{10.23}$$

$$G(\theta) = \begin{bmatrix} m_1 \, L_{cm1} \, g \, \cos\left(\theta_1\right) + m_2 \, g \, \left(L_{CM2} \, \cos\left(\theta_1 + \theta_2\right) + L_1 \, \cos\left(\theta_1\right)\right) \\ m_2 \, g \, L_{CM2} \, \cos\left(\theta_1 + \theta_2\right) \end{bmatrix} \tag{10.24}$$

where m_1, m_2 are the masses of the first and second segments, respectively; I_1, I_2 are the moments of inertia with respect to the center of mass of the first and second segments, respectively; L_1, L_2 are the lengths of the first and second segments, respectively; L_{CM1}, L_{CM2} are the lengths from the proximal joint to the segment center of mass for the first and second segments, respectively; and g is the acceleration due to gravity.

The grouping of all the terms including the moments of inertia gives the inertia matrix. This matrix is symmetric and positive definite, meaning that it is invertible (Golub and Van Loan 1983). The grouping of all terms involving angular velocities, which all relate to centripetal and Coriolis forces, gives the vector of centrifugal/Coriolis terms. Finally, grouping of all terms relating to gravity produces the vector of gravity terms. Terms from adjacent links occur in the equations for a link; therefore the equations are coupled. For each joint in the system, there is a second-order non-linear differential equation describing the relationship between the moments and the resulting angular motion.

The complexity of the equations in this form even for a planar two-link system makes the equations of more complex systems daunting to derive and therefore more so to program. Fortunately, there are packages which automatically generate the equations of motion and generate the appropriate computer code (e.g., Kurz et al. 2010; Lot and Da Lio 2004). Inverse dynamics starts from the motion of the body and determines the forces and moments causing the motion, while direct dynamics starts from the forces and moments acting on a body and determines the motion arising from these forces and moments. In simulation modeling direct dynamics modeling is commonly used and requires inversion of the inertia matrix and the solution of coupled ordinary differential equations.

10.1.6 Induced Acceleration Analysis

A muscle produces moments at the joints it crosses, but these moments can also cause accelerations at joints not crossed by the muscle (Zajac and Gordon 1989). This concept of moments generated at one joint accelerating another joint, not necessarily an adjacent joint, will be referred to as an induced acceleration.

To perform an induced acceleration analysis, the state-space representation of inverse dynamics can be used:

$$T = M(\theta)\ddot{\theta} + v(\theta, \dot{\theta}) + G(\theta) \tag{10.25}$$

where T is a $n \times 1$ vector of joint moments, $M(\theta)$ is the $n \times n$ inertia matrix, $v(\theta, \dot{\theta})$ is an $n \times 1$ vector of centrifugal/Coriolis terms, $G(\theta)$ is a $n \times 1$ vector of gravity terms, and $\theta, \dot{\theta}$, and $\ddot{\theta}$ are the generalized coordinates and their first and second derivatives. In an induced acceleration analysis, the accelerations caused by the centrifugal, Coriolis, and gravity terms are ignored. The analysis focuses on the accelerations due to the muscular moments acting at the joints, so these factors are ignored, which gives the following equation:

$$\ddot{\theta} \approx M(\theta)^{-1}T \tag{10.26}$$

where inversion of the inertia matrix is feasible because the matrix is positive definite (Craig 2005). It should be pointed out that the contributions of the other original terms in the equation can be important for fast movements (Putnam 1993). If in T the moments are set to zero except for the joint(s) of interest, the acceleration can be computed at the other (inactive) joints; these are the induced accelerations. In reality the joints are likely to all have active muscle moments simultaneously and therefore potentially effect one another simultaneously, but to decompose the influence of one joint or joints on other joints for this analysis the assumed active and inactive joints are systematically varied. Kepple et al. (1997) experimentally computed the resultant joint moments during human gait and then using induced accelerations computed the relative contributions of these moments to whole body center of mass support and forward progression. It is feasible to further decompose the moments to the muscles producing these moments, so the acceleration induced by an individual muscle can be computed. For example, Neptune et al. (2001) used a simulation model which estimated the forces produced by the muscles to generate gait. The accelerations induced by these muscle forces were computed to examine their contributions to whole body center of mass support and forward progression.

Challis (2011) derived an induced acceleration index (*IAI*). This index indicates, for a given system, how the orientation and inertial properties of that system influence the ability of any joint in the system to accelerate other joints in the system. The previous equation can be rewritten as:

$$\begin{bmatrix} T_A \\ [0] \end{bmatrix} = [M(\theta)] \cdot \begin{bmatrix} \ddot{\theta}_A \\ \ddot{\theta}_I \end{bmatrix} = \begin{bmatrix} M_{AA} & M_{IA} \\ M_{AI} & M_{II} \end{bmatrix} \cdot \begin{bmatrix} \ddot{\theta}_A \\ \ddot{\theta}_I \end{bmatrix} \tag{10.27}$$

where T_A is the vector of moments applied at active joints, [0] is a vector of zeros for the inactive joints, $\ddot{\theta}_A$ is the vector of accelerations at active joints, $\ddot{\theta}_I$ is the vector of accelerations induced at inactive joints, M_{AA} is a sub-matrix relating active moments to active joints' accelerations, and M_{IA} is the sub-matrix relating inactive moments to active joints' accelerations, and the other sub-matrices are similarly named. As the inertia matrix is positive definite, its sub-matrices must also have this property, so are they are invertible (Golub and Van Loan 1983). The equation can be written for any permutation of active and inactive joints.

Rearrangement of the equation gives:

$$M_{AI}.\ddot{\theta}_A + M_{II}.\ddot{\theta}_I = [0] \tag{10.28}$$

Therefore:

$$\ddot{\theta}_I = -M_{II}^{-1}.M_{AI}.\ddot{\theta}_A \tag{10.29}$$

So a matrix (M_{IAI}) can be defined which dictates the induced accelerations:

$$M_{IAI} = \left| M_{II}^{-1}.M_{AI} \right| \tag{10.30}$$

The matrix M_{IAI} indicates how the active joints in a system can induce acceleration of inactive joints in the system. The components of this matrix are functions of the segment lengths, center of mass locations, masses, moments of inertia, and orientations. For most human activities, all of these factors are constant for a given subject, other than segment orientations. Therefore, the variation in M_{IAI} is caused by changes in segment orientation. If there are m active joints and n inactive joints, then the matrix M_{IAI} will be a n by m matrix. From matrix M_{IAI}, an index can be extracted which indicates, for a given system, how the orientation and inertial properties of that system influence the ability of a joint in the system to accelerate other joints in the system. This index is called the induced acceleration index (*IAI*). For a simple two-joint system, with one joint active and the other inactive, the matrix M_{IAI} is actually a scalar. For more complex systems, M_{IAI} is either a vector or a matrix; in these cases, a scalar can be obtained from:

$$IAI = \|M_{IAI}\|_F \tag{10.31}$$

where $\|A\|_F$ is the Frobenius norm of A. The Frobenius norm is the square root of the sum of the squares of the singular values of a matrix (Lawson and Hanson 1974). In the present context, the singular values can be considered as the gains of the matrix M_{IAI} and thus provide an index of the potential for induced accelerations.

10.1.7 Signal Filtering

Typically, when an inverse dynamics analysis is performed data from a motion analysis system are combined with data from a force plate. The former typically have poorer signal to noise ratio so a common approach is to low-pass filter the force plate data with a higher cut-off frequency than the motion analysis data. Bisseling and Hof (2006) showed that in a landing task the selection of the cut-off frequencies for the motion analysis data and the force plate had a large influence on the computed resultant joint moments. If the cut-off frequency is high for the force plate data and by comparison the cut-off is low for the motion analysis data, some frequency content of the kinematics of impacts are retained in the force data but not in the motion analysis data. This discrepancy between the two signals creates an artifact in the computed moments, where the force and motion data have to be combined. Due to its signal to noise ratio the cut-off frequency for the motion analysis data cannot be too high. This is an area which requires more research, but the best current recommendation is to filter the motion analysis and force plate data with the same filter cut-off.

10.1.8 Other Approaches

Kuo (1998) presented a weighted least-squares approach to estimate resultant joint moments. Two sets of equations were formulated: the first relating joint moments to joint angular accelerations and the second relating these moments to the ground reaction forces. The resulting equations are overdetermined and can be solved using a pseudo-inverse (see Appendix A). His evaluation indicated that this approach produced more accurate joint moments than standard inverse dynamics. One disadvantage of this approach is that the foot had to be in contact with the ground and ground reaction forces measured. The approach of Kuo (1998) was extended by van den Bogert and Su (2008). Their revised approach could handle a flight phase, and incomplete ground reaction force data for a contact phase. These least-squares approaches are attractive but have not yet been generally adopted. They require analysis of the whole body which may create an experimental burden if only, for example, the moments at the wrist, elbow, and shoulder are required. A novel approach is to use a machine learning technique where given the body kinematics the resultant joint moments can be computed (Lv et al. 2016). Such an approach warrants further investigation.

10.2 Energetics

One way of analyzing movement is to compute the energy associated with recorded motion. From this energy analysis work and power can also be determined. These approaches have potential, but problems arise in their implementation, sometimes giving paradoxical results. In the following, the energetics of a particle, rigid body, and system of articulated rigid bodies and then some of the paradoxes which arise from their computation and analysis will be outlined.

10.2.1 General Principles

Energy is recognized in many forms, including thermal, chemical, electric, nuclear, and sound energies. These various forms are simply potential and kinetic energy, where potential energy is the energy possessed by an object because of its position relative to other objects, its electric charge, or stresses within itself. The focus here will be the gravitational potential energy which depends on the distance the object is from a planet. Kinetic energy is a function of the movement of an object. Here the units of energy are joules (J).

There are three laws of thermodynamics; the most relevant to the analysis of human movement is the first. The first law of thermodynamics can be expressed in many forms but proposes three principles:

1. The conservation of energy – energy cannot be created nor destroyed. Energy can change forms.
2. The flow of energy is a transfer of energy – such transfer does not imply a net loss of energy.
3. Work is a form of energy transfer – the work done on a body, or system of bodies, is equal to the change in mechanical energy of that body or system.

This law means that it should be possible to track the change and flow of energies in a system. Consider a simple pendulum (see Fig. 10.7). Why does a pendulum not oscillate forever? What are the sources of energy loss? Most of the loss of pendulum energy is due to friction, so it could be determined and the energy of system tracked. One problem which arises in the analysis of human movement is that there are so many sources of energy, with, for example, the muscles adding energy to the system as can strain in the tendons (Alexander and Bennet-Clark 1977). The problem of tracking all energy flows in human movement is akin to the famous quote from Pierre-Simon Laplace (1749–1827):

> An intellect knowing at any given instant of time, all forces acting in nature, as well as the momentary positions of all things of which the universe consist, would be able to comprehend the motions of the largest bodies of the world and those of the smallest atoms in one single formula, provided it were sufficiently powerful to subject all the data to analysis.

which of course is not feasible; similarly accurately identifying and tracking all energy sources and flows in the human body is problematic.

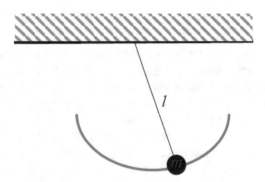

Fig. 10.7 An oscillating simple pendulum

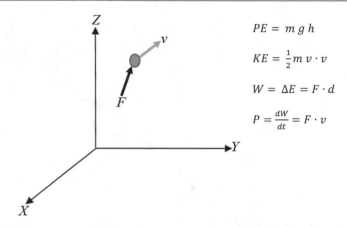

$$PE = m\,g\,h$$

$$KE = \tfrac{1}{2}m\,v \cdot v$$

$$W = \Delta E = F \cdot d$$

$$P = \frac{dW}{dt} = F \cdot v$$

Fig. 10.8 A point mass traveling with a velocity (v), acted on by a force (F). Note the height of the point mass (h) is the Z coordinate of the point mass

10.2.2 A Single Particle

For a single particle in space, it is possible to describe its potential and kinetic energy along with the work and power associated with any force applied to it (see Fig. 10.8).

The gravitational potential energy (PE) is due to work done on a point to counteract the effect of another force in this case gravity; it is the product of the object's mass (m), acceleration due to gravity (g), and the point mass height (h), in this case the position in the Z direction.

The kinetic energy is the capacity of a particle to do work due to its motion; it is the product of half, the mass of the point mass, and its velocity squared. The velocity squared is computed from the dot product of the velocity with itself.

The work done (W) on the point mass is equal to its change in mechanical energy. It describes the extent to which a force can move the point mass. It can be computed from the dot product of the applied force and the distance (d) over which that force acts. See Appendix L for details of the dot product.

Power (P) is the rate of doing work with respect to time, in which case the power is computed by differentiating the work done with respect to time. For a point mass, the power can be computed from the dot product of the applied force and its velocity.

The human body is not a point mass, but studies have computed the power generated during human movement by considering the whole body center of mass and taking the product of the ground force and the velocity of the center of mass (e.g., Cavagna et al. 1964; Davies and Rennie 1968). But in the definition of power for a point mass, the force acts directly on the point mass, but the ground reaction force acts at the foot not on the whole body center of mass. In most situations, the line of action of the ground reaction force vector does not intersect on the center of mass. When power is computed from the product of the ground reaction force and the center of mass, it assumed that this force is responsible for the generation of all power.

10.2.3 A Rigid Body

For a rigid body in space, it is possible to describe its potential and kinetic energy along with the work and power associated with any force applied to it (see Fig. 10.9).

The gravitational potential energy (PE) is due to work done on a body to counteract the effect of another force in this case gravity; it is the product of the object's mass (m), acceleration due to gravity (g), and height of the center of mass of the segment (h), in this case the position in the Z direction.

The kinetic energy is the capacity of a body to do work due to its motion. For a rigid body, there are two components. The first is its linear component which is computed from the product of half, the rigid body mass, and its center of mass velocity squared. The second is its angular component which is computed from the product of half, the inertia tensor (J) of the body about its principal axes, and its angular velocity squared.

The work done (W) on the rigid body is equal to its change in mechanical energy in both a linear and angular sense. It describes the extent to which a force and moment can move the body. It has two components: linear and angular. The linear

Fig. 10.9 A rigid body traveling with a center of mass velocity (v_{CM}), with an angular velocity (ω), acted on by a force (F), and moment (T). Note the height of the center of mass (h) is the Z coordinate of the center of mass

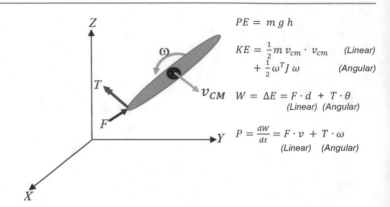

$$PE = mgh$$

$$KE = \tfrac{1}{2}m\,v_{cm} \cdot v_{cm} \quad \text{(Linear)}$$
$$+ \tfrac{1}{2}\omega^T J\,\omega \quad \text{(Angular)}$$

$$W = \Delta E = F \cdot d + T \cdot \theta$$
$$\quad\quad\quad\;\;\text{(Linear) (Angular)}$$

$$P = \tfrac{dW}{dt} = F \cdot v + T \cdot \omega$$
$$\quad\quad\quad\quad\text{(Linear) (Angular)}$$

component can be computed from the dot product of the applied force and the distance (d) over which that force acts. The angular component can be computed from the dot product of the applied moment and the angular distance (θ) over which that moment acts.

Power (P) is the rate of doing work with respect to time, in which case the power is computed by differentiating the work done with respect to time. For a rigid body, power has two components: linear and angular. The linear component can be computed from the dot product of the applied force and its velocity. The angular component can be computed from the dot product of the applied moment and its angular velocity.

Winter (1979) computed the sum of the kinetic and potential energies for each segment of the human body during human locomotion. To compute total body mechanical energy, the segment energies were computed:

$$E_{Segment} = m_{Seg}\, g\, h_{Seg} + \frac{1}{2} m_{Seg}\, v_{Seg} \cdot v_{Seg} + \frac{1}{2} \omega_{Seg} J_{Seg}\, \omega_{Seg} \tag{10.32}$$

The total body mechanical energy was then computed from:

$$E_{Body} = \sum_{i=1}^{ns} \left| E_{Segment,i} \right| \tag{10.33}$$

where ns is the number of body segments. This approach was used to examine the metabolic cost of locomotion relative to the mechanical costs.

10.2.4 A System of Rigid Bodies

For a system of rigid bodies in space, it is possible to describe its potential and kinetic energy along with the work and power associated with its motion (see Fig. 10.10).

For the system, the total gravitational potential energy (PE) is the sum of the work done on each body to counteract the effect of another force in this case gravity.

For a system of rigid bodies, the total kinetic energy is computed using König's theorem. The theorem states that the kinetic energy of a system of rigid bodies is the sum of the kinetic energy associated with the movement of the center of mass of the system and the kinetic energy associated with the movement of the bodies relative to the center of mass. Therefore, there are two components of kinetic energy: the kinetic energy due to the motion of the whole body center of mass which is the product of half, the whole body mass (m_{WB}), and the velocity of the whole body center of mass ($v_{WB,\ CM}$) squared. Then there is the peripheral energy component which is the sum of the linear kinetic energies of the segments computed from the product of half, the segment mass (m_i), and the velocity of the segment center of mass ($v_{R-CM,\ i}$) squared where the velocity of the segment is relative to the whole body center of mass. The other peripheral energy component is the sum of the angular kinetic energies of the segments computed from the product of half, the segment moment of inertias (I_i), and the angular velocity of the segment (ω_i) squared.

Fig. 10.10 Two jointed rigid bodies each with a center of mass velocity ($v_{CM,\,i}$), with an angular velocity (ω_i), and with both a force (F_J) and moment (T_J) acting at the joint. Note the height of each center of mass (h_i) is the Z coordinate of each center of mass

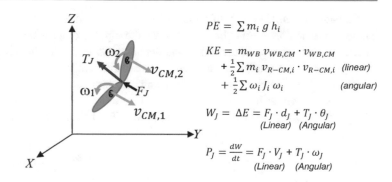

$$PE = \sum m_i\, g\, h_i$$

$$KE = m_{WB}\, v_{WB,CM} \cdot v_{WB,CM}$$
$$+ \tfrac{1}{2}\sum m_i\, v_{R-CM,i} \cdot v_{R-CM,i} \quad \text{(linear)}$$
$$+ \tfrac{1}{2}\sum \omega_i\, J_i\, \omega_i \qquad\qquad \text{(angular)}$$

$$W_J = \Delta E = \underset{\text{(Linear)}}{F_J \cdot d_J} + \underset{\text{(Angular)}}{T_J \cdot \theta_J}$$

$$P_J = \tfrac{dW}{dt} = \underset{\text{(Linear)}}{F_J \cdot V_J} + \underset{\text{(Angular)}}{T_J \cdot \omega_J}$$

For two segments with a common joint, the work done (W) on the system is computed with linear and angular components. The linear component can be computed from the dot product of the joint reaction force and the displacement of the joint (d). The angular component can be computed from the dot product of the joint moment and the angular distance (θ) over which that moment acts.

Power (P) is the rate of doing work with respect to time, in which case the power is computed by differentiating the work done with respect to time. For a rigid body system comprising two segments with a common joint, the power has two components: linear and angular. The linear component can be computed from the dot product of the joint reaction force and its velocity. The angular component can be computed from the dot product of the joint moment and the joints' angular velocity.

Positive work at a joint is considered to indicate that the net effect of the muscles crossing the particular joints is generating mechanical energy, or if there is negative work at a joint, the net effect of the muscles crossing that joint is dissipating mechanical energy.

10.2.5 Paradoxes

The analysis of human movement using energetics is an interesting proposition, and while there have been studies which have used this approach and provided insights to human movement, the theoretical basis of the analysis is not currently completely described. The following examples illustrate some of the issues with energetic analysis.

Devita et al. (2007) examined the energetics of human walking. They had 34 subjects walk at a horizontal center of mass velocity of 1.5 m.s^{-1} and computed the work associated with the support phase of the gait cycle. Theirs was a sagittal plane analysis with joint power computed for the ankle, knee, and hip joints from the product of their respective resultant joint moments and their angular velocities. Their results could be criticized because they are sagittal plane only, but Eng and Winter (1995) have provided evidence that this assumption probably only makes a difference at the hip joint. Indeed, analysis of the power curves of Eng and Winter (1995) gives similar results to Devita et al. (2007). Devita et al. (2007) highlighted that the absolute value of negative work during the stance phase did not match the positive work, yet in steady-state locomotion, these work values should match. Zelik et al. (2015) were able to find a greater correspondence between negative and positive work during steady-state walking, with a less than 1% difference. Their analysis took into account not only the rotational powers ($T_J \cdot \omega_J$) at the ankle, knee, and hip joints but also the linear powers ($F_J \cdot V_J$). The remaining negative work, which dissipates energy, is likely due to soft tissue motion during the impacts, which has been shown to effectively dissipate energy (Pain and Challis 2002).

In isolated muscle preparations, approximately 25% of chemical energy used produces mechanical work by the muscles; the majority of the chemical energy generates heat (Woledge et al. 1985). An issue which has intrigued researchers is the relationship in vivo between chemical energy expenditure and the mechanical energy produced during gait. Such analysis gets at concepts such as efficiency (the ratio of work done to energy expenditure), and should permit assessment of efficient and inefficient movement patterns, and how efficiency may perhaps be improved (Cavanagh and Kram 1985). Unfortunately, there is no agreed-upon method for measuring mechanical energy expenditure. Williams and Cavanagh (1983) compared a number of different methods for estimating mechanical energies for steady state running and produced estimated mechanical power values which ranged from 273 to 1775 W. Concurrent measurements of metabolic cost, using analysis of the volume of oxygen in inspired and expired air, produce efficiency ratios from 0.31 to 1.97. The approach of Zelik et al. (2015) has promise for resolving some of these issues, but to perform a joint-by-joint analysis of both linear and angular power is problematic, particularly for the trunk. Then the factors which are ignored, such as the contribution of soft tissue motion and the energetics of footwear, present additional problems.

Consider a simple system of a subject standing still but flapping their arms up and down, but out of phase with one another. Therefore, when one arm is heading toward the hip, the other is heading toward pointing at the sky. If the right arm is being raised, the joint powers at the right shoulder are positive, and in the left arm which is being lowered the shoulder has negative joint power. If the motion of the two limbs is sufficiently synchronized then the sum of the shoulder joint powers is zero, which is hard to match to the corresponding metabolic cost for the motions. The nature of the movement means there is no motion of the whole body center of mass. Therefore, we have motion but the total power is zero. Van Ingen Schenau and Cavanagh (1990) proposed that the total power (P_{Tot}) should be computed from:

$$P_{Tot} = \sum_{i=1}^{nj} P_i \qquad (10.34)$$

where nj is the number of body segments and P_i is the power at an individual joint. In contrast Blickhan and Full (1992) took the absolute values of the joint powers:

$$P_{Tot} = \sum_{i=1}^{nj} |P_i| \qquad (10.35)$$

This approach has an intuitive appeal because at least there is power present in the arm flapping example. But the sign of the power values indicates how power is distributed through the body, but it is then ignored in summing absolute joint powers.

One aspect of power which is interesting is the effort in the SI system to preserve an old unit of power output. In the English system of units one measure of power was the horsepower (hp), a unit first promulgated by James Watt designed to show how his engines were more powerful than a horse. In the English system of units one horsepower is equal to 745.7 watts, but the SI unit of one horsepower is equal to 735.499 watts. Indeed Watt underestimated the power-producing capabilities of a horse (Stevenson and Wassersug 1993), whether this was an error in his computations or an advertising ploy is not clear.

Independent of debates on how to compute and quantify human body energetics, the issues which arise with regard to the ability to experimentally measure pertinent system kinematics are important factors to consider. The experimental challenges include the following:

1. The power at a joint is typically computed from the dot product of the moment at the joint and the joints' angular velocity. The common presumption is that the moment is due to muscular activity. This net moment reflects the activity of the agonist to produce the moment, but these muscles also overcome any antagonist activity. Such co-contractions are ubiquitous part of muscle activity during everyday tasks (e.g., da Fonseca et al. 2005; Meulenbroek et al. 2005).
2. It is common when computing joint powers to ignore the linear component of the joint power (dot product of the joint reaction force and joint velocity). It was been shown including this component is important when fully accounting for joint energetics (Zelik et al. 2015). In normal locomotion, the feet are normally shod, meaning measuring the variables required for the computation of the both linear and joint powers of the joints of the feet is not readily measurable.
3. The soft tissue associated with a body segment moves independently of the underlying bone. This soft tissue motion makes a contribution to the total segment energy and seems to be effective for the dissipation of energy (e.g., Pain and Challis 2002).
4. There are tissues in the human body which can store and return mechanical energy (Alexander and Bennet-Clark 1977). This elastic energy can have an impact on the energetics of human movement (Wager and Challis 2016).
5. During breathing, there is motion of the thorax as its volume changes (Massaroni et al. 2018). Such motion requires energy but is not accounted for in typical energetic analysis of human movement.

10.2.6 Conclusions

In light of the problems of adequately defining and measuring the energetics of human movement, this approach has some limitations and should be used with caution. The mechanical power associated with human movement is not necessarily indicative of performance. The estimated mechanical power is not a surrogate for the power generated by the muscles. Finally, mechanical work does directly relate to the metabolic energy costs. Further research is needed to able to measure and track energy sources in human movement.

10.3   Review Questions

1. Write the equation which describes the relationship between all the forces applied to a segment and the acceleration of that segment center of mass and the equation which describes the relationship between all the moments applied to a segment and the angular acceleration of that segment about its center of mass.
2. The resultant joint moment is a result of a number of different forces applied to a segment; what are these forces?
3. Define a parameter and variable, and then list the parameters and variables required to compute a resultant joint moment. How can each of the parameters and variables be computed?
4. What are the assumptions made when computing resultant joint moments?
5. How can the distal forces and moments be determined?
6. Define and give the units for kinetic energy, potential energy, work, and power.
7. For the human body, it is hard to compute the total energy, explain why.
8. What are the problems when trying to compute the efficiency of human movement?

References

Cited References

Alexander, R. M., & Bennet-Clark, H. C. (1977). Storage of elastic strain energy in muscle and other tissues. *Nature, 265*, 114–117.

Bell, A. L., Brand, R. A., & Pedersen, R. (1989). Prediction of hip joint Centre location from external landmarks. *Human Movement Science, 8*(1), 3–16.

Bergmann, G., Deuretzbacher, G., Heller, M., Graichen, F., Rohlmann, A., Strauss, J., & Duda, G. N. (2001). Hip contact forces and gait patterns from routine activities. *Journal of Biomechanics, 34*(7), 859–871.

Bisseling, R. W., & Hof, A. L. (2006). Handling of impact forces in inverse dynamics. *Journal of Biomechanics, 39*(13), 2438–2444.

Blankevoort, L., Huiskes, R., & de Lange, A. (1990). Helical axes of passive knee joint motions. *Journal of Biomechanics, 23*(12), 1219–1229.

Blickhan, R., & Full, R. J. (1992). Mechanical work in terrestrial locomotion. In A. A. Biewener (Ed.), *Biomechanics - structures and systems: A practical approach* (pp. 75–95). New York: IRL Press.

Cavagna, G. A., Saibene, F., & Maragaria, R. (1964). Mechanical work in running. *Journal of Applied Physiology, 19*(2), 249–256.

Cavanagh, P. R., & Kram, R. (1985). Mechanical and muscular factors affecting the efficiency of human movement. *Medicine and Science in Sports and Exercise, 17*(3), 326-331.

Challis, J. H. (2011). An induced acceleration index for examining joint couplings. *Journal of Biomechanics, 44*(12), 2320–2322.

Craig, J. J. (2005). *Introduction to robotics: Mechanics and control* (3rd ed.). Reading, MA: Addison-Wesley Publishing Company, Inc.

da Fonseca, S. T., Vaz, D. V., de Aquino, C. F., & Bricio, R. S. (2005). Muscular co-contraction during walking and landing from a jump: Comparison between genders and influence of activity level. *Journal of Electromyography and Kinesiology, 16*(3), 273–280.

Davies, C. T., & Rennie, R. (1968). Human power output. *Nature, 217*(5130), 770–771.

DeVita, P., Helseth, J., & Hortobagyi, T. (2007). Muscles do more positive than negative work in human locomotion. *Journal of Experimental Biology, 210*(Pt 19), 3361–3373.

Elftman, H. (1939). The function of muscles in locomotion. *American Journal of Physiology, 125*(2), 357–366.

Eng, J. J., & Winter, D. A. (1995). Kinetic analysis of the lower limbs during walking: What information can be gained from a three-dimensional model? *Journal of Biomechanics, 28*(6), 753–758.

Golub, G. H., & Van Loan, C. F. (1983). *Matrix computations*. Oxford: North Oxford Academic Publishing Company Ltd.

Kane, T. R., & Levinson, D. A. (1985). *Dynamics: Theory and applications*. New York: McGraw-Hill.

Karduna, A. R., Williams, G. R., Williams, J. L., & Iannotti, J. P. (1997). Glenohumeral joint translations before and after total shoulder arthroplasty. A study in cadavera. *Journal of Bone and Joint Surgery, A79*(8), 1166–1174.

Kepple, T. M., Siegel, K. L., & Stanhope, S. J. (1997). Relative contributions of the lower extremity joint moments to forward progression and support during gait. *Gait & Posture, 6*(1), 1–8.

Kirkwood, R. N., Culham, E. G., & Costigan, P. (1999). Radiographic and non-invasive determination of the hip joint center location: Effect on hip joint moments. *Clinical biomechanics, 14*(4), 227–235.

Kuo, A. D. (1998). A least-squares estimation approach to improving the precision of inverse dynamics computations. *Journal of Biomechanical Engineering, 120*(1), 148–159.

Kurz, T., Eberhard, P., Henninger, C., & Schiehlen, W. (2010). From Neweul to Neweul-M2: Symbolical equations of motion for multibody system analysis and synthesis. *Multibody System Dynamics, 24*(1), 25–41.

Lawson, C. L., & Hanson, R. J. (1974). *Solving least squares problems*. New York: Prentice-Hall, Inc.

Lewis, F. L., Abdallah, C. T., & Dawson, D. M. (1993). *Control of robot manipulators*. New York: Macmillan Publishing Company.

Lot, R., & Da Lio, M. (2004). A symbolic approach for automatic generation of the equations of motion of multibody systems. *Multibody System Dynamics, 12*(2), 147–172.

Luh, J. Y. S., Walker, M. W., & Paul, R. P. C. (1980). On-line computational scheme for mechanical manipulators. *Journal of Dynamic Systems, Measurement, and Control, 102*(2), 69–76.

Lv, X., Chai, J., & Xia, S. (2016). Data-driven inverse dynamics for human motion. *ACM Transactions on Graphics, 35*(6), 163.

Massaroni, C., Piaia Silvatti, A., Levai, I. K., Dickinson, J., Winter, S., Schena, E., & Silvestri, S. (2018). Comparison of marker models for the analysis of the volume variation and thoracoabdominal motion pattern in untrained and trained participants. *Journal of Biomechanics, 76*, 247–252.

Meulenbroek, R. G., Van Galen, G. P., Hulstijn, M., Hulstijn, W., & Bloemsaat, G. (2005). Muscular co-contraction covaries with task load to control the flow of motion in fine motor tasks. *Biological Psychology, 68*(3), 331–352.

Morrison, J. B. (1968). Bioengineering analysis of force actions transmitted by the knee joint. *Bio-medical Engineering, 11*(5), 164–170.

Neptune, R. R., Kautz, S. A., & Zajac, F. E. (2001). Contributions of the individual ankle plantar flexors to support, forward progression and swing initiation during walking. *Journal of Biomechanics, 34*(11), 1387–1398.

Pain, M. T. G., & Challis, J. H. (2001). A high resolution technique for determining body segment inertial parameters and their variation due to soft tissue motion. *Journal of Applied Biomechanics, 17*(4), 326–334.

Pain, M. T. G., & Challis, J. H. (2002). Soft tissue motion during impacts: Their potential contributions to energy dissipation. *Journal of Applied Biomechanics, 18*(3), 231–242.

Pain, M. T. G., & Challis, J. H. (2006). The influence of soft tissue movement on ground reaction forces, joint torques and joint reaction forces in drop landings. *Journal of Biomechanics, 39*(1), 119–124.

Paul, J. P. (1978). Torques produce torsion. *Journal of Biomechanics, 11*(1), 87.

Piazza, S. J., Erdemir, A., Okita, N., & Cavanagh, P. R. (2004). Assessment of the functional method of hip joint center location subject to reduced range of hip motion. *Journal of Biomechanics, 37*(3), 349–356.

Putnam, C. A. (1993). Sequential motions of body segments in striking and throwing skills: Descriptions and explanations. *Journal of Biomechanics, 26*(Suppl 1), 125–135.

Scott, S. H., & Winter, D. A. (1990). Internal forces at chronic running injury sites. *Medicine and Science in Sports and Exercise, 22*(3), 357-369.

Silver, W. M. (1982). On the equivalence of Lagrangian and Newton-Euler dynamics for manipulators. *The International Journal of Robotics Research, 1*(2), 60–70.

Stevenson, R. D., & Wassersug, R. J. (1993). Horsepower from a horse. *Nature, 364*, 195.

van den Bogert, A. J., & Su, A. (2008). A weighted least squares method for inverse dynamic analysis. *Computer Methods in Biomechanics & Biomedical Engineering, 11*(1), 3–9.

Van Ingen Schenau, G. J., & Cavanagh, P. R. (1990). Power equations in endurance sports. *Journal of Biomechanics, 23*(9), 865–882.

Vaughan, C. L. (1996). Are joint torques the holy grail of human gait analysis? *Human Movement Science, 15*, 423–443.

Wager, J. C., & Challis, J. H. (2016). Elastic energy within the human plantar aponeurosis contributes to arch shortening during the push-off phase of running. *Journal of Biomechanics, 49*(5), 704–709.

Weinhandl, J. T., & O'Connor, K. M. (2010). Assessment of a greater trochanter-based method of locating the hip joint center. *Journal of Biomechanics, 43*(13), 2633–2636.

Williams, K. R., & Cavanagh, P. R. (1983). A model for the calculation of mechanical power during distance running. *Journal of Biomechanics, 16*(2), 115–128.

Winter, D. A. (1979). A new definition of mechanical work done in human movement. *Journal of Applied Physiology, 46*(1), 79–83.

Winter, D. A. (1980). Overall principle of lower limb support during stance phase of gait. *Journal of Biomechanics, 13*(11), 923–927.

Woledge, R. C., Curtin, N. A., & Homsher, E. (1985). *Energetic aspects of muscle contraction*. London: Academic Press.

Zajac, F. E., & Gordon, M. E. (1989). Determining muscle's force and action in multi-articular movement. *Exercise and Sport Sciences Reviews, 17*, 187–230.

Zelik, K. E., Takahashi, K. Z., & Sawicki, G. S. (2015). Six degree-of-freedom analysis of hip, knee, ankle and foot provides updated understanding of biomechanical work during human walking. *The Journal of Experimental Biology, 218*(6), 876–886.

Useful References

Andrews, J. G. (1982). On the relationship between resultant joint torques and muscular activity. *Medicine and Science in Sports and Exercise, 14*(5), 361-367.

Mommaerts, W. F. H. M. (1969). Energetics of muscular contraction. *Physiological Reviews, 49*, 427–508.

Winter, D. A., & Eng, P. (1995). Kinetics: Our window into the goals and strategies of the central nervous system. *Behavioural Brain Research, 67*, 111–120.

Winter, D. A., Patla, A. E., Frank, J. S., & Walt, S. E. (1990). Biomechanical walking pattern changes in the fit and healthy elderly. *Physical Therapy, 70*(6), 340–347.

Winter, D. A., & Robertson, D. G. E. (1978). Joint torque and energy patterns in normal gait. *Biological Cybernetics, 29*, 137–142.

Error Analysis

Error Analysis **11**

Overview

The English author G.K. Chesteron (1874–1936) wrote,

> It is human to err; and the only final and deadly error, among all our errors, is denying that we have ever erred.

Any data collected will be corrupted by errors; it is important to quantify these errors as the magnitude of the errors will influence the interpretation of the data. Errors arise in all four stages of the experimental process:

Calibration
Acquisition
Data analysis
Data combination

Calibration is required for any measurement device. Calibration is the determination of the coefficients of the model of the measurement device. The model of the experimental device might very simply comprise the stiffness coefficient for a spring being used to measure a force but might consist of multiple coefficients for modeling a series of cameras for close-range three-dimensional photogrammetry. Errors due to calibration can arise from two sources: the measurement system model, and errors in calibration data. The model of the measurement system, as with all models, will be not be perfect so introduces some errors. In 1982 Dapena et al. proposed a model for close-range three-dimensional photogrammetry; in the write-up of the model there were errors in four equations; paradoxically when Dapena (1985) compared the corrected model with the incorrect model the results were more accurate for the incorrect model! To perform calibration requires a standard with which to "tune" the model calibration coefficients. No perfect standard for calibration exists; for example, with a force plate system the weights that might be used for calibration are not known to infinite accuracy. In each stage of the experimental process, it is important to minimize errors; efforts to minimize these errors starts with appropriate calibration.

Data acquisition errors arise from a number of sources. For example, if the data are not collected at a sufficiently high sample rate, then aliasing will occur and the collected data will be corrupted. Markers are often placed on the skin to track underlying bone motion, but due to muscular actions and impact-generated shock waves these markers can move relative to the bones whose motion they are designed to track (e.g., Fuller et al. 1997). Errors that arise during data acquisition occur either because of incorrect decisions in designing data collection (e.g., too low a sample rate) or because conditions change compared to when calibration was performed.

Data analysis is the processing of the sampled data to have the data in a more usable form. This might include data filtering to reduce the noise that might corrupt sampled data, which in theory should reduce the errors in the sampled data. Although the veracity of the data can be improved during data analysis, imprudent decisions, for example, under or overfiltering data, can cause a reduction in data quality.

In data combination, many mechanical quantities require the combination parameters and variables. For example, the computation of resultant joint moments requires the combination of data from a variety of sources, including anthropometry, motion analysis, and data differentiation. Each of the parameters and variables will contain errors, and as these are combined, the errors are propagated.

11

In the following sections, it is the intention to review the ways in which errors can be analyzed and assessed, and examples will be presented for specific biomechanical analyses.

The content of this chapter requires a working knowledge of binary representations of numbers (see Appendix C) and partial differentiation (See Appendix J).

11.1   Accuracy, Precision, and Resolution

When referring to a measurement, reference is often made to the accuracy, precision, and resolution of the measurement. Here these terms will be defined, and their quantification explained.

Accuracy   Accuracy is quantified as the difference between a true value and an observed value. Accuracy quantifies the bias in a measure caused by measurement error. High accuracy would be a measure with a small deviation from the true value, and low accuracy would be a measure a large deviation from the true value. Accuracy is quantified by measuring the deviation between measures and a criterion; as such this is really a measure of inaccuracy. To illustrate the quantification of accuracy consider the data in Table 11.1.

So one commonly adopted approach given such data is to compute the mean error ($\bar{x}$):

$$\bar{x} = \frac{1}{n} \sum_{i=1}^{n} \Delta x_i \tag{11.1}$$

where n is the number of measures and Δx_i is difference between a measure and the criterion. For the data in Table 11.1, this gives an accuracy estimate of -2.9 N. This value informs us that the force plate system tends to underestimate the true value of the vertical ground reaction force. But a cursory look at the deviations suggests that this estimate of accuracy is low relative to the individual error values. This occurs because the negative error values can be canceled by positive error values during averaging. Therefore a better criterion could be the absolute mean error ($\bar{x}_{Abs}$):

$$\bar{x}_{Abs} = \frac{1}{n} \sum_{i=1}^{n} |\Delta x_i| \tag{11.2}$$

which for these data gives an accuracy value of 8.1 N, which reflects the magnitude of the recorded errors more appropriately than the mean error. Another criterion to consider is the root mean square error (RMSE), computed using:

$$\text{RMSE} = \sqrt{\frac{\sum (\Delta x_i)^2}{n}} \tag{11.3}$$

which for the data in Table 11.1 gives a RMSE of 9.9 N. As the least conservative estimate, the RMSE is a good way of estimating accuracy, but remember that the mean error has some utility as it does indicate the direction of any bias. Whenever accuracy is assessed, there is a problem of finding a suitable criterion or an absolute standard with which to compare measured values (Crease 2011).

Table 11.1 Ten measures of a reference weight measured by a force plate system; here is the criterion and measured vertical ground reaction force

Measure #	Criterion (N)	Measure (N)	Error (N)	Absolute error (N)
1	500	489	−11	11
2	500	502	2	2
3	500	497	−3	3
4	500	481	−19	19
5	500	510	10	10
6	500	502	2	2
7	500	491	−9	9
8	500	511	11	11
9	500	501	1	1
10	500	487	−13	13

Table 11.2 Ten measures of the mass of a human thigh

Measurement #	1	2	3	4	5	6	7	8	9	10
Thigh mass (kg)	10.08	9.67	9.78	9.29	9.41	9.45	9.72	9.44	9.64	9.09

Precision If a system measures something multiple times, then the results should be invariant; any deviations in the measured values would indicate lack of precision. Therefore precision is the difference between an observed and an expected mean value. To illustrate the quantification of precision, consider the data in Table 11.2.

Precision of these measures of thigh mass can be determined by computing the standard deviation (σ):

$$\sigma = \sqrt{\frac{1}{n-1}\sum(x_i - \bar{x})^2} \tag{11.4}$$

As the true value of thigh mass is unknown and is therefore estimated from the mean of the data, one degree of freedom is lost, so $n - 1$ is used. For the thigh mass estimates, the precision is 0.28 kg. This precision could also be expressed as the coefficient of variation:

$$CV = 100\,\frac{\sigma}{\bar{x}}\% \tag{11.5}$$

which for these data is 2.9%.

In some cases there maybe multiple estimates of precision, for example, if multiple loads were applied multiple times to a force plate or if multiple thighs were measured multiple times. For a group of m precision, estimates the overall precision (σ_{Group}) can be computed from:

$$\sigma_{Group} = \sqrt{\frac{1}{m}\sum_{i=1}^{m}\sigma_i^2} \tag{11.6}$$

Note the use of variance here for averaging rather than standard deviation; this is because the variance gives a better estimate of a Gaussian distribution. Then the group coefficient of variation is computed from:

$$CV_{Group} = 100\left(\sigma_{Group}\frac{1}{\frac{1}{m}\sum_{i=1}^{m}\bar{x}_i}\right)\% \tag{11.7}$$

Precision estimates in theory have no upper limit but have a lower limit as the precision estimates cannot be negative. Therefore if the confidence interval for precision estimates is to be estimated, it is not appropriate to take ± 2 times the standard deviation and get the 95% confidence interval. The confidence interval can be estimated using the chi-square (χ^2) distribution. For example, if the 95% confidence interval is required, then:

$$\sigma^2\,\frac{df}{\chi_{1-\frac{a}{2},df}^2} < \sigma^2 < \sigma^2\,\frac{df}{\chi_{\frac{a}{2},df}^2} \tag{11.8}$$

where df are the degrees of freedom which is total number of measures minus 1, α represents the confidence interval which for 95% would be equal to 0.05, and $\chi_{\alpha,df}^2$ is the critical chi-square value for a given level and degrees of freedom (standard statistical tables provide these values). For example, for the data in Table 11.2, and a 95% confidence interval:

$$0.28^2\,\frac{9}{\chi_{0.975,9}^2} < 0.28^2 < 0.28^2\,\frac{9}{\chi_{0.025,9}^2} \tag{11.9}$$

where $\chi_{0.975,9}^2 = 16.92$, and $\chi_{0.025,9}^2 = 2.70$ therefore,

$$0.20 < 0.28 < 0.51 \; kg \tag{11.10}$$

Thus a confidence interval is computed; notice that it is asymmetrical. The question arises: what is the appropriate level of precision? Generally this is a function of the study being performed and the nature of the comparisons to be made.

The terms precision and accuracy are sometimes used interchangeably, but they are distinct quantities. An example helps illustrate this difference; in an experiment you are measuring the point of peak pressure beneath the feet during standing. If a pedobarograph consists of one sensor only, the system could be very precise but accuracy in the identification of pressure distributions will be very poor.

Resolution Resolution is the "fineness" with which a measure can be made. It reflects for any measurement instrument the limit on the measured quantity which produces a change in the output of the instrument. For data collection equipment, resolution is expressed as a value or percentage of the instruments full measuring range. For example, for a force plate a sandbag is placed on the force plate, and the measured ground reaction force is its weight; if another grain of sand is added to the sand-bag a change in the output from the force plate equal to weight of a grain of sand would reflect the systems resolution. Given that a typical grain of sand is ~0.7 mg, its weight would be tiny and likely below the resolution of a force plate. It is important to appreciate that the output from a measurement instrument may include a large number of digits, but this does not necessarily reflect high resolution as a component of these values may reflect system noise.

11.2 Uncertainty

Airy (1875) introduced the concept of measurement uncertainty, where the uncertainty reflects the possible value that an error may have. All measurements contain potential deviations from the true value due to measurement system inaccuracy and system imprecision; therefore uncertainty can be expressed as:

$$\text{measured value} \pm \text{uncertainty} \tag{11.11}$$

In certain circumstances by basic probability theory the measured value might be the same as the true value, but this is seldom the case. The question becomes how to estimate uncertainty, particularly given measured values which will have a lack of accuracy and precision. For many measurement protocols, the accuracy of measurement is much greater than the precision, in which case the uncertainty is reflected predominantly by the precision.

11.3 General Sources of Errors

It is reasonable to expect variation in any collected data, or datum, between the true values of that data and that measured. The difference is referred to as an error. One definition of error is that a mistake has been made, but errors in measurements should generally not be considered as mistakes but differences that arises due to the measurement equipment and measurement process. There is one exception to this caveat and that is gross errors. Gross errors occur when a mistake or blunder has been made in the process of the experiment; often the data from such experiments have to be discarded. The sources of errors discussed in the remainder of this section will focus on those errors sources which can be described as noise.

A sampled value can be considered to consist of two parts: the true value and the noise that corrupts it:

$$\text{measured value} = \text{true value} + \text{noise} \tag{11.12}$$

The noise can be considered as the undesirable part of the signal, and careful experimental protocols will help to minimize the noise but never fully eliminate it. There are various types of noise, most of which will contaminate any measurement, for example, noise that arises from using electronic devices (Milburn and Sun 1998); the predominant types of noise are described in the following.

Systematic Noise This source of noise varies systematically and is correlated in some way with the measurement process or the thing being measured. If, for example, a motion analysis system was calibrated using a rod with markers on either end assumed to be 1 meter apart but the markers were actually 0.99 m apart this would produce a systematic error associated with the measurement process. If markers were placed on the subject but those markers move with respect to the underlying bones, they are designed to track this and can introduce systematic noise as the motion of the markers independent of the bone motion would be in part dependent on the task being analyzed. For example, the markers move relative to the underlying the bones due to impacts (e.g., Pain and Challis 2002). Many sources of systematic noise can be modeled during the calibration process, and removed from the sampled signal.

White Noise This is random in the sense that the noise signal is stationary and has a mean value of zero and a flat power spectrum (but across all of the spectrum). White light contains all the wavelengths of the visible spectrum, and this parallels white noise which has frequency components all across the spectrum. Quantification of the variations in human movement has indicated that some movement signals can also have white noise like properties (e.g., Winter and Challis 2017), and white noise can also arise from the electronic devices used for data collection due to thermal and shot noise.

Pink Noise This is the noise that varies inversely with frequency, therefore is often referred to as 1/f noise (Marinari et al. 1983) and sometimes as flicker noise. As pink noise reduces in amplitude with increasing frequency, its influence is typically greater at the lower end of the frequency spectrum. It was first noticed in vacuum tubes but then identified in semiconductors and occurs in many electronic devices (Keshner 1982).

Brown Noise This has a power spectrum where the signal power is inversely proportional to the frequency squared ($1/f^2$); it has much greater amplitude at low frequencies than pink noise. The patterns in Brown noise have been described as Brownian motion; it is also described as a random walk and is found in biology, for example, in the motion of single-celled organisms (Berg 1983). It was first identified by the Scottish botanist Robert Brown (1773–1858), thus the name Brown noise. Brown noise can be produced by integrating white noise.

Thermal Noise This is caused by the random thermal motion of charge carriers, typically electrons, inside an electrical conductor. This noise arises independent of the applied voltage. Thermal noise is a close approximation to white noise. It is sometimes called Johnson or Nyquist noise.

Shot Noise Electric current is carried by electrons, which have discrete arrival times. In each interval the number of particles in motion fluctuates; thus shot noise arises. The smaller the current, the larger the fluctuations. Shot noise has a flat spectrum, so manifests as white noise. Shot noise in electronic devices arises from the unavoidable random statistical fluctuations in the electric current when the electrons, carrying the charge, traverse a gap.

In electrical circuits there are five common noise sources:

- Pink noise
- Thermal noise
- Shot noise
- Burst noise
- Avalanche noise

Both burst and avalanche noise do not normally produce problems in the quality of sampled data.

11.4 Error Propagation

Many mechanical variables require two or more variables or parameters to be combined, for example, body mass and velocity to compute momentum, so the derived mechanical variable has two error sources. Error propagation quantifies how the errors in derived variables are influenced by the parameters and variables used to compute the derived variables.

For all parameters and state variables there will be an errors associated with their estimation, and there will be propagation of these errors in any value determined from their combination. There are two approaches for the examining the propagation of errors: use of the error propagation equations, and sensitivity analysis. In the following each of these will be described.

The error in any value determined from the combination of parameters or state variables can be expressed mathematically (Barford 1985). If the required state variable is P, which is a function of n state variables and or parameters (X_i), then:

$$P = f(X_1, X_2, \ldots, X_n) \tag{11.13}$$

There are errors associated with each of the variables/parameters (δX_i); the influence of each of these errors on the derived variable is determined by taking the partial derivative of the function with respect to the variable (see Appendix J for details on taking partial derivatives). Therefore the overall error associated with the derived variable (δP) can be computed from:

$$(\delta P)^2 = \sum_{i=1}^{n} \left[\frac{\partial P}{\partial X_i} \delta X_i \right]^2 \tag{11.14}$$

So when combining two variables, the error in the derived variable is not computed by simply adding or taking the mean of the errors in two variables. For example, for simple addition or subtraction:

$$P = f(x, y) = x \pm y \tag{11.15}$$

$$\frac{\partial P}{\partial x} = \frac{\partial P}{\partial y} = 1 \tag{11.16}$$

$$(\delta P)^2 = \left(\frac{\partial P}{\partial x} \delta x \right)^2 + \left(\frac{\partial P}{\partial y} \delta y \right)^2 = \delta x^2 + \delta y^2 \tag{11.17}$$

For the multiplication of two variables:

$$P = f(x, y) = x\, y \tag{11.18}$$

$$(\delta P)^2 = (y\, \delta x)^2 + (x\, \delta y)^2 \tag{11.19}$$

Performing such an analysis assumes the errors for each of the variables are random and uncorrelated, which may not strictly always be the case. These error propagation equations can easily be employed for simple biomechanical analyses, but for more complex biomechanical analyses the resulting system of equations can become unwieldy.

Another way to assess the influence of uncertainties, which does not require the same assumptions and a potentially simpler one is to perform a sensitivity analysis, where the parameters and state variables are changed (perturbated) by amounts relating to the estimated uncertainty or error with which they were measured. The change in the system output as a consequence of these perturbations is then quantified. The major problem associated with assessing uncertainty in this way is how to allow for all the possible combinations of potential errors. This approach can also be computationally expensive. The major advantage is that it can be easier to implement compared with the error propagation formulae. To some extent this problem can be viewed as a programming problem, as appropriately written computer code can easily investigate a reasonable subset of the options. Writing software for this purpose can be problematic, and many commercial software packages are not written with this level of flexibility.

Figure 11.1 is designed to allow visualization of the potential combination of errors when there are three variables combined to produce another variable. The rectangular parallelepiped contains all possible combinations of the errors. There are an infinite number of combinations of errors in the three variables contained in the rectangular parallelepiped. A compromise is to run calculations for nine of the possible error options which are presented in the figure by the original data set (the origin in Fig. 11.1), and each of the corners of the rectangular parallelepiped. Such an approach would represent

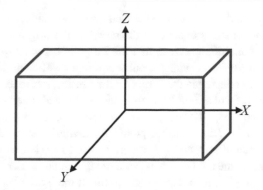

Fig. 11.1 The rectangular parallelepiped which encompasses all possible error combinations in variables X, Y, and Z

the largest errors in the variables. The root mean square difference between the perturbated result of the analysis and the original result can then be computed.

Irrespective of how the uncertainty is quantified, it is necessary to have estimates of the error in the variables and parameters. The following sub-sections will provide some examples for estimating errors in common biomechanical variables. Selection of values from the literature is an option but should be done prudently by matching the selected published study to the study at hand as closely as possible.

11.5 Variability of Human Movement and Outliers

When processing human motion data, there are two issues to consider: the inherent variability of human movement data and the handling of outliers in the collected data.

Measurements of human movement are variable for two reasons. The first is intrinsic variability of the subject performing the task at hand. For example, Winter (1995) quantified for a group of subjects the variability in human gait by collecting for each subject ten walking trials. The second is the variability introduced by the process of measurement (e.g., a force plate vibrating at its natural frequency), extrinsic noise. Ideally, these two sources of variability can be separated so that the signal of interest, human generated, can be analyzed. As has been highlighted, the electronic equipment used for data collection introduces noise into sampled data, and the properties of this noise often have similar characteristics to the variability of human movement, for example, Brown noise (Collins and De Luca 1994) and 1/f noise (Farrell et al. 2006). If the magnitude of the extrinsic variability is small relative to the intrinsic variability, then the analysis of the human movement generated signal is viable; otherwise subtle aspects of movement dynamics will be washed away in data processing. It should also be appreciated that data processing choices can change the nature of the noise contained in a sampled signal, highlighting the importance of careful data collection and processing.

In statistics, an outlier is an observation which is outside of the overall pattern of distribution of the collected data (observations). Sometimes outliers are easy to identify and useful. For example, in a study of athletes performing the clean and jerk one unfortunate participant ruptured their patellar tendon during the jerk phase of the lift, which of course gave the authors the opportunity to examine the loads at the joints associated with the rupture (Zernicke et al. 1977). In other cases outliers can cause problems with the appropriate application of statistics, as it can cause a violation of the data properties required to apply various statistics analyses. These outliers can either be for individual data points, for example, the peak vertical ground reaction force in a vertical jump, or for a time series, for example, the vertical ground reaction force profile during walking. Checking for outliers is an important step once the data has been collected in case some error occurred during data collection, independent of some odd performance of a task by a subject.

A data set consisting of individual data points (e.g., peak force, thigh mass) may contain outliers. One recommended practice is to assume any data points which lie outside of a range specified by the mean ± 3 standard deviations are outliers (e.g., Howell 1998). The main flaw with this approach is that the computations of the mean and standard deviation are both influenced by the data set being examined; another is that this approach is less likely to work with small data sets. A more robust method to use is the median of the absolute deviation. The median is a measure of central tendency which is insensitive to outliers. The absolute deviation for each data point from the median is computed, and then the median of this time series is computed. Any data point with a value which is outside of the range of the original median $\pm$ 1.4826 $\times$ median absolute deviation is considered an outlier. The analysis assumes that the range is between one and three quarters of the standard normal cumulative distribution function (different ranges would require a coefficient different to 1.4826).

Much of the data in biomechanics are collected in the form of a time series, so are there ways of identifying a time series in a set of time series which is an outlier? If there are multiple time series, these all need to be temporally aligned, for example, from 0 to 100% of movement time (see Chap. 12 for methods of temporal alignment). Imagine there are 10 data sets all temporally aligned, with, for example, each having 101 data points. A moving window along the time series can be used to test for outliers on a time instant by time instant basis, and see if there is an outlier (Brownlees and Gallo 2006). If for each time instant the mean and standard deviation of the data points is computed, then an outlier could be classified if a data point is $\pm$ three standard deviations above or below the mean. As already outlined there are reasons why a median absolute deviation might be a superior way of identifying an outlier in a time series. There are other methods for identifying outliers in a time series, for example, the outliergram (Arribas-Gil and Romo 2014), but their utility has yet to be explored in biomechanics.

11.6   Quantization Errors

Most equipment used for data collection use analog-to-digital converters (ADC). They convert the analog voltage signal into a numerical format, most commonly digital. Each ADC has a limit to the number of digits that are used to represent each component of the sampled signal; the data values are quantized. Most ADC quantize with a resolution of 8 to 32 bit. The number of values that can be represented by different n bit values is calculated from:

$$\text{Number of unique values} = 2^n \tag{11.20}$$

Given this relationship it is easy to compute the number of values available from each bit resolution (see Table 11.3) (see Appendix C for details on binary arithmetic).

The digitized values are each a single packet of information (word), represented in binary form, so the decimal value 4 is represented by 0100 in binary where the word contains 4 bits of information. Figure 11.2 shows the quantization steps of a 4 bit ADC. With 4 bits the integers from 0 to 15 can be represented, so in the figure each integer is accurately represented but not any numbers in between. Clearly the more bits of information, the higher the resolution of the system. If the input signal has a range from 0 to 10 volts and has an 8 bit ADC, then we can represent numbers from 0 to 255, so 00000000 would represent 0 volts, and 11111111 represents 10 volts. The resolution is:

$$\frac{10}{255} = 0.0392 \text{ volts} \tag{11.21}$$

If, for example, this voltage was from a force plate and full-scale range of the vertical ground and reaction force was 3000 N (= 10 volts), then the resolution would correspond to just under 12 N. It is important to set the gain so that as much of the 10 volts range as possible is used; otherwise the opportunity for higher resolution is being lost. Of course it is not always feasible to know a priori what range of forces, or other metric, will arise during data capture.

Different ADC can have different bit resolutions. For example, if a 2 Hz sine wave is sampled using a 1 bit, 2 bit, 4 bit, or 16 bit ADC, the resulting sampled version of the signal varies (see Fig. 11.3). For some applications a 1 bit ADC may be

Table 11.3 The number of bits and the number of unique values that correspond to when representing a numerical value

Number of bits	Number of unique values
2	4
4	16
6	64
8	256
10	1024
12	4096
14	16,384
16	65,536

Fig. 11.2 Quantization steps for a 4 bit ADC, for values from 0 to 9

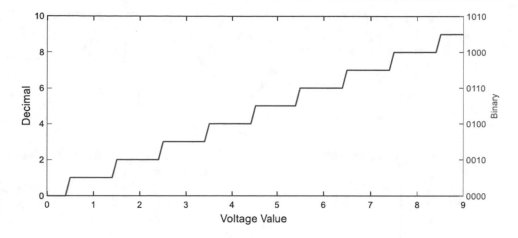

Fig. 11.3 Quantization of a 2 Hz sine wave, sampled with a 1, 2, 4, and 16 bit ADC

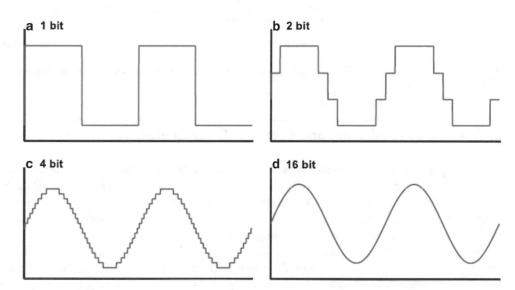

sufficient, for example, if measuring periods of contact. If the change in a signal to be measured is anticipated to be small, a higher-resolution ADC may be appropriate.

The distance between adjacent quantization steps is the least significant bit (LSB); this value reflects the resolution of the system. Figure 11.4 illustrates the original signal which is then sampled and then digitized. The difference between the sampled and digitized signal indicates the errors which are added to the signal due to digitization. Notice the quantization error has a magnitude which is ± 0.5 LSB and is randomly either positive or negative. This quantization error is random and has a mean value of zero and a standard deviation equal to $\frac{1}{LSB\sqrt{12}}$, so varies with the bit resolution of the ADC:

$$8 \text{ bit} \qquad \sigma = \frac{1}{255\sqrt{12}} \approx \frac{1}{900} \tag{11.22}$$

$$16 \text{ bit} \qquad \sigma = \frac{1}{65536\sqrt{12}} \approx \frac{1}{227000} \tag{11.23}$$

The noise which corrupts a sampled signal using an ADC is a close approximation to white noise, as the absolute amplitude of the noise is theoretically constant and is present across all of the power spectrum.

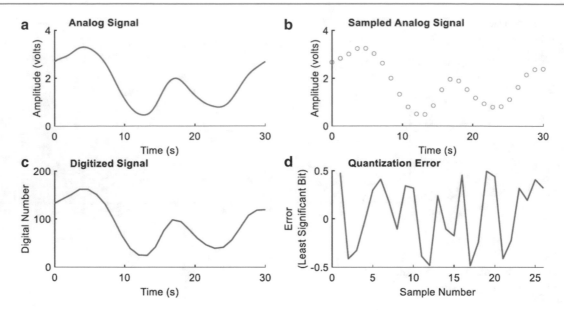

Fig. 11.4 A signal in its (**a**) analog form, (**b**) sampled, (**c**) digitized, and (**d**) the error introduced by quantization (the difference between graphs **b** and **c**)

11.7   Low-Pass Filtering and Computation of Derivatives

For many biomechanical analyses, it is necessary to reduce the white noise which contaminates the sampled data by low-pass filtering the data, prior to the computation of signal derivatives. There are a variety of methods used in biomechanics for low-pass filtering data (e.g., low-pass digital filter, generalized cross-validated splines); to a large extent all of these approaches are equivalent (Craven and Wahba 1979). This section presents analyses of the influence of filtering and differentiation on the errors corrupting a sampled signal.

In the frequency domain signal differentiation is equivalent to signal amplification, with increasing amplification with increasing frequency. A simple example illustrates the influence of data differentiation on signal components. If the true signal is a sine wave with an amplitude of *1*, which is corrupted by noise across the spectrum, this example will focus on one noise frequency component with an amplitude of *0.001*. The signal to noise ratio can be computed for the zero, first, and second order derivatives:

Displacement data	$x(t) = sin(t)$
Noise data	$noise(t) = 0.001\ sin(50\ t)$
Signal to noise ratio	1: 0.001
Velocity data	$v(t) = cos(t)$
Noise data	$noise(t) = 0.05\ cos(50\ t)$
Signal to noise ratio	1: 0.05
Acceleration data	$a(t) = -sin(t)$
Noise data	$noise(t) = -2.5\ sin(50\ t)$
Signal to noise ratio	1: 2.5

This example illustrates how important it is to low-pass filter data prior to data differentiation, to reduce the high-frequency noise components of the sampled signal. In the example the noise was only present at one frequency, but in reality with white noise, there would be noise across the frequency spectrum.

In 1982 Lanshammar presented a series of formulae that estimate the noise which can be expected to remain in a noisy signal after filtering and or differentiation. The basic format of the formula is:

$$\sigma_k^2 \geq \frac{\sigma^2\ \tau\ (2\ \pi\ \omega_b)^{(2k+1)}}{\pi(2k + 1)} \tag{11.24}$$

where σ_k^2 is the minimum variance of the noise affecting the kth derivative after processing, σ^2 is the variance of the additive white noise, τ is the sample interval, ω_b is the bandwidth of the signal in radians per second ($\omega_b = 2\ \pi f_b$), and f_b is the sample bandwidth in Hertz. Notice the equation only provides at best a minimum estimate of the errors.

There are three major assumptions implicit with the formulae of Lanshammar (1982):

1. It is assumed that the signal is band-limited, which means all frequencies above the Nyquist frequency are ignored. But from a mathematical point of view, no signal is band-limited (Slepian 1976); after all a higher sample rate may reveal frequencies which were not observable with a lower sample rate.
2. The noise contaminating the signal is assumed to be white; given the many sources of error that can corrupt collected data, it is unrealistic to assume that the noise corrupting a signal is perfectly white.
3. The formulae assumes the frequency response of the low-pass filter is ideal, similarly for the differentiator. For a low-pass filter to be ideal, it assumes the input signal passes the signal unattenuated up to the specified filter cut-off, and after that cut-off removes all of the signal. Such a low-pass filter is not practically realizable. Similarly for an ideal differentiator, all signal up to a cut-off is suitably amplified and then completely attenuated after that cut-off.

If the initial focus is on the zero-order derivative, displacement, the equation becomes:

$$\sigma_0^2 \geq 2\,\sigma^2\,\tau\,\omega_b \tag{11.25}$$

As would be expected to decrease the error in the zero-order derivative, the original error in the data should be as small as possible. The sample interval should also be small, which means having a high sample rate. Finally the bandwidth of the signal should be as small as possible; this is task-dependent and indicates that tasks with narrower bandwidths will have smaller error variance after data processing compared with broader bandwidth tasks with all other aspects of data collection being equal. For a motion analysis system, there are two components which specify the system's spatial and temporal resolution, a figure of merit for comparing different analysis systems which is the product the measure of spatial resolution (σ) and square root of the temporal resolution ($\sqrt{\tau}$). It allows comparison of the relative balance between a system which produces little noise at a low sample rate with one which produces noisier data but at a higher sample rate.

To move the focus on to the first-order derivative, velocity, the equation becomes:

$$\sigma_1^2 \geq \frac{\sigma^2\,\tau\,(2\,\pi\,\omega_b)^3}{3\,\pi} \tag{11.26}$$

This version of the formulae highlight that derivatives will always be more noisy than the data representing the previous derivative, with this problem growing with increasing order derivatives. If the task is to reduce the error variance in the computed derivative, the variance of the noise and the sample interval should be decreased and a task selected with a narrow bandwidth. The process of differentiation amplifies any noise in the signal; therefore it is essential that as much noise as possible is removed from the signal prior to differentiation.

To use the formulae from Lanshammar (1982), the variance of the additive white noise is required. One approach available to compute the variance of the noise is to make repeat measures of the sample; this approach is feasible if, for example, motion analysis data are collected via manual digitization. Another option with automatic motion analysis data collection is to compute the variance of the positions of markers when they are stationary. For example, Winter et al. (1974) estimated noise levels by quantifying the variation in location of a marker on the foot when the foot was stationary during the stance phase of gait.

11.8 Segment Orientation and Joint Angles

Given the locations of landmarks on a body segment, the orientation of the segment can be determined, and the relative orientation of one segment relative to an adjacent segment (joint angles) can be determined. In two dimensions, segment orientations are typically computed by measuring the locations of landmarks at the distal and proximal ends of the segment; the segment orientation is then the angle between a line joining these landmarks and the horizontal. The variance of an angle (σ_ϕ^2) computed in this way can be calculated from:

$$\sigma_\phi^2 = \frac{2\,\sigma^2}{l^2} \tag{11.27}$$

where σ^2 is the error variance of noise affecting coordinates and l is the distance between markers or body landmarks.

To increase the accuracy of the angle computation, the equation indicates that the noise affecting the coordinates should be minimized. In addition the distance between the markers or landmarks should be maximized. There is a limit to the distance between markers on a body segment, and marker locations selection must also consider factors such as minimizing skin movement and ensuring the markers are visible to cameras.

The definition of the orientation of a segment or a joint in three dimensions is more complex than their definition in two dimensions and so therefore is the quantification of the errors in their computation. In three dimensions, orientations are typically described by three angles extracted from the 3 by 3 attitude matrix. To examine the influence of errors on this vector of angles, it is necessary to form a covariance matrix. The covariance matrix for a 3-element vector is a 3×3 symmetric matrix, where the diagonal elements of the matrix are the variance of the components of the vector and the off diagonal terms are the scalar covariances between the vector components. This matrix is positive semi-definitive.

For Cardan angles the attitude matrix can be described in equation form:

$$R = R(\gamma)\, R(\beta)\, R(\alpha) \tag{11.28}$$

where $R(\alpha)$, $R(\beta)$, and $R(\gamma)$ represent rotations about different axes, which might, for example, be associated with flexion/extension, abduction/adduction, and internal/external rotation. Given the vector of angles $\vartheta = (\alpha,\ \beta,\ \gamma)$, a covariance matrix can be determined. Using an error model based on Woltring et al. (1985) where errors ($\Delta\varphi$) are multiplicative with an isotropic distribution, then the error in the attitude matrix ($\widehat{R}$) is:

$$\widehat{R} = (I + A(\Delta\varphi))\, R \tag{11.29}$$

where I is the identity matrix and $A(\Delta\varphi)$ is a skew-symmetric matrix:

$$A(\Delta\varphi) = \begin{bmatrix} 0 & -\Delta\varphi_z & \Delta\varphi_y \\ \Delta\varphi_y & 0 & -\Delta\varphi_x \\ -\Delta\varphi_y & \Delta\varphi_x & 0 \end{bmatrix}$$

The error vector $\Delta\varphi$ refers to small rotational errors about the reference frame affixed to the body of interest, which have an error variance of σ_φ^2. The covariance matrix (ϑ) for the three Cardan angles is:

$$\vartheta = \sigma_\varphi^2 \begin{bmatrix} \dfrac{1}{\cos^2\beta} & 0 & \dfrac{-\sin\beta}{\cos^2\beta} \\ 0 & 1 & 0 \\ \dfrac{-\sin\beta}{\cos^2\beta} & 0 & \dfrac{1}{\cos^2\beta} \end{bmatrix} \tag{11.30}$$

The errors in the three angles are correlated and cannot be considered independent when analyzing the influence of signal noise on the angles. Note that the influence of noise on the terminal angles (α, γ) becomes greater when the middle angle approaches $\pm(2n + 1)\,\pi/2$, where n is any integer. Using a Cardanic angle sequence at a middle angle of $\pm(2n + 1)\,\pi/2$, the other two (terminal) angles are undefined, and the angle system has a singularity. The Euler angle convention has similar properties.

Sometimes helical or screw axes are used to quantify the motion at the joints, with the finite helical axis (FHA) used to describe the motion of a rigid body from one position to another (see Chap. 9 for more details). The helical axis parameters describing the FHAs are the angle of rotation (θ) about the axis, the unit direction vector of the helical axis (e), the amount of translation (u) along the axis, and the location of a point (p) on the helical axis. Woltring et al. (1985) presented formulae for the estimation of the error variance associated with estimating these parameters; they are:

$$\sigma_\theta^2 = \frac{2\,\sigma^2}{m\,\rho^2} \tag{11.31}$$

$$\sigma_{\underline{e}}^2 = \frac{\sigma^2}{m\,\rho^2\,\sin^2\frac{\theta}{2}}$$

(11.32)

$$\sigma_u^2 = \frac{2\,\sigma^2}{m}\left[1 + \frac{|s - p|^2}{\rho^2\cos^2\frac{\theta}{2}}\right]$$

(11.33)

where m is the number of markers, ρ is the marker distribution radius with $\rho^2 = \frac{2}{3}\,r^2$, r^2 is the mean squared distance of the markers to the marker centroid, s is the position vector of the projection onto the helical axis of p, and p is the position vector of the midpoint of the marker centroid before and after the motion.

As would be expected, the FHA parameters are more accurate if the coordinate data have less noise ($\downarrow\sigma$). Similarly the more markers, the better the accuracy ($\uparrow m$, where $m \geq 3$), and the further apart the markers ($\uparrow r$), the greater the accuracy. The greater the amount of rotation about the helical axis ($\uparrow\theta$), the better the accuracy, but in doing this it reduces the ability of the FHA to serve as an approximation to the instantaneous helical axis. The analysis using the formulae assumes the following:

1. The markers are isotropically distributed, which means the markers would be on the vertices of a regular polyhedron (e.g., triangular pyramid ($m = 4$), or cube ($m = 8$)).
2. All markers must have the same zero-mean uncorrelated isotropic errors.
3. The FHA angle of rotation is less than or equal to 1 radian.

Violation of these assumptions mean the formulae would represent a fortuitous best case scenario, but the formulae do indicate steps which must be taken experimentally to give the best estimate of the FHA.

The error analyses presented in this section assume that the errors are isotropic. Isotropic errors do not vary with respect to direction of measurement, as opposed to anisotropic errors. Image-based motion analysis does not always produce isotropic errors, for example, with a two-camera system, the depth axis (axis perpendicular to a line between the principal points of the two cameras) is measured with lower accuracy than for the other two axes. Given anisotropic noise, error effects will be increased. When defining angles in three dimensions a reference frame must be defined for each body segment being analyzed; doing so accurately is challenging to achieve (Brennan et al. 2011).

11.9   Conclusions

Good experimental practice is to seek to minimize the errors at their source, and once this has been performed it does not matter what the source of the error is, but it is important to quantify the error. This chapter has outlined the terminology used in error analyses, and provided examples of how errors can be determined and minimized.

It is not possible to set acceptable levels of uncertainty as these will vary from study to study. The purpose of the study dictates what the acceptable levels of uncertainty are, and a pedantic search for increased accuracy with a particular measuring device may be redundant for the purpose of one study and yet be essential for another. Some of the factors which affect uncertainty levels cannot be known a priori, for example, bandwidth of the sampled signal. When the measurement uncertainty cannot be sufficiently well estimated before data collection, a strict adherence to good experimental protocols followed by an error analysis is the experimenter's only recourse.

11.10   Review Questions

1. Define accuracy, precision, and resolution. How might you numerically quantify accuracy, precision, and resolution?
2. If trying to compute derivatives from displacement data, what measures should be taken to ensure these estimates are as accurate as possible?
3. What do we mean by error propagation? Why is it important to consider the influence of error propagation when analyzing human movement?
4. What do we mean by spatial temporal resolution?

References

Cited References

Airy, G. B. (1875). *On the Algebraical and numerical theory of errors of observations and the combination of observations*. London: MacMillan and Co.

Arribas-Gil, A., & Romo, J. (2014). Shape outlier detection and visualization for functional data: The outliergram. *Biostatistics, 15*(4), 603–619.

Berg, H. C. (1983). *Random walks in biology*. Princeton, NJ: Princeton University Press.

Brennan, A., Deluzio, K., & Li, Q. (2011). Assessment of anatomical frame variation effect on joint angles: A linear perturbation approach. *Journal of Biomechanics, 44*(16), 2838–2842.

Brownlees, C. T., & Gallo, G. M. (2006). Financial econometric analysis at ultra-high frequency: Data handling concerns. *Computational Statistics & Data Analysis, 51*(4), 2232–2245.

Collins, J. J., & De Luca, C. J. (1994). Random walking during quiet standing. *Physical Review Letters, 73*(5), 764–767.

Craven, P., & Wahba, G. (1979). Smoothing noisy data with spline functions: Estimating the correct degree of smoothing by the method of generalised cross-validation. *Numerische Mathematik, 31*(4), 377–403.

Crease, R. P. (2011). *World in the balance: The historic quest for an absolute system of measurement*. New York: W.W. Norton & Company.

Dapena, J. (1985). Correction for 'Three-dimensional cinematography with control object of unknown shape. *Journal of Biomechanics, 18*, 163.

Dapena, J., Harman, E. A., & Miller, J. A. (1982). Three-dimensional cinematography with control object of unknown shape. *Journal of Biomechanics, 15*(1), 11–19.

Farrell, S., Wagenmakers, E. J., & Ratcliff, R. (2006). 1/f noise in human cognition: Is it ubiquitous, and what does it mean? *Psychonomic Bulletin & Review, 13*(4), 737–741.

Fuller, J., Liu, L. J., Murphy, M. C., & Mann, R. W. (1997). A comparison of lower-extremity skeletal kinematics measured using skin- and pin-mounted markers. *Human Movement Science, 16*, 219–242.

Howell, D. C. (1998). *Statistical methods for psychology*. New York: Wadsworth.

Keshner, M. S. (1982). 1/f noise. *Proceedings of the IEEE, 70*(3), 212–218.

Lanshammar, H. (1982). On precision limits for derivatives numerically calculated from noisy data. *Journal of Biomechanics, 15*(6), 459–470.

Marinari, E., Parisi, G., Ruelle, D., & Windey, P. (1983). On the interpretation of 1/f noise. *Communications in Mathematical Physics, 89*(1), 1–12.

Milburn, G. J., & Sun, H. B. I. (1998). Classical and quantum noise in electronic systems. *Contemporary Physics, 39*(1), 67–79.

Pain, M. T. G., & Challis, J. H. (2002). Soft tissue motion during impacts: Their potential contributions to energy dissipation. *Journal of Applied Biomechanics, 18*(3), 231–242.

Slepian, D. (1976). On bandwidth. *Proceedings of the IEEE, 64*, 292–300.

Winter, D. A. (1995). Human balance and posture control during standing and walking. *Gait & Posture, 3*(4), 193–214.

Winter, D. A., Sidwall, H. G., & Hobson, D. A. (1974). Measurement and reduction of noise in kinematics of locomotion. *Journal of Biomechanics, 7*(2), 157–159.

Winter, S. L., & Challis, J. H. (2017). Classifying the variability in impact and active peak vertical ground reaction forces during running using DFA and ARFIMA models. *Human Movement Science, 51*, 153–160.

Woltring, H. J., Huiskes, R., de Lange, A., & Veldpaus, F. E. (1985). Finite centroid and helical axis estimation from noisy landmark measurements in the study of human joint kinematics. *Journal of Biomechanics, 18*(5), 379–389.

Zernicke, R. F., Garhammer, J., & Jobe, F. W. (1977). Human patellar-tendon rupture. A kinetic analysis. *Journal of Bone and Joint Surgery, A59*(2), 179–183.

Useful References

Barford, N. C. (1985). *Experimental measurements: Precision, error and truth*. New York: John Wiley.

Bates, B. T. L., Osternig, L. R., Sawhill, J. A., & James, S. L. (1983). An assessment of subject variability, subject-shoe interaction, and the evaluation of running shoes using ground reaction force data. *Journal of Biomechanics, 16*(3), 181–191.

Gupta, S. V. (2012). *Measurement uncertainties: Physical parameters and calibration of instruments*. Berlin: Springer.

Rabinovich, S. G. (2005). *Measurement errors and uncertainties theory and practice*. Berlin: Springer.

Taylor, J. R. (1982). *An introduction to error analysis: The study of uncertainties in physical measurements*. Mill Valley, CA: University Science Books.

Scaling

12

12.1 Overview

Biology has used methods of scaling for many purposes, for example, understanding the influence of size on biological function (Schmidt-Nielsen 1984). Ontogenetic scaling examines the growth relationships during development, either between two traits within an organism, or between one trait and the whole organism (e.g., Huxley 1932). Phylogenetic scaling examines the size of relationships of organisms among various species, often with a focus on evolution (Gould 1971). All of these applications can apply to biomechanics (e.g., Alexander 1989), but often in biomechanics the interest in using scaling is either to understand the function of a system or to normalize collected data to account for the variation in the data due to subjects of different sizes. The focus of this chapter will be the latter, although this will also provide some insight to the former.

Galileo (1564–1642) in Galilei 1637 discussed the necessary strength and size of the bones of large animals. With increasing body size, body mass increases as do limb lengths; therefore, there should be a corresponding increase in limb and bone cross-sectional area to support this increased mass. The classic figure (Fig. 12.1) reveals an error in Galileo's thinking. The large bone has a threefold increase in length and a ninefold increase in diameter. Basic scaling principles suggest with this increase in mass, the increase in diameter would be proportional to $3^{3/2}$, not 3^2.

12.2 Why Scale?

Many biomechanical variables can be influenced by body size, so how can scaling help in these cases? There are two ways in which scaling can be useful: normalization of data to account for body size, and providing insight into the relationship between size and movement outcome (e.g., Pennycuick 1992).

Normalization to account for body size can be important if using inferential statistics on collected data. With many parametric statistical tests, it is assumed that the data demonstrate normality and homogeneity of variance. If in the originally collected data this is not the case, then scaling to account for body size may produce data that do satisfy these criteria. Data with normal distributions are symmetric around the center, the mean, and have a bell-shaped distribution. Homogeneity of variance means that there are equal variances across different samples or groups.

12.3 Geometric Scaling

Geometric scaling occurs if for bodies the size varies but not the shape. Under this assumption then:

length $\propto$ length	*area $\propto$ area*	*volume $\propto$ volume*
length $\propto$ area$^{1/2}$	*area $\propto$ length^2*	*volume $\propto$ length^3*
length $\propto$ volume$^{1/3}$	*area $\propto$ volume$^{2/3}$*	*volume $\propto$ area$^{3/2}$*
length $\propto$ mass$^{1/3}$	*area $\propto$ mass$^{2/3}$*	*volume $\propto$ mass*

© The Editor(s) (if applicable) and The Author(s), under exclusive license to Springer Nature Switzerland AG 2021
J. H. Challis, *Experimental Methods in Biomechanics*, https://doi.org/10.1007/978-3-030-52256-8_12

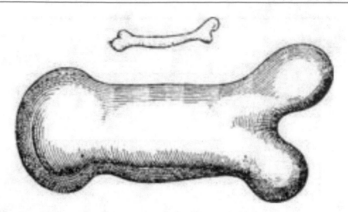

Fig. 12.1 Galileo's illustration of the changes in dimensions of the bones of animals of different sizes, with an error in their scale

Table 12.1 The length, surface area, and volume for three cubes of different sizes. As the cubes are the same shape but different sizes, they demonstrate the general principles of geometric scaling

Length	1	2	3	4
Area Side	1	4	9	16
Volume	1	8	27	64

So lengths are any characteristic length, for example, standing height or leg length. Areas can be any area, for example, body surface area, plantar surface area, or muscle cross-sectional area. Volume can be any area, for example, lung volume or heart volume. Mass and volume are assumed to be linearly related to one another. Mass could, for example, be total body mass, skeletal mass, or muscle mass. These general principles are illustrated in Table 12.1.

Here are a few examples illustrating how geometric scaling principles apply in man. In humans, an analysis of height and mass data produces a scaling exponent equal to 2.9 ($height^{2.9} \propto mass$), which is close to the exponent of 3 predicted by geometric scaling (McMahon and Bonner 1983). Cook and Hamann (1961) examined the relationship between standing height and lung volume in man, which should scale relative to height cubed ($volume \propto height^3$). The data for 106 males gave an exponent slightly more than the theory (3.04) and for 65 females a scaling exponent less than the theoretical value (2.91). McMahon and Bonner (1983) demonstrated that upper limb length scaled with body height with a scaling exponent of 1.0, as would be predicted by geometric scaling.

Allometry broadly refers to the study of size relationships between different features of an organism, although often considered to be the examination of features that do not scale geometrically. A common example of this is the scaling of metabolic rate, and it was thought to scale in proportion to surface area ($\propto mass^{2/3}$), but Kleiber (1932) showed that it actually scales $\propto mass^{3/4}$. There have been a variety of theories for why this may be the case, and one of the most compelling is that of Banavar et al. (1999) who posit that it arises due to the distributed networks found in both the cardiovascular and respiratory systems.

12.4 Scaling of Key Biomechanical Variables

For many biomechanical variables, their dimensions can be expressed in terms of length *(L)*, mass *(M)*, and time *(T)*. When scaling these variables, lengths are scaled using characteristics length (e.g., standing height), so:

l_B – representative body length (e.g., leg length, height)

Mass is typically scaled using whole-body mass:

m_B – body mass

Table 12.2 Some common physical quantities measured in biomechanics and suggestions on their appropriate scaling. Scaled values have a circumflex (e.g., $\hat{m}, \hat{I}$)

Physical quantity	Symbol	Dimension	Dimensionless number
Mass	M	M	$\hat{m} = \dfrac{m}{m_B}$
Moment of inertia	I	M.L^2	$\hat{I} = \dfrac{I}{m_B l_B^2}$
Length/displacement	l, s	L	$\hat{l} = \dfrac{l}{l_B}$
Velocity	V	L.T^{-1}	$\hat{v} = \dfrac{v}{\sqrt{gl_B}}$
Acceleration	A	L.T^{-2}	$\hat{a} = \dfrac{a}{g}$
Angular displacement	θ	–	Already dimensionless
Angular velocity	ω	T^{-1}	$\hat{\omega} = \dfrac{\omega}{\sqrt{g/l_B}}$
Angular acceleration	α	T^{-2}	$\hat{\alpha} = \dfrac{\alpha}{g/l_B}$
Force	F	M.L.T^{-2}	$\hat{F} = \dfrac{F}{m_B g}$
Moment	T	M.L^2.T^{-2}	$\hat{T} = \dfrac{T}{m_B g l_B}$
Time	T	T	$\hat{t} = \dfrac{t}{\sqrt{l_B/g}}$
Frequency	F	T^{-1}	$\hat{f} = \dfrac{f}{\sqrt{g/l_B}}$
Work/energy	E	M.L^2.T^{-2}	$\hat{W} = \dfrac{W}{m_B g l_B}$
Power	P	M.L^2.T^{-3}	$\hat{P} = \dfrac{P}{m_B g^{1/2} l_B^{3/2}}$

Expressions containing the units of force can often use the acceleration due to gravity for scaling:

g – acceleration due to gravity

Table 12.2 summarizes how key biomechanical variables can be scaled. Some of these will be further elaborated on in the following paragraphs.

The scaling of segmental masses can be scaled relative to body mass and segmental moments of inertia with respect to the product of body mass and height squared. Chapter 8 provides further details how segmental inertial properties can be scaled.

William Froude (1810–1879) examined the wave patterns of ships with hulls of different sizes (l) moving at different velocities (V). He identified a characteristic number for which ships of different sizes have the same wave patterns, the Froude number (F_r):

$$F_r = \frac{V}{\sqrt{gl}} \tag{12.1}$$

The number has found utility in scaling the velocity of locomotion where it can be considered to be ratios of the kinetic energy ($\frac{1}{2}mv^2$) to potential energy (mgh) during the support phase of locomotion. In quadrupeds ranging in size from the cat to the rhino, the transition from a walk to trotting occurs at the same Froude number (l = leg length) and at another constant Froude number for the transition from a trot to a gallop (Alexander and Jayes 1983). The Froude number also accounts for the scaling of walking velocity in passive dynamic walkers (McGeer 1992). Challis (2018) examined the relationship between stride length and walking velocity in children and adults. Stride lengths were scaled relative to standing height and the walking velocities scaled using Froude number, and with appropriate scaling, the children and adults had the same relationship between walking speed and stride length.

Ground reaction forces during locomotion are typically scaled to body weight under the assumption that the magnitude of these forces is proportional to body weight. There is experimental evidence that during walking the ground reaction forces scale in proportion to body weight (Browning and Kram 2007), similarly for running (Frederick and Hagy 1986).

In the literature, varieties of approaches have been used to normalize resultant joint moments. These moments can be considered to be the result of a force acting about a moment arm relative to the joint. The basic math suggests that resultant joint moments should be scaled using factors related to a force and a length. For example, Besier et al. (2001) in a study of running normalized the resultant joint moments using body weight but did not include a length scaling factor, while McClay and Manal (1999) used the product of body weight and height, and Kadaba et al. (1989) examining walking used the product of body weight and leg length. There is no way to fully assess the superiority of one method over another, but Moisio et al. (2003) showed there was a greater reduction in the variability of joint moments when normalizing with respect to the product of body weight and standing height compared with body weight only.

12.5   Scaling Time

If a task is repeated n times by the same subject, it might be of value to take the average of these repeat trials. One problem is that each trial will not necessarily take the same time period, so if the data set is aligned at the beginning (start time is the same), the end time will vary between trials. Figure 12.2 presents data for 12 running trials and shows how each footfall does not last for the same period of time.

The simplest option for scaling time is:

1. For each trial, identify times of the start and end of movement.
2. Divide each trial's movement time into 101 equidistant time values.
3. Interpolate each trial over its 101 value time base.
4. Now consider that the 101 data points represent the data as percentage of movement time.

Figure 12.3 illustrates for the 12-footfall data set, from Fig. 12.2, when the data has been synchronized. Given the data over a common time base (percentage of movement time), it is feasible to compute the mean of the signal and its standard deviation for each instant of the moment.

Fig. 12.2 The vertical ground reaction forces for 12 footfalls during running. Note the force is expressed in units of body weight (BW)

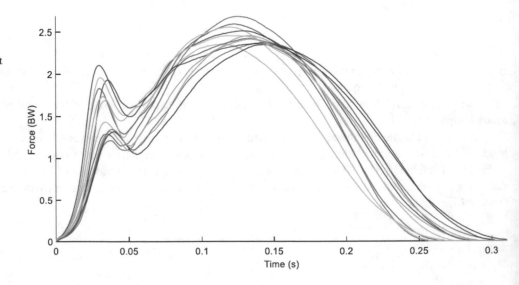

Fig. 12.3 The vertical ground reaction forces for 12 footfalls during running, where time is expressed as a percentage of stance time, where (**a**) is each of the 12 footfalls and (**b**) is the mean of the 12 footfalls, and one standard deviation either side of the mean. Note the force is expressed in units of body weight (BW)

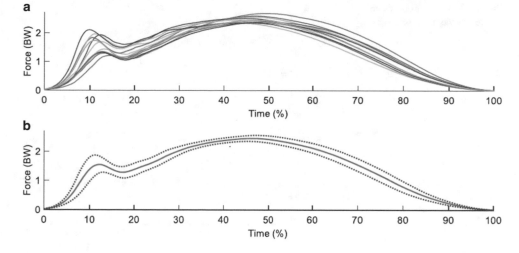

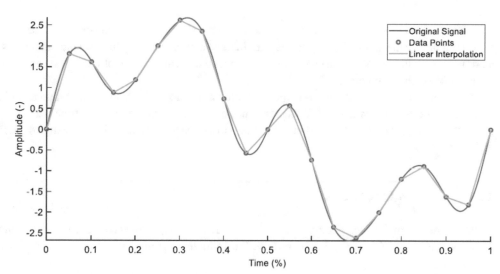

Fig. 12.4 Illustration of linear interpolation of a signal

There are many options for interpolating data. The simplest is linear interpolation, which fits a straight line between data points and can then predict the dependent variable values at time instants along that line. If the signal of interests has low-frequency components and is sampled at a high sample rate, then this type of interpolation may be adequate. Figure 12.4 illustrates a situation where linear interpolation may not be appropriate.

The other common technique is the fitting of an interpolating cubic spline to the data (de Boor 1978). Splines are piecewise polynomials, where the pieces join at knots. The polynomials must fulfill continuity conditions for the function and its derivatives. For cubic splines, the polynomials must be continuous at the knots up to and including the second derivative. For each interval, between the knots, the fitted cubic polynomial can be used to estimate function values not necessarily measured. Figure 12.5 shows that the interpolating cubic spline does a good job of capturing the pattern underlying signal, with the only apparent deviations being around 0.08 seconds and 0.94 seconds.

There are other methods available for temporally aligning data. The method examined here focused on converting time to percentages of the movement time. But there are other options, which form a class of methods often called time warping (Ramsay and Silverman 2002). In dynamic time warping, the signal time axis can be compressed or expanded to match individual data sets with a target data set (Sakoe and Chiba 1978). An alternative is to perform time axis compression or expansion on the original signal but using its first derivative to identify the appropriate strategy, and this gives rise to derivative dynamic time warping (Keogh and Pazzani 2001). Other time-scaling methods use a variety of methods to segment the time series into subphases and then temporally align each of the subphases individually (Ramsay and Li 1998). This time scaling can be performed on the collected data or its derivatives. These methods rely on the identification of landmarks (events) within the time series to identify phases, and often these events are more plentiful in the first derivative of a time series.

Fig. 12.5 Illustration of cubic spline interpolation of a signal

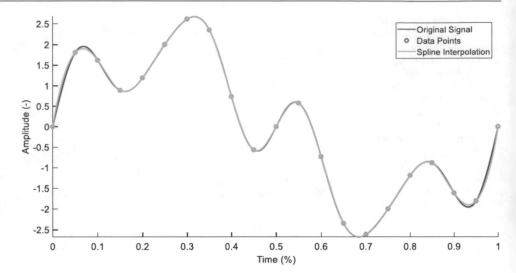

When various time-scaling methods are used, the data should meet certain criteria. The curves to be aligned should have the same structural pattern. Ideally, the curves should have common clearly identifiable landmarks. The order of appearance of the landmarks should also be consistent. Time-scaling methods enforce monotonicity that ensures the preservation of the order of appearance of signal landmarks.

It is an open question if the time-warping techniques are appropriate to use on biomechanical data. It could be that a part(s) of the time series being warped is functionally important in its original form. If time warping is performed for repeat trials for the same subject, this may be justified, and it is more questionable if such time warping should be performed for data from different subjects.

12.6   Review Questions

1. For an object that scales geometrically, state the relationships between lengths, areas, volumes, and masses.
2. What scaling might be used to normalize the following: ground reaction forces, resultant joint moments, and walking velocity?
3. Why might the scaling of time be useful? How could this be achieved?

References

Cited References

Alexander, R. M. (1989). *Dynamics of dinosaurs and other extinct giants*. New York: Columbia University Press.
Alexander, R. M., & Jayes, A. S. (1983). A dynamic similarity hypothesis for gaits of quadrupedal mammals. *Journal of Zoology, 201*, 135–152.
Banavar, J. R., Maritan, A., & Rinaldo, A. (1999). Size and form in efficient transportation networks. *Nature, 399*, 130–132.
Besier, T. F., Lloyd, D. G., Cochrane, J. L., & Ackland, T. R. (2001). External loading of the knee joint during running and cutting maneuvers. *Medicine & Science in Sports & Exercise, 33*(7), 1168–1175.
Browning, R. C., & Kram, R. (2007). Effects of obesity on the biomechanics of walking at different speeds. *Medicine & Science in Sports & Exercise, 39*(9), 1632–1641.
Challis, J. H. (2018). Body size and movement. *Kinesiology Review, 7*(1), 88–93.
Cook, C. D., & Hamann, J. F. (1961). Relation of lung volumes to height in healthy persons between the ages of 5 and 38 years. *The Journal of Pediatrics, 59*(5), 710–714.
de Boor, C. (1978). *A practical guide to splines*. New York: Springer.
Frederick, E. C., & Hagy, J. L. (1986). The prediction of vertical impact force during running. *International Journal of Sports Biomechanics, 2*(1), 41–49.
Galilei, G. (1637). *Dialogues concerning two new sciences*. (H. Crew & A. DeSalvio, Trans. New York: Macmillan.
Gould, S. J. (1971). Geometric similarity in allometric growth: A contribution to the problem of scaling in the evolution of size. *The American Naturalist, 105*(942), 113–136.

Huxley, J. (1932). *Problems of relative growth.* New York: L. MacVeagh, The Dial Press.

Kadaba, M. P., Ramakrishnan, H. K., Wootten, M. E., Gainey, J., Gorton, G., & Cochran, G. V. (1989). Repeatability of kinematic, kinetic, and electromyographic data in normal adult gait. *Journal of Orthopaedic Research, 7*(6), 849–860.

Keogh, E. J., & Pazzani, M. J. (2001). Derivative dynamic time warping. In *Paper presented at the first SIAM international conference on data mining (SDM'2001).* Chicago, IL, USA.

Kleiber, M. (1932). Body size and metabolism. *Hilgardia, 6,* 315–353.

McClay, I., & Manal, K. (1999). Three-dimensional kinetic analysis of running: Significance of secondary planes of motion. *Medicine & Science in Sports & Exercise, 31*(11), 1629–1637.

McGeer, T. (1992). Principles of walking and running. In R. M. Alexander (Ed.), *Advances in comparative and environmental physiology* (Vol. 11, pp. 113–139). Berlin: Springer.

McMahon, T. A., & Bonner, J. T. (1983). *On size and life.* New York: Scientific American Library : Distributed by W.H. Freeman.

Moisio, K. C., Sumner, D. R., Shott, S., & Hurwitz, D. E. (2003). Normalization of joint moments during gait: A comparison of two techniques. *Journal of Biomechanics, 36*(4), 599–603.

Pennycuick, C. J. (1992). *Newton rules biology: A physical approach to biological problems.* Oxford, UK: Oxford University Press.

Ramsay, J. O., & Li, X. (1998). Curve registration. *Journal of the Royal Statistical Society: Series B: Statistical Methodology, 60*(2), 351–363.

Ramsay, J. O., & Silverman, B. W. (2002). *Applied functional data analysis: Methods and case studies.* New York: Springer.

Sakoe, H., & Chiba, S. (1978). Dynamic programming algorithm optimization for spoken word recognition. *IEEE Transactions on Acoustics, Speech, and Signal Processing, 26*(1), 43–49.

Schmidt-Nielsen, K. (1984). *Scaling: Why is animal size so important?* Cambridge: Cambridge University of Press.

Useful References

Alexander, R. M. (1971). *Size and shape.* London: Edward Arnold (Publishers) Limited.

Delattre, N., Lafortune, M. A., & Moretto, P. (2009). Dynamic similarity during human running: About Froude and Strouhal dimensionless numbers. *Journal of Biomechanics, 42*(3), 312–318.

Hill, A. V. (1950). The dimensions of animals and their muscular dimensions. *Science Progress, 38*(150), 209–230.

Llenart, J., Salat, J., & Torres, G. T. (2000). Removing allometric effects of body size in morphological analysis. *Journal of Theoretical Biology, 205,* 85–93.

Thompson, D. W. (1992). *On growth and form.* (Abridged ed. Cambridge: Cambridge University Press.

Vaughan, C. L., & O'Malley, M. J. (2005). Froude and the contribution of naval architecture to our understanding of bipedal locomotion. *Gait & Posture, 21*(3), 350–362.

West, G. B., Woodruff, W. H., & Brown, J. H. (2002). Allometric scaling of metabolic rate from molecules and mitochondria to cells and mammals. *Proceedings of the National Academy of Sciences, 99*(Suppl 1), 2473–2478.

Appendix A Matrices

A.1 What Is a Matrix?

A matrix is just a rectangular array of numbers or symbols organized into rows and columns, for example:

$$A = \begin{bmatrix} \cos(\theta) & -\sin(\theta) \\ \sin(\theta) & \cos(\theta) \end{bmatrix} \qquad B = \begin{bmatrix} 0.25 & 0.33 \\ 0.125 & 0.666 \end{bmatrix}$$

A matrix C of m rows and n columns has the following form:

$$C = \begin{bmatrix} c_{1,1} & c_{1,2} & \cdots & c_{1,n} \\ c_{2,1} & c_{2,2} & \cdots & c_{m,n} \\ \cdot & \cdot & \cdot & \cdot \\ c_{m,1} & c_{m,2} & \cdots & c_{m,n} \end{bmatrix} \begin{matrix} \leftarrow & 1^{st}\text{row} \\ \leftarrow & 2^{nd}\text{row} \\ \\ \leftarrow & m^{th}\text{row} \end{matrix}$$

$$\begin{matrix} \uparrow & \uparrow & & \uparrow \\ 1^{st} & 2^{nd} & & n^{th}\text{ column} \end{matrix}$$

An individual element in a matrix can be referred to by listing its row and column, for example, $b_{1,1} = 0.25$, and $b_{2,1} = 0.125$. Both matrices A and B have two rows and two columns, so are 2×2 matrices – these are the dimensions of the matrix. A matrix with an equal number of rows and columns is called a square matrix. The number of rows and columns does not have to be equal:

$$D = \begin{bmatrix} 0.1 & 0.7 & 0.0 \\ 0.5 & 0.2 & 0.2 \end{bmatrix}$$

So matrix D is a 2×3 matrix.

A matrix can consist of a single row or single column, in which case they can be described as row and column vectors, respectively. For example:

$$a = \begin{bmatrix} 02 & 27 & 88 \end{bmatrix} \qquad b = \begin{bmatrix} 6 \\ 5 \\ 2 \end{bmatrix}$$

A.2 Basic Matrix Algebra

A matrix (A) can be multiplied by a constant (k):

J. H. Challis, *Experimental Methods in Biomechanics*, https://doi.org/10.1007/978-3-030-52256-8

$$kA = k.\begin{bmatrix} a & b \\ c & d \end{bmatrix} = \begin{bmatrix} k.a & k.b \\ k.c & k.d \end{bmatrix}$$

Two matrices, A and E, can be added or subtracted as long as they have the same dimensions:

$$A \pm E = \begin{bmatrix} a & b \\ c & d \end{bmatrix} \pm \begin{bmatrix} e & f \\ g & h \end{bmatrix} = \begin{bmatrix} a \pm e & b \pm f \\ c \pm g & d \pm h \end{bmatrix}$$

Two matrices can be multiplied:

$$C = AB$$

For this operation to be valid, the number of columns of matrix A must equal the number of rows of matrix B. The result of this multiplication, matrix C, will have the same number of rows as matrix A and the same number of columns as matrix B. Therefore if A is a 2×3, and B a 3×4, then C will be a 2×4 matrix. The elements of the matrix C are computed from:

$$c_{ij} = \sum_{k=1}^{n} a_{ik}b_{kj} \qquad \begin{aligned} i &= 1, 2, \ldots, \text{number of rows in } A \\ j &= 1, 2, \ldots, \text{number of columns in } B \end{aligned}$$

Therefore, for example:

$$AB = \begin{bmatrix} a & b \\ c & d \end{bmatrix} . \begin{bmatrix} e & f \\ g & h \end{bmatrix} = \begin{bmatrix} a.e + b.g & a.f + b.h \\ c.e + d.g & c.f + d.h \end{bmatrix}$$

$$\begin{bmatrix} 6 & 1 \\ 2 & 3 \end{bmatrix}\begin{bmatrix} 2 & 5 \\ 3 & 1 \end{bmatrix} = \begin{bmatrix} 15 & 31 \\ 13 & 13 \end{bmatrix}$$

Matrix multiplication can be illustrated using Falk's scheme:

$$A = \begin{bmatrix} 1 & 3 & 7 \\ 2 & -1 & 4 \\ -1 & 0 & 1 \end{bmatrix} \qquad B = \begin{bmatrix} 3 & 2 \\ -5 & 1 \\ 0 & 3 \end{bmatrix}$$

So for the highlighted areas:

$$(2 \times 2) + (-1 \times 1) + (4 \times 3) = 4 - 1 + 12 = 15$$

Matrix multiplication is associative so:

$$A(BC) = (AB)C$$

is distributive so:

$$A(B + C) = AB + AC$$

but not commutative:

$$AB \neq BA$$

For example:

$$A = \begin{bmatrix} 1 & 2 \\ 3 & 4 \end{bmatrix} \qquad B = \begin{bmatrix} 1 & 3 \\ 2 & 4 \end{bmatrix}$$

$$AB = \begin{bmatrix} 1 & 2 \\ 3 & 4 \end{bmatrix}\begin{bmatrix} 1 & 3 \\ 2 & 4 \end{bmatrix} = \begin{bmatrix} 5 & 11 \\ 11 & 25 \end{bmatrix} \qquad BA = \begin{bmatrix} 1 & 3 \\ 2 & 4 \end{bmatrix}\begin{bmatrix} 1 & 2 \\ 3 & 4 \end{bmatrix} = \begin{bmatrix} 10 & 14 \\ 14 & 20 \end{bmatrix}$$

$$AB \neq BA$$

A column vector can be multiplied by a matrix:

$$Ax = \begin{bmatrix} a & b \\ c & d \end{bmatrix}\begin{bmatrix} x_1 \\ x_2 \end{bmatrix} = \begin{bmatrix} ax_1 + bx_2 \\ cx_1 + dx_2 \end{bmatrix}$$

Matrix division is not defined, but you can multiply by the inverse of a matrix to achieve the same effect as division. For example:

$$A x = b \qquad A^{-1} b = x$$

Therefore you need to compute the inverse of a matrix, which is outlined later.

A matrix can be transposed, where the rows and columns are switched. The transpose operation is indicated by a superscript T:

$$\begin{bmatrix} a & b \\ c & d \end{bmatrix}^T = \begin{bmatrix} a & c \\ b & d \end{bmatrix}$$

$$\begin{bmatrix} 1 & 2 \\ 3 & 4 \\ 5 & 6 \end{bmatrix}^T = \begin{bmatrix} 1 & 3 & 5 \\ 2 & 4 & 5 \end{bmatrix}$$

The trace of a matrix is a value associated with every square matrix (equal number of rows and columns); it is the sum of the elements on the main diagonal (top left to bottom right). The trace of a matrix is computed; thus:

$$A = \begin{bmatrix} a & b \\ c & d \end{bmatrix} \quad tr(A) = a + d$$

or for a 3 x 3:

$$A = \begin{bmatrix} a & b & c \\ d & e & f \\ g & h & i \end{bmatrix} \quad tr(A) = a + e + i$$

The trace has some useful properties, including the following:

$$tr(AB) = tr(A) + tr(B)$$
$$tr(A) = tr\left(A^T\right)$$
$$tr\left(A^T B\right) = tr\left(AB^T\right) = tr\left(B^T A\right) = tr\left(BA^T\right) = \sum_{i,j} a_{ij} b_{ij}$$

A.3 Types of Matrices

There are a number of special matrices which are important. For example, I is the identity matrix, the matrix equivalent of 1 so multiplication by an identity matrix has no effect. For example, for a 2×2 matrix:

$$I = \begin{bmatrix} 1 & 0 \\ 0 & 1 \end{bmatrix}$$

$$IB = \begin{bmatrix} 1 & 0 \\ 0 & 1 \end{bmatrix} \begin{bmatrix} 0.25 & 0.33 \\ 0.125 & 0.666 \end{bmatrix} = \begin{bmatrix} 0.25 & 0.33 \\ 0.125 & 0.666 \end{bmatrix}$$

A null or zero matrix, $\mathbf{0}$, is a matrix where all elements are zero:

$$0 = \begin{bmatrix} 0 & 0 \\ 0 & 0 \end{bmatrix}$$

Matrices with only nonzero elements on its diagonal running from the upper left to the lower right of the matrix (leading diagonal) are diagonal matrices. For example:

$$A = \begin{bmatrix} 8 & 0 & 0 \\ 0 & 11 & 0 \\ 0 & 0 & 65 \end{bmatrix}$$

A symmetric matrix is a square matrix which is unchanged in its transpose:

$$A^T = A$$

For example:

$$\begin{bmatrix} 1 & 2 & 3 \\ 2 & 4 & 5 \\ 3 & 5 & 6 \end{bmatrix}^T = \begin{bmatrix} 1 & 2 & 3 \\ 2 & 4 & 5 \\ 3 & 5 & 6 \end{bmatrix}$$

A skew-symmetric matrix is a square matrix, with zero elements on the leading diagonal, and the transpose is its negative:

$$A^T = -A$$

For example:

$$\begin{bmatrix} 0 & 9 & 7 \\ -9 & 0 & -12 \\ -7 & 12 & 0 \end{bmatrix}^T = \begin{bmatrix} 0 & -9 & -7 \\ 9 & 0 & 12 \\ 7 & -12 & 0 \end{bmatrix}$$

So the operator $S\{\}$ generates a skew-symmetric matrix from vector $a = (a_1, a_2, a_3)$:

$$S\{a\} = \begin{bmatrix} 0 & -a_3 & a_2 \\ a_3 & 0 & -a_1 \\ -a_2 & a_1 & 0 \end{bmatrix}$$

A matrix is called non-singular, for example, matrix A, if its inverse exists (A^{-1}), so:

$$A^{-1}A = I$$

A matrix is orthogonal if its transpose is equal to its inverse; therefore:

$$A^T = A^{-1}$$

and therefore:

$$AA^T = AA^{-1} = I$$

$$\begin{bmatrix} 0.5253 & -0.8509 \\ 0.8509 & 0.5253 \end{bmatrix} \begin{bmatrix} 0.5253 & 0.8509 \\ -0.8509 & 0.5253 \end{bmatrix} = \begin{bmatrix} 1 & 0 \\ 0 & 1 \end{bmatrix}$$

An orthogonal matrix is a proper orthogonal matrix if its determinant is 1.

A.4 Matrix Properties

The rank of a matrix is the dimension of the vector space generated by its columns. It informs if the columns are linearly independent of one another; if this is the case, we would say the matrix A is non-singular, so the inverse can be computed.

The determinant is a value associated with every square matrix (equal number of rows and columns), for example, for a 2×2 matrix, it is computed; thus:

$$A = \begin{bmatrix} a & b \\ c & d \end{bmatrix} \quad \det(A) = a.d - b.c$$

or for a 3×3:

$$D = \begin{bmatrix} a & b & c \\ d & e & f \\ g & h & i \end{bmatrix} \quad \det(A) = aei + bfg + cdh - ceg - bdi - afh$$

Determinants can be used for a variety of purposes, for example, they can be used to solve a system of equations or identify the properties of a matrix. If the determinant of a matrix is nonzero, then the matrix is invertible; if two of the rows, or columns, are identical, then the determinant is zero.

Other properties of the determinant include the following:

$$\det(I) = 1$$
$$\det(A^T) = \det(A)$$
$$\det(AB) = \det(A)\det(B)$$
$$\det(A^{-1}) = \frac{1}{\det(A)}$$

A.5 Matrix Eigenvalues and Eigenvectors

Consider the following equation,

$$Av = \lambda v$$

Where A is a square matrix, v is non-zero vector, and λ is a real number. The problem for a given square matrix A is identifying the scaling number λ, and vector v which characterize this matrix. An $n \times n$ square matrix will not have more than n scaling numbers and vectors. Formally these numbers and vectors are referred to as the eigenvalues and eigenvectors of a matrix respectively. The eigenvalues and eigenvectors of a matrix have important applications, for example in determining the principal moments of inertia and principal axes of a rigid body, in stability analyses, and vibration analysis.

The eigenvalues can be determined by solving the following equation, called the characteristic equation:

$$\det(A - \lambda I) = 0$$

The eigenvectors are then computed from,

$$(A - \lambda I) = 0$$

For the following matrix:

$$A = \begin{bmatrix} 1 & 2 \\ 2 & 1 \end{bmatrix}$$

the eigenvalues are:

$$\lambda_1 = -1 \quad \lambda_2 = 3$$

and the eigenvalues are:

$$v_1 = \begin{bmatrix} 1 \\ -1 \end{bmatrix} \quad v_2 = \begin{bmatrix} 1 \\ 1 \end{bmatrix}$$

An important property for an $n \times n$ matrix A with the eigenvalues $\lambda_1, \ldots, \lambda_n$ is,

$$tr(A) = \lambda_1 + \ldots + \lambda_n$$

A.6 Matrix Inversion

If A is a matrix with an equal number of rows and columns, then:

$$A^{-1}A = AA^{-1} = I$$

where I is the identity matrix.

If a matrix is not invertible, it is said to be **singular** (it exists on its own), in which case its determinant would be zero.

If a matrix has m rows and m columns, and the matrix has a determinant which is nonzero then the inverse exists. For example:

$$B = \begin{bmatrix} 0.25 & 0.33 \\ 0.125 & 0.666 \end{bmatrix} \quad B^{-1} = \begin{bmatrix} 5.3174 & -2.6347 \\ -0.9980 & 1.9960 \end{bmatrix}$$

$$B^{-1}B = \begin{bmatrix} 1 & 0 \\ 0 & 1 \end{bmatrix} = I$$

There are a variety of techniques for the numerical inversion of a matrix, including:

- Gaussian elimination (Gauss-Seidel, LU factorization)
- Cholesky factorization (symmetric definite matrices)
- Singular value decomposition (see Appendix H)

If a matrix has m rows and n columns, the system is over-determined if $m > n$ and undetermined if $m < n$. In both of these cases, an approximate solution can be obtained using a pseudo-inverse. The most popular of these is the Moore-Penrose inverse, which is designated by A^+. The Moore-Penrose inverse has the following properties:

$$A A^+ A = A$$
$$A^+ A A^+ = A^+$$
$$(A A^+)^T = A A^+$$
$$(A^+ A)^T = A^+ A$$

For example:

$$A x = b \qquad \begin{bmatrix} 1 & 2 \\ 2 & 0 \\ 0 & 2 \end{bmatrix} x = \begin{bmatrix} 1 \\ 2 \\ 3 \end{bmatrix}$$

$$A^+ b = x \qquad A^+ = \frac{1}{9} \begin{bmatrix} 1 & 4 & -1 \\ 2 & -1 & 2.5 \end{bmatrix}$$

$$x = \frac{1}{6} \begin{bmatrix} 4 \\ 5 \end{bmatrix}$$

If the system is over-determined, that is, if $m > n$, then the Moore-Penrose inverse gives a solution which is closest in a least-square sense to the desired solution vector, while if $m < n$, the Moore-Penrose inverse gives the solution whose vector 2 norm is minimized.

The Moore-Penrose inverse can be computed using the singular value decomposition (see Appendix H).

A.7 Matrix Norms

Use of matrices often requires some quantification of the distance of the space of the matrix – norms provide this quantification. For a matrix A, its norm is $\|A\|$. If a matrix A is $m \times n$, then its norm can be computed from:

$$A_P = \left\{ \sum_{i=1}^m \sum_{j=1}^n |a_{ij}|^p \right\}^{1/p}$$

Common values for p are 1, 2, and ∞. When $p = 2$, then this is the Frobenius norm, which has some other interesting properties:

$$\|A\|_2 = \|A\|_F = \sqrt{\sum_{i=1}^m \sum_{j=1}^n |a_{ij}|^2} = \sqrt{tr(A^T A)} = \sqrt{\sum_{i=1}^r \lambda_i} = \sqrt{\sum_{i=1}^r \sigma_i^2}$$

where r is the rank of matrix A, λ_i is the ith nonzero eigenvalue of $[A]^T[A]$, and σ_i^2 is the ith singular value of A.

If we have an orthogonal matrix R ($n \times n$), then its Frobenius norm has the following properties:

$$\|R\|_F = \sqrt{tr(R^{-1} R)} = \sqrt{tr(I)} = \sqrt{n}$$

Matrix norms have the following key properties:

1. $\|A\| > 0$, if $A \neq 0$
2. $\|kA\| = |k| \|A\|$
3. $\|A + B\| \leq \|A\| + \|B\|$
4. $\|AB\| \leq \|A\| \|B\|$

A.8 Useful References

Ben-Israel, A., & Greville, T. N. E. (2003). *Generalized inverses: Theory and applications* (2nd ed.). New York: Springer.

Bhatia, R. (1997). *Matrix analysis*. New York: Springer.

Golub, G. H., & Van Loan, C. F. (1983). *Matrix computations*. Oxford: North Oxford Academic Publishing Company Ltd.

Lütkepohl, H. (1996). *Handbook of matrices*. Chichester; New York: Wiley.

Appendix B SI Units and Quantities Used in Biomechanics

B.1 SI System

The standard units and abbreviations for units and quantities for measurements are derived from Le Système International d'Unités (SI). The majority of quantities defined in biomechanics can be derived from the three basic quantities of length, mass, and time.

Length (L) – measured in meters, which is are defined as the length traveled by light in a vacuum in 1/299,792,458 of a second. SI unit is the meter (metre in Europe), the symbol for which is "m" (lower case).

Mass (M) – measured in kilograms, which is defined relative to the Planck constant, an invariant constant from quantum mechanics. SI unit is the kilogram, the symbol for which is "kg" (lower case).

Time (T) – unit is the second, which is defined by the characteristic frequency of a cesium clock. The symbol for second(s) is "s" (lower case).

B.2 Prefixes

There are a number of standard prefixes which can be used to designate units of measurement which are either larger or smaller than the base SI unit. The table presents the most common prefixes used in biomechanics.

Prefix	Multiplier	Symbol	Prefix	Multiplier	Symbol
Tera	10^{12}	T	Deci	10^{-1}	D
Giga	10^9	G	Centi	10^{-2}	C
Mega	10^6	M	Milli	10^{-3}	M
Kilo	10^3	k	Micro	10^{-6}	μ
Hecto	10^2	h	Nano	10^{-9}	N
Deca	10^1	da	Pico	10^{-12}	P

Therefore a centimeter is:

$$1 \text{ m} \times 10^{-2} = 0.01 \text{ m} \equiv 1 \text{ cm}$$

Alternatively a kilometer is:

$$1 \text{ m} \times 10^3 = 1000 \text{ m} \equiv 1 \text{ km}$$

Note that the SI unit for mass is the kilogram (1000 g):

$$1 \text{ g} \times 10^3 = 1000 \text{ g} \equiv 1 \text{ kg}$$

© The Editor(s) (if applicable) and The Author(s), under exclusive license to Springer Nature Switzerland AG 2021
J. H. Challis, *Experimental Methods in Biomechanics*, https://doi.org/10.1007/978-3-030-52256-8

Table B.1 Example of using SI prefixes for the SI unit for length

Unit	Equivalents	Examples
Nanometer	1 nm $= 0.001$ μm $= 10^{-9}$m	15 nm myosin filament diameter in muscle
Micrometer	1 μm $= 0.001$ mm $= 10^{-6}$m	1.25–3.65 μm sarcomere range of operation
Millimeter	1 mm $= 0.1$ cm $= 0.001$ m	0.1 mm typical thickness of human hair
Centimeter	1 cm $= 0.01$ m	1 cm approximate width of little finger nail
Meter	1 m	1.62 m mean of height of mature US woman
Kilometer	1 km $= 1000$ m	6 city blocks or 0.62 miles
Megameter	1 Mm $= 1000$ km $= 10^{6}$ m	40 Mm perimeter of earth, around poles
Gigameter	1 Gm $= 1000$ Mm $= 10^{9}$ m	0.38 Gm mean distance from earth to moon

Table B.2 Names for some derived SI units, with special names for the units

Quantity	Name	SI symbol	SI units
Frequency	Hertz	Hz	s^{-1}
Force	Newton	N	$kg.m.s^{-2}$
Pressure	Pascal	Pa	$N. m^{-2} \equiv kg. m^{-1}$
Energy	Joule	J	$N. m \equiv kg. m^{2}. s^{-2}$
Power	Watt	W	$J. s^{-1} \equiv kg. m^{2}. s^{-1}$

Table B.1 gives examples of the uses of the SI prefixes for the SI unit for length (meter), along with examples of their use for defining lengths.

B.3 Compound Units

A compound unit can be formed by the multiplication of two or more units; in these cases, the compound unit should be written with either a space between the symbols or a period (.), for example,

Units for velocity $m.s^{-1}$ or $m\ s^{-1}$, but not ms^{-1}

Note that when one unit is divided by another, there are two acceptable formats, e.g.,

Units for velocity $m.s^{-1}$ or m/s

Some of these compound units have special names; see Table B.2.

B.4 Common Quantities Used in Biomechanics

Table B.3 presents the common quantities used in biomechanics, including the common symbol used for their designations, their dimensions, units, and a definition. Dimensions for these quantities are expressed in terms of Length, Mass, and Time.

B.5 Conversion Factors

On occasions where we have to convert between units, Table B.4 will be of use.

Table B.3 Common biomechanical related quantities derived from the SI system

Physical quantity	Symbol	Dimension	SI unit	Definition
Linear displacement	s	L	m	The change in location of a point, change in position vector $\underline{r}$
Velocity	v	$L.T^{-1}$	$m.s^{-1}$	The time rate of change of location
Acceleration	a	$L.T^{-2}$	$m.s^{-2}$	The time rate of change of velocity
Gravity	g	$L.T^{-2}$	$m.s^{-2}$	Acceleration of a freely falling body due to gravity. (At sea level $g = 9.80665$ $m.s^{-2}$)
Angular displacement	θ	–	radian (rad)	Change in orientation of a line segment, 2D motion plane angle between initial and final orientation
Angular velocity	ω	T^{-1}	$rad.s^{-1}$	Time rate of change of orientation
Angular acceleration	α	T^{-2}	$rad.s^{-2}$	Time rate of change of angular velocity
Mass	M	M	kg	A measure of the amount of matter comprising an object
Weight	G	$M.L.T^{-2}$	N	Force acting on a body due to gravitational attraction; equal to product of body mass and acceleration due to gravity (wt = m.g)
Moment of inertia	I	$M.L^2$	$kg.m^2$	Measure of a body's resistance to accelerated angular motion about a given axis. ($I = \Sigma m.r^2$)
Force	F	$M.L.T^{-2}$	newton (N) $(kg.m.s^{-2})$	Effect of one body on another which causes the bodies to accelerate relative to an inertial reference frame
Moment of force	T	$M.L^2.T^{-2}$	N.m	Turning effect of a force about a defined axis; equal to the product of the magnitude of the force and the perpendicular distance (moment arm) from the line of action of the force to the axis about which the motion is defined
Linear momentum	P	$M.L.T^{-1}$	N.s $(kg.m.s^{-1})$	Quantity possessed by a moving body; equal to the product of its mass and the velocity of its mass center
Angular momentum	H	$M.L^2.T^{-1}$	$kg.m^2.s^{-1}$	The moment of the linear momentum of a body about a point.
Impulse	J	$M.L.T^{-1}$	N.s $(kg.m.s^{-1})$	The integral of a force over a time interval
Area	–	L^2	m^2	Portion of a surface
Volume	–	L^3	m^3	Space encompassed
Density	ρ	$M.L^{-3}$	$kg.m^{-3}$	Concentration of matter in an object
Specific gravity	D	–	–	The ratio of the density of a substance to the density of water; if reference is made to a substance other than water, then it is called relative density
Frequency	f	T^{-1}	hertz (Hz)	Number of cycles/repetitions of a periodic event in a second (e.g., 1 Hz = 1 cycle per second)
Pressure/ stress	p/σ	$M.L^{-1}.T^{-2}$	pascal (Pa)	Force distributed over a surface, the force per unit area. (1 Pa= 1 N.m^{-2})
Linear strain	ε	–	–	Deformation due to a stress, defined by change in length of object
Energy	E	$M.L^2.T^{-2}$	joule (J)	Capacity to do work, quantified by the sum of potential and kinetic energies
Power	P	$M.L^2.T^{-3}$	watt (W)	Rate at which energy is expended or work is done

Table B.4 Common conversion factors

Physical quantity	SI unit	Alternate unit	Conversion
	meter	centimeter	1 cm = 0.01 m
Linear	meter	kilometer	1 km = 1000 m
Displacement	meter	inch	1 inch = 0.0254 m
	meter	foot	1 foot = 0.3048 ml
	meter	mile	1 mile = 1609 m
Acceleration	$m.s^{-2}$	$ft.s^{-2}$	1 $ft.s^{-2}$ = 0.3048 $m.s^{-2}$
Angular	radian	degrees	1 degree = 0.01745 radians
Displacement	radian	revolution	1 rev. = 360 degrees = 6.283 rad
Mass	kg	pounds	1 pound = 0.453592 kg
Moment of inertia	$kg.m^2$	$lb.ft^2$	1 $lb.ft^2$ = 0.04241 $kg.m^2$

Appendix C Binary Arithmetic

Overview

The most common numeral system is the decimal system (base 10), but there are other base systems, for example, the Maya civilization used base 20 (Blume, 2011). The concept of binary arithmetic was first proposed by Gottfried Wilhelm Leibniz (1646–1716). Its name is derived from the Latin word *binaries* which means "two at a time." Binary arithmetic is important because of its use in analog-to-digital converters and in computers.

C.1 Concept of Binary Numbers

The idea of binary representation of quantities can be illustrated using the following example. *Question*: you have a balance scale and need to purchase the weights so you can measures all masses from 1 g to 1000 g in 1 g increments, what mass weights would you need?

For 1 g you need a 1 g weight, for 2 g you would need a 2 g weight, and for 3 g you could use the 1 g and 2 g weights. For up to 16 g the following weights are needed 1, 2, 4, 8, and 16 g. The combinations can be presented in tabular form,

Mass to be weighed	Weights needed
1 g	**1 g**
2 g	**2 g**
3 g	1 g +2 g
4 g	**4 g**
5 g	4 g +1 g
6 g	4 g +2 g
7 g	4 g +2 g +1 g
8 g	**8 g**
9 g	8 g +1 g
10 g	8 g +2 g
11 g	8 g +2 g +1 g
12 g	8 g +4 g
13 g	8 g +4 g +1 g
14 g	8 g +4 g +2 g
15 g	8 g +4 g +2 g +1 g
16 g	**16 g**

So to weigh up to 16 grams requires only five weights (1, 2, 4, 8, and 16 g). To weigh up to 1000 g (1 kg) requires the following weights 1, 2, 4, 8, 16, 32, 64, 128, 256, and 512 g. This provides a clue to an alternative to base ten for numbers. All of these weights are powers of 2,

J. H. Challis, *Experimental Methods in Biomechanics*, https://doi.org/10.1007/978-3-030-52256-8

Number	Power of 2
1	2^0
2	2^1
4	2^2
8	2^3
16	2^4
32	2^5
64	2^6
128	2^7
256	2^8
512	2^9

Therefore rather than working in the base of 10 (decimal system), it is possible to work in the base of 2 (binary). In this system only ones and zeros are used; the position of the number refers to a power of 2. The following table illustrates how this works:

Decimal number	Binary form	Binary breakdown
1	1	(1×2^0)
2	10	$(1 \times 2^1) + (0 \times 2^0)$
3	11	$(1 \times 2^1) + (1 \times 2^0)$
4	100	$(1 \times 2^2) + (0 \times 2^1) + (0 \times 2^0)$
5	101	$(1 \times 2^2) + (0 \times 2^1) + (1 \times 2^0)$
6	110	$(1 \times 2^2) + (1 \times 2^1) + (0 \times 2^0)$
7	111	$(1 \times 2^2) + (1 \times 2^1) + (1 \times 2^0)$
8	1000	$(1 \times 2^3) + (0 \times 2^2) + (0 \times 2^1) + (0 \times 2^0)$

The following table illustrates how this works for fractions:

Decimal number	Binary form	Binary breakdown
0.5	0.1	(1×2^{-1})
0.25	0.01	(1×2^{-2})
0.125	0.001	(1×2^{-3})
0.0625	0.0001	(1×2^{-4})

C.2 Binary Representation

Any decimal number can be represented in binary. To indicate the base, the base in subscript can be used; therefore:

$$(74)_{10} = (1001010)_2$$
$$(756)_{10} = (1011110100)_2$$
$$(0.25)_{10} = (0.01)_2$$
$$(0.75)_{10} = (0.11)_2$$

Within the decimal system, there are irrational numbers which cannot be precisely represented, for example $\frac{1}{3}$. Similar problems exist in the binary system, for example, $\frac{1}{10}$.

Within a computer, numbers are typically represented in binary form in the following format:

$$-1^s \times (1.f)_2 \times 2^{e-E}$$

where s is used to designate the sign of the number, f represents the bits in the significand, e is an integer exponent, and E is the bias exponent (127 for 32 bits). Notice that 1 is automatically used so the significand does not need include the first 1. Similarly when data is sampled, the analog-to-digital converter converts the sampled data into binary form.

C.3 Binary Arithmetic

The basic manipulations of addition and multiplication are presented below:

Addition	Multiplication
$0 + 0 = 0$	$0 \times 0 = 0$
$1 + 0 = 1$	$1 \times 0 = 0$
$0 + 1 = 1$	$0 \times 1 = 0$
$1 + 1 = 10$	$1 \times 1 = 1$
$1 + 10 = 11$	$10 \times 10 = 100$
$11 + 11 = 110$	$11 \times 11 = 1001$

The following example illustrates both multiplication and addition:

```
              1  0  1
        1  1  0  1  x
        ──────────────
              1  0  1
           0  0  0  0
        1  0  1  0  0
     1  0  1  0  0  0
  ───────────────────
  1  0  0  0  0  0  1
```

C.4 Reference

Blume, A. (2011). Maya concepts of zero. *Proceedings of the American Philosophical Society, 155*(1), 51–88.

Appendix D Trigonometry

Overview

Trigonometry has many applications in biomechanics: the following reviews the basics of trigonometry.

D.1 Basic Functions

The trigonometric functions are sometimes called the trigonometric ratios because they relate to the ratios of the lengths of the sides of right-angle triangles. For a right-angle triangle, these functions express the relationships between the ratio of the lengths of the sides and the angle between those sides.

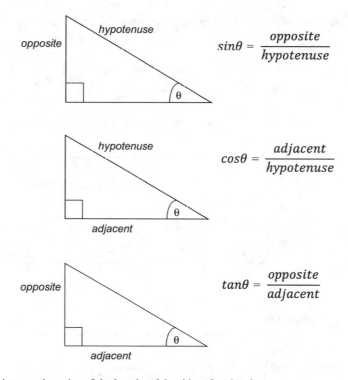

$$sin\theta = \frac{opposite}{hypotenuse}$$

$$cos\theta = \frac{adjacent}{hypotenuse}$$

$$tan\theta = \frac{opposite}{adjacent}$$

Fig. D.1 The trigonometric functions, as the ratios of the lengths of the sides of a triangle

J. H. Challis, *Experimental Methods in Biomechanics*, https://doi.org/10.1007/978-3-030-52256-8

D.2 Visualizing the Trigonometric Functions

Angles from $0°$ to $360°$ (2π radians) can be visualized by examining the motion of a line of length r rotating about the origin of a reference frame.

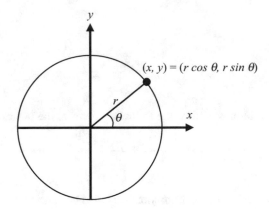

Fig. D.2 Illustration of how the trigonometric functions can be used to compute the end point of a line rotating about the origin of an axis system, thus defining the trigonometric functions from 0 to 360 degrees

Therefore, each of the three trigonometric functions for all angles from $0°$ to $360°$ can be plotted, see following figure.

Fig. D.3 The values of the sine, cosine, and tangent of angle, θ, from 0 to 360 degrees

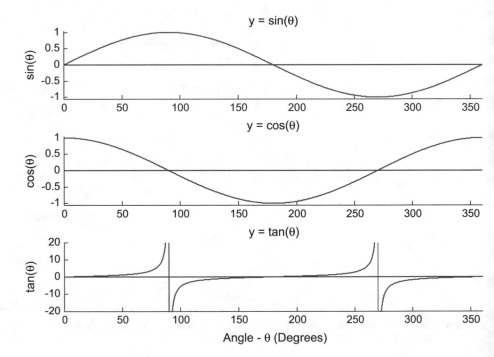

D.3 Amplitude, Frequency, and Phase

The following figure gives the basic anatomy of a sine wave, but the same principles apply to a cosine wave.

Fig. D.4 Sine wave, with period and amplitude identified

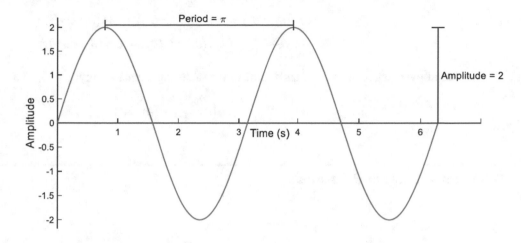

For a function, f, with values varying with time, t, a sine wave can be described by:

$$f(t) = b \, \sin(\omega_0 \, t)$$

where b is the amplitude and ω_0 is angular frequency (radians/second). The angular frequency is related to the period of the function (T) and the frequency of the signal (f_0):

$$f_0 = \frac{1}{T} = \frac{\omega_0}{2\pi}$$

In addition, the phase of a sine wave can shift the wave along the horizontal axis; therefore:

$$f(t) = b \, \sin(\omega_0 \, t - \phi)$$

where ϕ is the phase.

D.4 Trigonometric Identities

The following identities exist due to the periodic nature of the sine and cosine functions:

$$\sin(\theta) = -\sin(-\theta) = -\cos\left(\theta + \frac{\pi}{2}\right) = \cos\left(\theta - \frac{\pi}{2}\right)$$

$$\cos(\theta) = -\cos(-\theta) = \sin\left(\theta + \frac{\pi}{2}\right) = -\sin\left(\theta - \frac{\pi}{2}\right)$$

Also:

$$\tan(\theta) = \frac{\sin(\theta)}{\cos(\theta)}$$

The sum and differences between the sine and cosine of two angles have the following relationships:

$$\cos(\theta_1 + \theta_2) = \cos(\theta_1)\cos(\theta_2) - \sin(\theta_1)\sin(\theta_2)$$
$$\sin(\theta_1 + \theta_2) = \cos(\theta_1)\sin(\theta_2) - \sin(\theta_1)\cos(\theta_2)$$
$$\cos(\theta_1 - \theta_2) = \cos(\theta_1)\cos(\theta_2) + \sin(\theta_1)\sin(\theta_2)$$
$$\sin(\theta_1 - \theta_2) = \sin(\theta_1)\cos(\theta_2) - \cos(\theta_1)\sin(\theta_2)$$

The sum of the squares of the sine and cosine of the same angle equals unity:

$$\cos(\theta)^2 + \sin(\theta)^2 = 1$$

D.5 Laws of Sines and Cosines

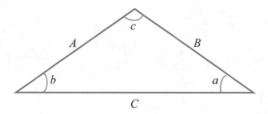

Fig. D.5 A triangle used to illustrate the Law of Sines, and Law of Cosines, with the sides A, B, and C and angles a, b, and c marked

The Law of Sines $\quad \dfrac{A}{\sin(a)} = \dfrac{B}{\sin(b)} = \dfrac{C}{\sin(c)}$

$$A^2 = B^2 + C^2 - 2BC\cos(a)$$
The Law of Cosines $\quad B^2 = A^2 + C^2 - 2AC\cos(b)$
$$C^2 = A^2 + B^2 - 2AB\cos(c)$$

D.6 Euler's Formula

The following relationship exists between the sine and cosine functions and the exponential function:

$$e^{\pm i\theta} = \cos(\theta) \pm \sin(\theta) \ \text{ where } \ i = \sqrt{-1}$$
$$\cos(\theta) = \frac{1}{2}\left(e^{i\theta} + e^{-i\theta}\right)$$
$$\sin(\theta) = \frac{1}{2i}\left(e^{i\theta} - e^{-i\theta}\right)$$

D.7 Rotation Matrices

The following matrices represent rotations about three orthogonal axes:

$$R_x(\alpha) = \begin{bmatrix} 1 & 0 & 0 \\ 0 & \cos(\alpha) & \sin(\alpha) \\ 0 & -\sin(\alpha) & \cos(\alpha) \end{bmatrix}$$

$$R_y(\beta) = \begin{bmatrix} \cos(\beta) & 0 & \sin(\beta) \\ 0 & 1 & 0 \\ -\sin(\beta) & 0 & \cos(\beta) \end{bmatrix}$$

$$R_Z(\gamma) = \begin{bmatrix} \cos(\gamma) & -\sin(\gamma) & 0 \\ \sin(\gamma) & \cos(\gamma) & 0 \\ 0 & 0 & 1 \end{bmatrix}$$

where α, β, and γ are angles of rotation about the X, Y, and Z axes, respectively.

D.8 Useful Equivalents

It can be useful to remember the following:

$$\tan\left(0^\circ\right) = \sin\left(0^\circ\right) = \cos\left(90^\circ\right) = 0$$

$$\tan\left(45^\circ\right) = \sin\left(90^\circ\right) = \cos\left(0^\circ\right) = 1$$

$$\sin\left(45^\circ\right) = \cos\left(45^\circ\right) = 1/\sqrt{2} \approx 0.707$$

$$\sin\left(30^\circ\right) = \cos\left(60^\circ\right) = 1/2$$

$$\sin\left(60^\circ\right) = \cos\left(30^\circ\right) = \sqrt{3}/2 \approx 0.866$$

$$\tan\left(30^\circ\right) = 1/\sqrt{3} \approx 0.577$$

As a reminder here are some equivalents between radians and degrees:

$$1 \text{ radian} \equiv 57.296 \text{ degrees}$$

$$\frac{\pi}{4} \text{ radians} \equiv 45 \text{ degrees}$$

$$\frac{\pi}{2} \text{ radians} \equiv 90 \text{ degrees}$$

$$\pi \text{ radians} \equiv 180 \text{ degrees}$$

$$2\pi \text{ radians} \equiv 360 \text{ degrees}$$

Appendix E Logarithms

Overview

Logarithms, or logs, were invented independently by the Scot John Napier (1550–1617) and the Swiss Joost Bürgi (1552–1632). Their target was to simplify mathematical calculations; so by using logs, multiplication is replaced by addition and division by subtraction. Before calculators and computers became generally available much arithmetic was performed using a table of logs; now that is not necessary but logs are still useful as ways of modeling data as many biological phenomena behave in a logarithmic way, for example,

- pH is the negative logarithm of the concentration of free hydrogen ions.
- We perceive sound as the logarithm of the sound intensity, so we measure sound on a log scale (decibels).
- We perceive the brightness of light as the logarithm of the actual light energy.
- Earthquake intensity is measured on a log scale (Richter).

In addition some problems in calculus are solved using logs (e.g., the area under the curve $\frac{1}{x}$ is provided by the natural logarithm of x).

E.1 Expressing Numbers on a Log Scale

All numbers can be expressed as a combination of one number serving as a base and the other serving as an exponent. For example, using *10* as the base:

$$0.1 = 10^{-1} \quad \text{or} \quad 100 = 10^2$$

Logs work on the principle of expressing a decimal number by the exponent expressed relative to a base. All logs have a base, and therefore any number can be represented by the exponent to which the base must be raised in order to produce that number. The two most common bases for logs are base e and *10*. e is an irrational number; here it is to the first *20* decimal places:

$$2.71828182845904523536$$

For convenience here we will either use e or *2.718*. The log of a number x to the base *2.718* is written ln x; these logs are referred to as the natural logs. The log of a number x to the base *10* is written $\log_{10}x$; these logs are referred to as the common logs. While we can determine logs we can also compute the inverse logs; the symbols used for logs and inverse (anti-) log are illustrated in Table E.1.

E.2 Common Logs

Table E.2 provides some examples of common logs, that is, logs to the base *10*.

Table E.1 The common symbols used for logarithms

Symbol	Meaning	Example
$\log_{10} x$	Common logarithm of x, log in base 10	$\log_{10}(100) = 2$
$\log_{10}^{-1} x$	Anti-logarithm of x, log in base 10	$\log_{10}^{-1}(2) = 100$
$\ln x$	Natural logarithm of x, log in base 2.718	$\ln(50) = 3.912$
e^x	Exponential of x, which is 2.718^x	$e^{3.912} = 50$

Table E.2 The common log of some numbers

x	$x = 10^y$	$\log_{10} x$
0.1	$0.1 = 10^{-1}$	-1
0.2	$0.2 = 10^{-0.69897}$	-0.69897
1	$\mathbf{1 = 10^0}$	**0**
2	$2 = 10^{0.30103}$	0.30103
10	$\mathbf{10 = 10^1}$	**1**
20	$20 = 10^{1.30103}$	1.30103
100	$100 = 10^2$	2
1000	$1000 = 10^3$	3

Table E.3 The natural log of some numbers

x	$x = e^y$	$\ln x$
0.5	$0.5 = e^{-0.69315}$	-0.69315
1	$\mathbf{1 = e^0}$	**0**
1.5	$1.5 = e^{0.40547}$	0.40547
2	$2 = e^{0.69315}$	0.69315
2.7183	$\mathbf{2.7183 = e^1}$	**1**
5	$5 = e^{1.6094}$	1.6094
10	$10 = e^{2.3026}$	2.3026
50	$50 = e^{3.912}$	3.912
100	$100 = e^{4.6052}$	4.6052
1000	$1000 = e^{6.9078}$	6.9078

Notice that:

$$\log_{10}(2 \times 10) = \log_{10}(2) + \log_{10}(10)$$

$$\log_{10}\left(\frac{1000}{100}\right) = \log_{10}(1000) - \log_{10}(100)$$

$$\log_{10}\left(10^2\right) = 2 \times \log_{10}(10)$$

E.3 Natural Logs

Table E.3 provides some examples of natural logs, that is, logs to the base e $(2.718\ldots)$.

Notice that:

$$\ln(5 \times 10) = \ln(5) + \ln(10)$$

$$\ln\left(\frac{1000}{100}\right) = \ln(1000) - \ln(100)$$

$$\ln\left(10^2\right) = 2 \times \ln(10)$$

E.4 Log Arithmetic

As logs are really exponents, arithmetic for logs is the same as the arithmetic of exponents. Some cases will be presented here to illustrate their use, the examples are presented in base *10*, common logs, but the same principles apply to all other logs.

Multiplication the log of a product of two numbers is equal to the sum of their logs.

$$\log_{10}(x.y) = \log_{10}(x) + \log_{10}(y)$$

For example, $$\log_{10}(2 \times 3) = \log_{10}(2) + \log_{10}(3)$$

Division the log of one number divided by another is equal to the difference between their logs.

$$\log_{10}\left(\frac{x}{y}\right) = \log_{10}(x) - \log_{10}(y)$$

For example, $$\log_{10}\left(\frac{2}{3}\right) = \log_{10}(2) - \log_{10}(3)$$

Raising to the power the log of a number raised to a power is equal to log of that number multiplied by the power.

$$\log_{10}(x^y) = y.\log_{10}(x)$$

For example, $$\log_{10}(2^3) = 3 \times \log_{10}(2)$$

Appendix F Numerical Data Differentiation

Overview

In many scenarios the derivative of a signal is required. There are many options for numerically computing these derivatives; a common method for determining signal derivatives is application of the finite difference equations. Forward finite difference equations can be used at the start of a data set where there are no previous signal values. The backward finite difference equations can be used at the end of a data set where there are no future signal values. The central difference equations work in the middle of the data set. These equations can have a varying number of terms. In this Appendix the forward, central, and backward finite difference equation coefficients are presented for the computation of first and second derivatives; in addition, the Lanczos equations for numerical differentiation are also presented.

F.1 Finite Difference Equations: First Derivative

Table F.1 presents the coefficients for the central finite difference equations with two, four, six, and eight terms. For example, for the central difference equations for estimating first derivatives at $\dot{f}_0$ with the two terms, the equation has one coefficient (½):

$$\dot{f}(0) = \frac{f_1 - f_{-1}}{2\Delta t}$$

Therefore, for the four term central difference equation, its format is:

$$\dot{f}_0 = \frac{-f_2 + 8 f_1 - 8 f_{-1} + f_{-2}}{12\Delta t}$$

Table F.2 presents the coefficients for the forward finite difference equations with two, three, four, and five terms.

Table F.1 Coefficients for central finite difference equations, first derivatives

Component	Number of terms			
	2	4	6	8
$f_1 - f_{-1}$	½	8/12	−45/60	672/840
$f_2 - f_{-2}$		−1/12	9/60	−168/840
$f_3 - f_{-3}$			−1/60	32/840
$f_4 - f_{-4}$				−3/840

N.B. – the fractions for each column use the same denominator even when they could be potentially simplified; this makes it easy to program the equations

Table F.2 Coefficients for forward finite difference equations, first derivatives

Component	Number of terms			
	2	3	4	5
f_0	−1	−3/2	−11/6	−25/12
f_1	1	4/2	18/6	48/12
f_2		−1/2	−9/6	−36/12
f_3			2/6	16/12
f_4				−3/12

© The Editor(s) (if applicable) and The Author(s), under exclusive license to Springer Nature Switzerland AG 2021
J. H. Challis, *Experimental Methods in Biomechanics*, https://doi.org/10.1007/978-3-030-52256-8

Table F.3 Coefficients for backward finite difference equations, first derivatives

Component	Number of terms			
	2	3	4	5
f_0	1	1/2	−2/6	3/12
f_{-1}	−1	−4/2	9/6	−16/12
f_{-2}		3/2	−18/6	36/12
f_{-3}			11/6	−48/12
f_{-4}				25/12

Table F.4 Coefficients for central finite difference equations, second derivatives

Component	Number of terms			
	3	5	7	9
f_0	−2	−30/12	−490/180	−14,350/5040
$f_1 + f_{-1}$	1	16/12	270/180	8064/5040
$f_2 + f_{-2}$		−1/12	−27/180	−1008/5040
$f_3 + f_{-3}$			2/180	128/5040
$f_4 + f_{-4}$				−9/5040

N.B. – the fractions for each column use the same denominator even when they could be potentially simplified; this makes it easy to program the equations

Therefore, for the two term forward difference equation, its format is:

$$\dot{f}_0 = \frac{f_1 - f_0}{\Delta t}$$

Table F.3 presents the coefficients for the backward finite difference equations with two, three, four, and five terms.

Therefore, for the two term backward difference equation, its format is:

$$\dot{f}_0 = \frac{f_0 - f_{-1}}{\Delta t}$$

F.2 Finite Difference Equations: Second Derivative

Table F.4 presents the coefficients for the central finite difference equations with three, five, seven, and nine terms for determining second derivatives.

Therefore, for the five term central difference equation, its format is:

$$\ddot{f}_0 = \frac{-f_2 + 16 f_1 - 30f_0 + 16 f_{-1} - f_{-2}}{12\Delta t^2}$$

Table F.5 presents the coefficients for the forward finite difference equations with three, four, five, and six terms for determining second derivatives.

Table F.6 presents the coefficients for the backward finite difference equations with three, four, five, and six terms for determining second derivatives.

Table F.5 Coefficients for forward finite difference equations, second derivatives

Component	Number of terms			
	3	4	5	6
f_0	1	2	35/12	45/12
f_1	−2	−5	−104/12	−154/12
f_2	1	4	114/12	215/12
f_3		−1	−56/12	−156/12
f_4			11/12	61/12
f_5				−10/12

Table F.6 Coefficients for backward finite difference equations, second derivatives

Component	Number of terms			
	3	4	5	6
f_0	1	2	35/12	45/12
f_{-1}	−2	−5	−104/12	−154/12
f_{-2}	1	4	114/12	215/12
f_{-3}		−1	−56/12	−156/12
f_{-4}			11/12	61/12
f_{-5}				−10/12

Table F.7 Coefficients for Lanczos (1967) central equations for estimating first derivatives

Component	Number of terms			
	4	4	6	8
$f_1 - f_{-1}$	1/10	1/28	1/60	1/110
$f_2 - f_{-2}$	2/10	2/28	2/60	2/110
$f_3 - f_{-3}$		3/28	3/60	3/110
$f_4 - f_{-4}$			4/60	4/110
$f_5 - f_{-5}$				5/110

F.3 Lanczos Equations: First Derivative

Others have presented variations on the finite difference method which use different coefficients and therefore produce a difference transfer function. For example, Lanczos (1967) presented the following equations (Table F.7):

$$\dot{f}_0 = \frac{2f_2 + f_1 - f_{-1} + 2f_{-2}}{10\Delta t} \quad \text{Central}$$

$$\dot{f}_0 = \frac{-21f_0 + 13f_1 + 17f_2 - 9f_3}{20\Delta t} \quad \text{Forward}$$

$$\dot{f}_0 = \frac{-21f_0 + 13f_{-1} + 17f_{-2} - 9f_{-3}}{20\Delta t} \quad \text{Backward}$$

The following graph presents the frequency magnitude response for the equations of Lanczos (1967). Note an ideal differentiator has an increasing influence on the signal with increasing signal frequency component; see Fig. F.1 for ideal response. With increasing number of coefficients in the equation, there is greater attenuation of the higher-frequency components of the signal.

Fig. F.1 The frequency magnitude response of the equations for estimating first derivatives from Lanczos (1967), with an increasing number of terms, where Nt is the number of terms in the formula. The dotted line represents the theoretical ideal differentiator

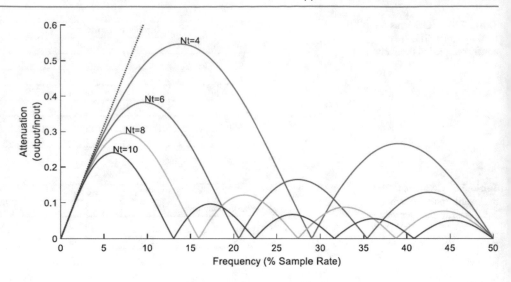

F.4 Reference

Lanczos, C. (1967). *Applied analysis*. London: Pitman and Sons.

Appendix G Data Sets for Testing Data Differentiation

Overview

To evaluate filtering and differentiation procedures, a number of different data sets have been proposed. Typically these data sets comprise a noisy zero-order derivative data and criterion second-order derivative data. The task is to process the noisy data and evaluate a filtering and differentiation procedure by comparing estimated acceleration values with the criterion values. The common numerical criterion used is the percentage root mean square difference (%RMSD) which is computed from:

$$\%\text{RMSD} = 100\sqrt{\frac{\frac{1}{n}\sum_{i=1}^{n}\left(x_{ci} - x_i\right)^2}{\frac{1}{n}\sum_{i=1}^{n}\left(x_{ci}\right)^2}}$$

where n is the number of data points, x_{ci} is the ith value of the criterion signal, and x_i is the estimated i^{th} value of the signal.

These test data sets can be divided into two broad categories: experimentally generated data sets and function generated data sets.

G.1 Experimentally Generated Data Sets

Pezzack et al. (1977) – this data set comprises angular displacement from film records and angular acceleration data from an accelerometer measured from limb motion. The data set comprises 142 data points sampled at 50 Hz. These data are presented in the paper (Fig. G.1).

Fig. G.1 The noisy angular displacement data and criterion angular acceleration data of Pezzack et al. (1977)

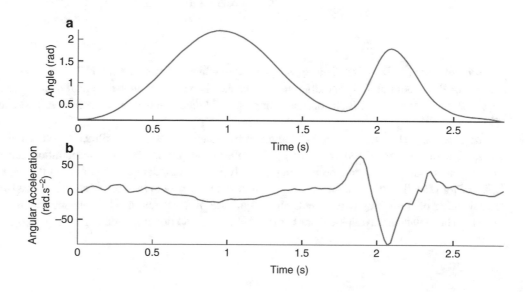

Fig. G.2 The noisy angular displacement data and criterion angular acceleration data of Lanshammar (1982)

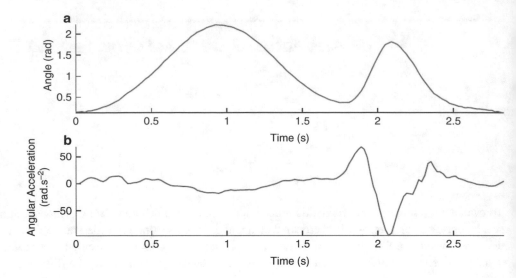

Fig. G.3 The noisy displacement data and criterion acceleration data of Vaughan (1982)

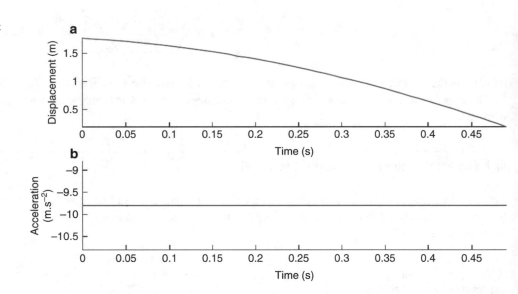

Lanshammar (1982) – this data set is based on the data of Pezzack et al. (1977). Lanshammar (1982) identified that the data set of Pezzack et al. (1977) did not have noise levels typical of biomechanical data, so added more noise to the original data of Pezzack et al. (1977). The data set comprises 142 data points sampled at 50 Hz. These data are presented in the paper (Fig. G.2).

Vaughan (1982) – this data set was presented by Vaughan (1982) who filmed a golf ball dropping under the influence of gravity. He then presented vertical (noisy) ball displacement data and the local acceleration due to gravity (-9.802418 m.s^{-2}). The data set comprises 50 data points sampled at 100 Hz. These data are presented in the paper (Fig. G.3).

Dowling (1985) – this data set was produced in a similar way to Pezzack et al. (1977) by Dowling (1985) but contained a greater range of frequency components and magnitude of acceleration. The data set comprises 600 data points sampled at 512 Hz. These data are available from https://isbweb.org/data/dowling/index.html (Fig. G.4).

Fig. G.4 The noisy angular displacement data and criterion angular acceleration data of Dowling (1985)

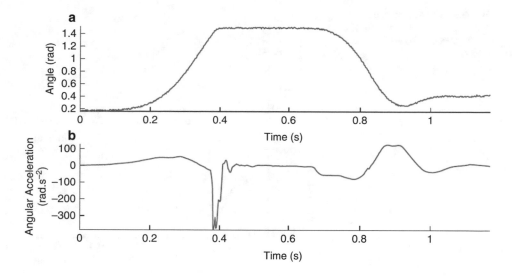

Table G.1 The 13 constant pairs for Data Set A

a_i	16.0	12.0	6.3	0.5	1.2	0.6	0.3	0.2	0.1	0.09	0.04	0.06	0.01
ϕ_i	0.99	0.23	−2.97	0.0	−3.13	−0.84	2.14	0.58	1.15	0.17	−1.13	0.76	−1.21

G.2 Function Generated Data Sets

A number of authors have presented functions which can generate displacement data and which can be analytically differentiated to produce criterion acceleration data. To make the process of estimating acceleration values more challenging, noise is added to the displacement data.

Data Set A This data set was derived from one of the functions presented by Andrews et al. (1982). It was designed to model the displacement of the iliac crest tubercle of an amputee during level walking. The function used to provide the criterion displacement data was:

$$f(t) = \sum_{i=1}^{13} a_i . \sin{(i.\omega_0.t + \phi_i)}$$

where:

t – time for the following range $0 \leq t \leq 1.0s$
a_i, ϕ_i – constants whose values are presented in Table G.1
ω_0 – equal to $2.\pi$

The function is generated for 91 samples. For a 1% root mean square difference between the true signal and the noisy signal, the standard deviation of the noise is 0.15, while at 20%, the standard deviation is 2.99 (Fig. G.5).

Data Set B This data set was generated from one of the functions presented by Andrews et al. (1982) and is an extension of the function presented for Data Set A. The function used to provide the criterion displacement data was:

$$f(t) = k\left(t - \frac{p}{2}\right)^2 + \sum_{i=1}^{13} a_i . \sin{(i.\omega_0.t + \phi_i)}$$

Fig. G.5 The displacement data and criterion acceleration for Data Set A, presented by Andrews et al. (1982)

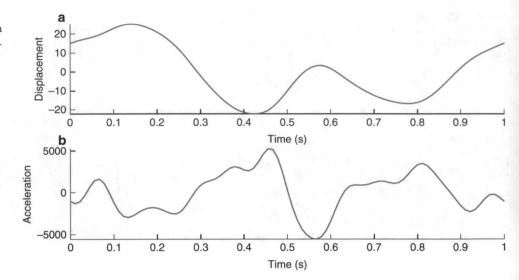

Fig. G.6 The displacement data and criterion acceleration for Data Set B, presented by Andrews et al. (1982)

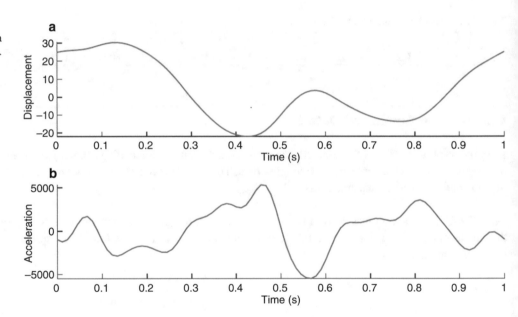

where:

t – time for the following range $0 \leq t \leq 1.0s$

k – constant equal to 40

$p = \Delta t(n-1) - \frac{\Delta t}{2}$

a_i, ϕ_i – constants (presented in Table G.1 for Data Set A)

ω_0 – equal to $2.\pi$

The function is generated for 91 samples. For a 1% root mean square difference between the true signal and the noisy signal, the standard deviation of the noise is 0.17, while at 20%, the standard deviation is 3.48 (Fig. G.6).

Data Set C This data set was derived from modeling the data of Felkel (1951) which described the displacement of the shank during walking. The function used to provide the criterion displacement data was:

$$f(t) = -55.1t + 427t^3 - 342t^4 \qquad t \leq 0.85s$$
$$f(t) = 579.97 - 304.32t - 241.77t^{-1} \qquad t > 0.85s$$

where t is time for the following range $0 \leq t \leq 1.0$.

The function is generated for 91 samples. For a 1% root mean square difference between the true signal and the noisy signal, the standard deviation of the noise is 0.21, while at 20%, the standard deviation is 4.16 (Fig. G.7).

Data Set D This data set is generated from a function proposed by Trujillo and Busby (1983) for testing techniques designed to differentiate noisy data. The function used to provide the criterion displacement data was:

$$f(t) = \frac{1}{\pi^2} \cdot \sin(\pi \cdot t) \qquad 0 \le t \le 4.0s$$

$$f(t) = \frac{1}{\pi}(t - 4) \qquad 4.0 \le t \le 5.0s$$

where t is time for the following range $0 \le t \le 5.0$.

The function is generated for 200 samples. For a 1% root mean square difference between the true signal and the noisy signal, the standard deviation of the noise is 0.001, while at 20%, the standard deviation is 0.020 (Fig. G.8).

Fig. G.7 The displacement data and criterion acceleration for Data Set C, presented by Felkel (1951)

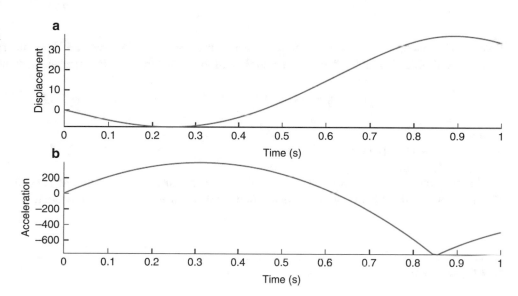

Fig. G.8 The displacement data and criterion acceleration for Data Set D, presented by Trujillo and Busby (1983)

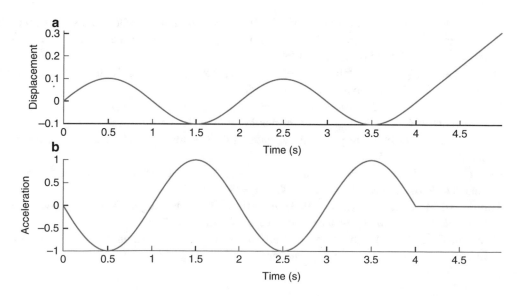

Fig. G.9 The displacement data and criterion acceleration for Data Set E, presented by Anderssen and Bloomfield (1974)

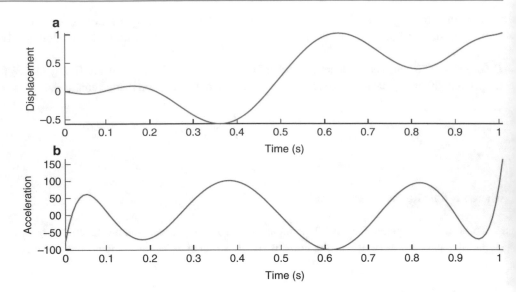

Data Set E This data set is generated from a function proposed by Anderssen and Bloomfield (1974) for testing techniques designed to differentiate noisy data. The function used to provide the criterion displacement data was:

$$f(t) = B.t^2.(t-0.1).(t-0.2).(t-0.5)(t-0.75).(t-0.95).(t-1)^2 + t^2$$

where:

B – constant equal to 10,000

t – time for the following range $0 \le t \le 1.0s$

The function is generated for 128 samples. For a 1% root mean square difference between the true signal and the noisy signal, the standard deviation of the noise is 0.006, while at 20%, the standard deviation is 0.115 (Fig. G.9).

G.3 References

Anderssen, R. S., & Bloomfield, P. (1974). Numerical differentiation procedures for non-exact data. *Numerische Mathematik, 22,* 157–182.

Andrews, B. J., Cappozzo, A., & Gazzani, F. (1982). A quantitative method for assessment of differentiation techniques used for locomotion analysis. In J. P. Paul, M. M. Jordan, M. W. Ferguson-Pell, & B. J. Andrews (Eds.), *Computing in medicine* (pp. 146–154). London: Macmillan Press.

Dowling, J. J. (1985). A modelling strategy for the smoothing of biomechanical data. In B. Jonsson (Ed.), *Biomechanics X-B* (pp. 1163–1167). Champaign: Human Kinetics Publishers.

Felkel, E. O. (1951). *Determination of acceleration from displacement-time data.* Institute of Engineering Research, University of California, Berkeley: Prosthetic Devices Research Project, Series 11, Volume 16.

Lanshammar, H. (1982). On practical evaluation of differentiation techniques for human gait analysis. *Journal of Biomechanics, 15*(2), 99–105.

Pezzack, J. C., Norman, R. W., & Winter, D. A. (1977). An assessment of derivative determining techniques used for motion analysis. *Journal of Biomechanics, 10*(5–6), 377–382.

Trujillo, D. M., & Busby, H. R. (1983). Investigation of a technique for the differentiation of empirical data. *Journal of Dynamic Systems, Measurement, and Control, 105*(3), 200–202.

Vaughan, C. L. (1982). Smoothing and differentiation of displacement-time data: An application of splines and digital filtering. *International Journal of Bio-Medical Computing, 13*(5), 375–385.

Appendix H Singular Value Decomposition

Overview

The singular value decomposition (SVD) is a popular method for analyzing data. As a method it has a long history (Stewart, 1993), but became popular when computer code became available for its fast implementation (Golub and Reinsch, 1971). Any real matrix, A, with dimension $m \times n$ can be decomposed as follows:

$$A = UDV^T$$

where U is an $m \times m$ orthogonal matrix consisting of vectors u_1, u_2, $\ldots u_m$, D is an $m \times n$ diagonal matrix, the non-negative real values are the singular values, and V is a $n \times n$ orthogonal matrix, consisting of vectors v_1, v_2, $\ldots v_n$.

The matrix D is diagonal, where the values along the leading diagonal are the ordered singular values so:

$$D = \text{diag}(\sigma_1, \sigma_2, \ldots, \sigma_n)$$

In D the singular values are ordered so that $\sigma_1 \geq \sigma_2 \ldots \sigma_n$.

The SVD of a matrix has great utility in computing various aspect of matrix algebra. A number will be outlined in the following.

H.1 Condition Number

The condition number is the ratio of the largest to smallest singular value. The matrix is singular if the condition number is infinite and is ill-conditioned if its reciprocal approaches machine precision.

H.2 Matrix Rank

The rank of a matrix is the dimension of the vector space generated by its columns. It informs if the columns are linearly independent of one another; if this is the case, we would say the matrix A is non-singular, so the inverse can be computed from its SVD. The singular values provide the rank of a matrix, with the number of singular values greater than zero (or machine precision) giving the rank.

H.3 Computing the Inverse of a Matrix

The inverse of a square matrix, A, exists only if all the singular values are not equal to zero (the rank is equal to the dimensions of the matrix). As the matrices U and V, from the SVD, are orthogonal, their inverses are simply their transposes. Therefore the inverse of matrix A can be calculated from:

$$A = UDV^T$$
$$A^{-1} = VD^{-1}U^T$$

where $D^{-1} = \text{diag}\left(\frac{1}{\sigma_1}, \frac{1}{\sigma_2}, \ldots, \frac{1}{\sigma_n}\right)$.

© The Editor(s) (if applicable) and The Author(s), under exclusive license to Springer Nature Switzerland AG 2021
J. H. Challis, *Experimental Methods in Biomechanics*, https://doi.org/10.1007/978-3-030-52256-8

The utility of the SVD includes that it can be used to provide an approximation of the inverse if the matrix is singular or ill-conditioned:

$$A^{-1} = VD^{-1}U^T$$

where $D^{-1} = \text{diag}\left(\frac{1}{\sigma_1}, \frac{1}{\sigma_2}, \ldots, \frac{1}{\sigma_n}\right)$ but if $\sigma_i > \varepsilon$ $D^{-1}(i, i) = 0$, where ε is a threshold (typically the precision of the computer).

H.4 Least-Squares Solution of Systems of Linear Equations

An over-determined system of equations can be expressed in equation form:

$$Ax = b$$

where,

A is an $n \times n$ matrix.
x is an unknown $n \times 1$ vector.
and b is an $n \times 1$ vector.

If both sides are multiplied by the inverse of A, the vector of unknowns can be determined,

$$A^{-1}Ax = A^{-1}b \quad \text{so} \quad x = A^{-1}b$$

Here is an example; the two simultaneous equations can be expressed in matrix form and then solved using an SVD generated inverse:

$$x - 2y = 4 \tag{1}$$

$$x + 3y = 9 \tag{2}$$

$$Ax = b \rightarrow \begin{bmatrix} 1 & -2 \\ 1 & 3 \end{bmatrix} \cdot \begin{bmatrix} x \\ y \end{bmatrix} = \begin{bmatrix} 4 \\ 9 \end{bmatrix}$$

$$A^{-1} = \begin{bmatrix} 0.6 & 0.4 \\ -0.2 & 0.2 \end{bmatrix}$$

$$\begin{bmatrix} 0.6 & 0.4 \\ -0.2 & 0.2 \end{bmatrix}\begin{bmatrix} 1 & -2 \\ 1 & 3 \end{bmatrix} \cdot \begin{bmatrix} x \\ y \end{bmatrix} = \begin{bmatrix} 0.6 & 0.4 \\ -0.2 & 0.2 \end{bmatrix}\begin{bmatrix} 4 \\ 9 \end{bmatrix}$$

$$\begin{bmatrix} 1 & 0 \\ 0 & 1 \end{bmatrix} \cdot \begin{bmatrix} x \\ y \end{bmatrix} = \begin{bmatrix} 0.6 & 0.4 \\ -0.2 & 0.2 \end{bmatrix}\begin{bmatrix} 4 \\ 9 \end{bmatrix} = \begin{bmatrix} 6 \\ 1 \end{bmatrix}$$

$$x = 6, y = 1$$

H.5 Moore-Penrose Pseudo-inverse

The SVD can be used to compute the pseudo-inverse of a matrix A, where the Moore-Penrose inverse is designated by A^+. The computation of the pseudo-inverse parallels the computation of the inverse of a matrix, so its SVD is computed:

$$A = UDV^T$$

Then:

$$A^+ = VD^{-1}U^T$$

where:

$$D^{-1} = \begin{bmatrix} \dfrac{1}{\sigma_1} & 0 & & 0 & 0 \\ & & \cdots & & \\ 0 & \dfrac{1}{\sigma_2} & & 0 & 0 \\ \vdots & & \ddots & & \vdots \\ 0 & 0 & & 0 & 0 \\ 0 & 0 & \cdots & \dfrac{1}{\sigma_m} & 0 \end{bmatrix}$$

If the system is over-determined, that is, if $m > n$, then the Moore-Penrose inverse gives a solution which is closest in a least-square sense to desired solution vector, while if $m < n$, the Moore-Penrose inverse gives the solution whose vector 2 norm is minimized.

H.6 Eigenvalues and Eigenvectors

The SVD of a matrix A is intrinsically linked with the eigenvalues and eigenvectors:

$$A = UDV^T$$

where U is an $m \times m$ matrix whose columns are the eigenvectors of the matrix AA^T, D is an $n \times n$ diagonal matrix where the squares of the singular values are the eigenvalues of the matrix A^TA, and V is an $n \times n$ matrix whose columns are the eigenvectors of the matrix A^TA.

H.7 References

Golub, G. H., & Reinsch, C. (1971). Singular value decomposition and least squares solutions. In J. H. Wilkinson & C. Reinsch (Eds.), *Handbook for automatic computations vol II: Linear algebra* (pp. 131–154). Heidelberg: Springer.
Stewart, G. W. (1993). On the early history of the singular-value decomposition. *SIAM Review, 35*(4), 551–566.

Appendix I Ellipses

Overview

In science the ellipse has many uses, for example, pupil tracking, cell segmentation, and astronomical modeling. Ellipse fitting is a first step for 3D analysis of circular or elliptical objects in computer vision applications, including marker-based automatic motion analysis systems used in biomechanics. In biomechanics another application is to compute an ellipse that fits the center of pressure stabilogram, and then the area of the ellipse can be used to quantify the center of pressure motion. In the following sections, the basic anatomy of ellipse will be described, the modeling of the ellipse outlined, and the process for fitting an ellipse to data described.

I.1 Basic Anatomy

The basic anatomy of an ellipse depends on the lengths of two axes (see Fig. I.1).

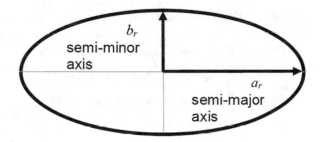

Fig. I.1 The axes of an ellipse, the semi-major axis being the longer of the two axes and the semi-minor axis the shorter

The eccentricity of the ellipse can be computed from:

$$\varepsilon = \frac{\sqrt{a_r^2 - b_r^2}}{a_r}$$

An eccentricity of *0* occurs for a circle, and a ratio of nearly *1* occurs for a long and narrow ellipse. The area of the ellipse (*A*) is computed from:

$$A = \pi\, a_r\, b_r$$

I.2 Modeling the Ellipse

The lengths of the two axes of the ellipse can be used to define the coordinates of the edges of the ellipse (*x*, *y*):

$$\frac{x^2}{a_r^2} + \frac{y^2}{b_r^2} = 1$$

J. H. Challis, *Experimental Methods in Biomechanics*, https://doi.org/10.1007/978-3-030-52256-8

If the center of the ellipse is not at the origin of the axis system, then the equation becomes:

$$\frac{(x - x_0)^2}{a_r^2} + \frac{(y - y_0)^2}{b_r^2} = 1$$

The more general way to present the equation describing an ellipse is the quadratic equation, which also allows the rotation of the ellipse so the major and minor axes are not aligned with the coordinate system:

$$ax^2 + b\,xy + cy^2 + dx + ey + f = 0$$

If for an ellipse $a = c$ and $b = 0$, then the equation describes a circle. The equation can be represented in matrix form:

$$X^T A X + Y^T X + f = 0$$

where $X = (x, y)^T$

$$A = \begin{bmatrix} a & b/2 \\ b/2 & c \end{bmatrix}$$

$Y = (d, e)^T$

The determinant of matrix A can be computed from:

$$\det(A) = a_{11}a_{22} - a_{21}a_{12}$$

Then determinant can be used to confirm the quadratic represents an ellipse:

$$\det(A) < 0 \quad \text{quadratic represents a hyperbola}$$
$$\det(A) = 0 \quad \text{quadratic represents a parabola}$$
$$\det(A) > 0 \quad \text{quadratic represents an ellipse}$$

I.3 Center of the Ellipse

The ellipse center is not necessarily the origin of the coordinate system. The center (x_c, y_c) can be computed from:

$$\begin{bmatrix} x_c \\ y_c \end{bmatrix} = A^{-1} \begin{bmatrix} -\dfrac{d}{2} \\ -\dfrac{e}{2} \end{bmatrix}$$

I.4 Angle of Major Axis

The major axis is not necessarily aligned with the axes of the coordinate system. The angle of the major axis with respect to the positive x axis (θ) can be computed from:

$$\theta = \operatorname{atan}\left(\frac{a_{21}}{a_{11}}\right)$$

I.5 Length of Axes

The lengths of the minor and major axes can be computed from the eigenvalues (λ_1, λ_2) of matrix A and the determinants of matrix A and B:

$$B = \begin{bmatrix} a & b/2 & d/2 \\ b/2 & c & e/2 \\ d/2 & e/2 & f \end{bmatrix}$$

$$k = -\frac{\det(B)}{\det(A)}$$

$$b_r^2 = \frac{k}{\lambda_1}$$

$$b_r^2 = \frac{k}{\lambda_2}$$

I.6 Fitting an Ellipse

There have been various algorithms proposed to fit the ellipse to a series of data points (e.g., Bookstein, 1979; Batschelet, 1981; Gander and Hřebíček, 1997), where the fit is a least-squares fit so not all data points are necessarily within the fitted ellipse. Most of the algorithms enforce some constraint on the parameters which described the ellipse. The ellipse parameters are:

$$e_p = (a, b, c, d, e, f)$$

For example, if the following constraint is enforced:

$$a + c = 1$$

then the following equation can be formulated (Gander et al., 1994):

$$\begin{bmatrix} 2x_1y_1 & y_1^2 - x_1^2 & x_1 & y_1 & 1 \\ \cdots & \cdots & \cdots \cdots & \cdots \\ 2x_ny_n & y_n^2 - x_n^2 & x_n & y_n & 1 \end{bmatrix} [u] = \begin{bmatrix} -x_1^2 \\ \cdots \\ -x_n^2 \end{bmatrix}$$

where n is the number of data points. The vector u can be computed by multiplying both sides of the equation by the inverse of the matrix. Then the ellipse parameters can be computed from:

$$e_p = (1 - u_2, u_1, u_2, u_3, u_4, u_5)$$

Example
Using the algorithm for the estimation of the parameters of an ellipse, the data in Table I.1 are used.

The algorithm is described in the following pseudo-code, and the results are illustrated in Fig. I.2.

Table I.1 Data used illustrate ellipse fitting

x	1	2	5	7	9	6	3	8
y	7	6	8	7	5	7	2	4

From Gander et al. (1994)

Pseudo-code: Computation of the Area of an Ellipse to Encompass x, y Data

Purpose – To illustrate the computation of the area of an ellipse encompassing a set of x, y coordinate data	
Inputs x – Corresponding x coordinates of data to be described with an ellipse y – Corresponding y coordinates of data to be described with an ellipse	
Outputs a – Set of six parameters describing an ellipse, fitted to model of form: $ax^2 + b\,xy + cy^2 + dx + ey + f = 0$ Area – Area of ellipse	
G ← [2 * x. * y, y.^2 - x.^2, x, y, ones(n, 1)]	
h ← -x.^2	
u ← G \ h'	Solve for vector u
a ← [1-u(2); u]	Vector of quadratic equation coefficients
A ← eig([a(1), a(2)/2; a(2)/2, a(3)])	Form two matrices, and compute eigenvectors
B ← eig([a(1), a(2)/2 a(4)/2; a(2)/2, a(3), a(5)/2; a(4)/2, a(5)/2, a(6)]	
[Q, D] ← eig(A)	Compute eigenvalues (D)
k ← - det(A)/det(B)	Ratio of determinants of two matrices
alen ← sqrt(k / D(1,1)) blen ← sqrt(k / D(2,2)) area ← pi * alen * blen	Length of two axes

Fig. I.2 An ellipse fitted to the example data. Note the fit does not encompass all of the data points

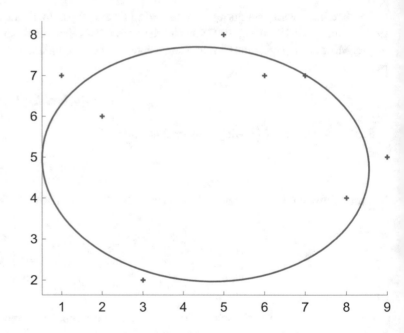

I.7 References

Batschelet, E. (1981). *Circular statistics in biology*. London; New York: Academic Press.

Bookstein, F. L. (1979). Fitting conic sections to scattered data. *Computer Graphics and Image Processing, 9*(1), 56–71.

Gander, W., Golub, G. H., & Strebel, R. (1994). Least-squares fitting of circles and ellipses. *BIT Numerical Mathematics, 34* (4), 558–578.

Gander, W., & Hřebíček, J. (1997). *Solving problems in scientific computing using maple and MATLAB* (4th, expanded and rev. ed.). New York: Springer.

Appendix J Calculus

Overview

Calculus is only the math of change; it uses two main techniques: differentiation and integration. Differentiation allows us to examine the rate of change, and integration relates to the area under a curve. It turns out that integration is the inverse process of differentiation.

The methods to be described here were initially developed in the seventeenth century by two famous scholars. Both Isaac Newton (1642–1727) and Gottfried Wilhelm Leibnitz (1646–1716) discovered the principles of calculus. They both made these discoveries independently, although there was dispute about who should be given credit for this important discovery. Since their work, these methods have been refined and applied in many contexts.

Both differentiation and integration can be performed analytically on a function or numerically given a data set. The focus in this Appendix is analytical differentiation and integration of functions; for numerical differentiation, see Appendix F, and for numerical integration, see relevant section in Chap. 7.

J.1 Derivatives

The derivative is the instantaneous rate of change of a function with respect to one of its variables. This is equivalent to finding the gradient of the tangent line to the function at a particular point. In many areas of science, the most commonly used variable is time, so the rate of change of a function is examined with respect to time. For example, if we have a function which describes the motion of a body as a function of time, then the value of the derivative of the function at a particular point would provide the velocity at that point.

There is common notation associated with derivatives, which was developed by Gottfried Wilhelm Leibnitz, the Leibnitz notation. If we have a function, for example, here are three example functions:

$$y = 2x^2 + x - 3$$
$$y = e^{2.x} - x + 4$$
$$y = 2.\sin(x) + 3.\cos^2(x)$$

In a generic sense, we can write that y is a function of x:

$$y = f(x)$$

Therefore, the derivative of y with respect to x can be written as:

$$\frac{dy}{dx}$$

This is described in words by stating "*dee y dee x.*" Alternatively as y is derived using a function f, we can also write:

$$\frac{df}{dx}$$

© The Editor(s) (if applicable) and The Author(s), under exclusive license to Springer Nature Switzerland AG 2021
J. H. Challis, *Experimental Methods in Biomechanics*, https://doi.org/10.1007/978-3-030-52256-8

Table J.1 Various representations of derivatives of the function $y = f(x)$

Derivative order	Notation		
	Leibnitz	Newton	Lagrange
First	$\frac{dy}{dx}$	$\dot{f}(x)$	$f'(x)$
Second	$\frac{d^2y}{dx^2}$	$\ddot{f}(x)$	$f''(x)$
Third	$\frac{d^3y}{dx^3}$	$\dddot{f}(x)$	$f'''(x)$
n^{th}	$\frac{d^ny}{dx^n}$		

Higher-order derivatives in the Leibnitz notation have the form:

$$\frac{d^ny}{dx^n}$$

For example, if $n = 2$, we have the second derivative of the function and so on for greater values of n.

There are three different styles of notation used for derivatives. The one that has been used here is due to the work of Gottfried Wilhelm Leibnitz, but the co-discover of calculus Isaac Newton also presented a notation, as did the Italian mathematician Joseph-Louis Lagrange (1736–1813). These three different styles of notation are all presented in Table J.1, so you will recognize them when you see them. High-order derivatives are most easily represented using the notation of Leibnitz.

The definition of a function, which can be differentiated, is:

$$\frac{dy}{dx} = \lim_{\Delta x \to 0} \frac{f(x + \Delta x) - f(x)}{\Delta x}$$

So we are taking a function and determining the change in the function ($f(x + \Delta x) - f(x)$), over a very small interval (Δx) as the interval shrinks to zero. We can use this definition of differentiable function to determine the derivative of a function. So as an example, we define a function:

$$y = f(x) = x^2$$

So:

$$f(x + \Delta x) = (x + \Delta x)^2 = x^2 + 2.x.\Delta x + \Delta x^2$$

We can now write out the full equation for the rate of change in the function with respect to x, which can be simplified:

$$\frac{f(x + \Delta x) - f(x)}{\Delta x} = \frac{x^2 + 2.x.\Delta x + \Delta x^2 - x^2}{\Delta x}$$

$$\frac{f(x + \Delta x) - f(x)}{\Delta x} = \frac{2.x.\Delta x + \Delta x^2}{\Delta x}$$

Now we divide by Δx:

$$\frac{f(x + \Delta x) - f(x)}{\Delta x} = 2.x + \Delta x$$

But the size of this interval (Δx) is shrinking toward zero so disappears in the equation. In which case we can write out the equation for the derivative of our original function:

$$y = f(x) = x^2$$

$$\frac{df}{dx} = 2.x$$

Table J.2 Derivatives of some common function of the form $y = f(x)$

Function	Derivative
$y = f(x)$	$\frac{dy}{dx}$ or $\frac{df(x)}{dx}$
Any constant k	0
x^n	nx^{n-1}
$k.f(x)$	$k.\frac{df}{dx}$
$\sin(x)$	$\cos(x)$
$\cos(x)$	$-\sin(x)$
$\tan(x)$	$\frac{1}{\cos^2(x)}$
$\ln(x)$	$\frac{1}{x}$
e^x	e^x

Table J.3 Formulae for computing derivatives of functions where the key variable appears more than once

Rule	Function	Derivative
The sum rule	$a(x) + b(x)$	$\frac{da}{dx} + \frac{db}{dx}$
The product rule	$a(x).b(x)$	$a(x).\frac{db}{dx} + b(x).\frac{da}{dx}$
The chain rule	$y(a(x))$	$\frac{dy}{da} \cdot \frac{da}{dx}$

So for this function, we do not need to recourse to numerical differentiation, but there is an equation which allows us to compute the derivative of the function directly. It turns out than many functions have partner functions which represent their derivative. In many cases, it is possible to determine the derivative (second order) of a derivative (first order) and so on. There follows a Table J.2 of basic derivatives; this is not a comprehensive list.

Using the information in Table J.3, we can write some functions, their derivatives, and their second derivatives. Here are two examples:

$$y = x^2 \qquad \frac{df}{dx} = 2.x \qquad \frac{d^2f}{dx^2} = 2$$

$$y = \sin(x) \qquad \frac{df}{dx} = \cos(x) \qquad \frac{d^2f}{dx^2} = -\sin(x)$$

If the function to be differentiated contains the key variable more than once, there are three rules which allow these functions to be differentiated.

There follow example for the application of each of these three rules.

Rule 1: Sum Rule

Example 1 $\qquad y = x^2 + x^3$

$$\frac{dy}{dx} = 2x + 3x^2$$

Example 2 $\qquad y = a.\sin(x) + b.\cos(x)$

$$\frac{dy}{dx} = a.\cos(x) - b.\sin(x)$$

Rule 2: Product Rule

Example $\qquad y = x^2.s(x)$

$$a(x) = x^2 \qquad\qquad \frac{da}{dx} = 2.x$$

$$b(x) = \sin(x) \qquad\qquad \frac{db}{dx} = \cos(x)$$

$$\frac{dy}{dx} = x^2\cos(x) + 2x\sin(x)$$

Rule 3: Chain Rule

Example

$$y = \sin^2(x)$$

$$a(x) = \sin(x) \frac{da}{dx} = \cos(x)$$

$$y(a) = a^2 \frac{dy}{da} = 2.a$$

$$\frac{dy}{dx} = 2.a.\cos(x) = 2.\sin(x).\cos(x)$$

J.2 Integrals

Integration is the reverse process of differentiation. For example, if we have a graph of time plotted against velocity, then integration of this curve gives displacement. For a function $f(x)$, the integral is the area under the curve of the plot of the function. If the curve is divided into small vertical strips, all Δx wide, then the integral is the sum of the areas of the strips. Formally this can be written as:

$$\lim_{\Delta x \to 0} \sum_{k=1}^{n} f(x_k).\Delta x$$

Therefore, the larger the n, the greater the number of strips the function has been divided into, and thus the smaller the Δx. The common notation used when computing the integral of a function $f(x)$ with respect to variable x is:

$$\int f(x)\, dx$$

where the elongated *"S"* $\int$ is called the summa and is the integral sign. This is described in words by stating *the integral, with respect to x, of the function f(x)*. If we are seeking the area under the curve for a specific section, for example, between points a and b, we would write:

$$\int_a^b f(x)\, dx$$

If we set the limits for the integration, then we determine the definitive integral; if we simply want to know the integral of a function, without any limits, then we compute the indefinite integral.

The function to be integrated is called the integrand, and the integrated function the integral ($F(x)$):

$$\int f(x)dx = F(x)$$

Integrand   Integral

The Fundamental Theorem of Calculus expresses the relationship between integration and differentiation, which in equation form is:

$$f(x) = \frac{dF(x)}{dx}$$

So the derivative of the integral is the original function; thus, integration is the reverse process of differentiation and vice versa (within certain limits). We can exploit this relationship to explore how to determine the integral of a function. For a simple function:

$$f(x) = x^m$$

The derivative is:

$$\frac{df}{dx} = \frac{d(x^m)}{dx} = mx^{m-1}$$

If $m = n + 1$, then:

$$\frac{d(x^{n+1})}{dx} = (n+1)x^n$$

and therefore:

$$x^n = \frac{d\left(\frac{x^{n+1}}{n+1}\right)}{dx}$$

If we have a function where:

$$x^n = \frac{dF(x)}{dx}$$

therefore:

$$F(x) = \int x^n dx = \frac{x^{n+1}}{n+1} + c$$

So for this function, we have a rule for its integration; therefore, we do not need to recourse to numerical integration as there is an equation which allows us to compute the integral of the function directly. It turns out that many functions have partner functions which represent their integrals. Notice that in the equation, we have added a constant c. The Fundamental Theorem of Calculus was expressed in the following form:

$$f(x) = \frac{dF(x)}{dx}$$

but can also be stated as:

$$\int \frac{df(x)}{dx} dx = f(x) + c$$

Here the constant c is required because the derivative of an expression such as $c . x$ is c. This constant is referred to as the constant of integration. Here is a simple example:

$$y = 4 - x \rightarrow \int (4 - x)\, dx = 4x - \frac{1}{2}x^2 + c$$

So the indefinite integral is:

$$F(x) = 4x - \frac{1}{2}x^2 + c$$

But if we compute the definitive integral of the same function:

$$\int_1^4 (4-x)\,dx = \left[4x - \frac{1}{2}x^2 + c\right]_1^4$$

$$= \left[(4 \times 4) - \frac{4^2}{2} + c\right] - \left[(1 \times 4) - \frac{1^2}{2} + c\right] = \frac{9}{2}$$

Notice that in this case the constant of integration cancels out.

The relationship between integration and differentiation can be illustrated by the following, which is just a way of stating the Fundamental Theorem of Calculus:

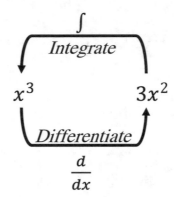

There follows a Table J.4 of integrals; this is not a comprehensive list.

Using the information in Table J.4, we can write some functions and their integrals; here are some examples:

$$y = x^2 \qquad \rightarrow \qquad \int f(x)\,dx = \frac{x^3}{3} + c$$

$$y = 4x^3 \qquad \rightarrow \qquad \int f(x)\,dx = x^4 + c$$

$$y = \sin(x) \qquad \rightarrow \qquad \int f(x)\,dx = -\cos(x) + c$$

$$y = \cos(x) \qquad \rightarrow \qquad \int f(x)\,dx = \sin(x) + c$$

Functions can be integrated twice, for example:

$$y = \cos(x) \quad \int \cos(x) = \sin(x) + c_1$$

$$\int \sin(x) + c_1 = c_1 x - \cos(x) + c_2$$

$$\int\int \cos(x)\,dx = c_1 x - \cos(x) + c_2$$

Table J.4 Indefinite integrals of some common function of the form $y = f(x)$

Function	Integral
$y = f(x)$	$\int f(x)dx$
x^n	$\frac{1}{n+1}x^{n+1}$
Any constant k	$k\,x$
$\sin(x)$	$-\cos(x)$
$\cos(x)$	$\sin(x)$
$\tan(x)$	$-\ln|\cos(x)|$
$\frac{1}{x}$	$\ln|x|$
e^x	e^x

Note – the constant of integration has been omitted here

There are some key properties of integrals which make the computation of integrals easier.

Homogenous Property

$$\int c.f(x)dx = c \int f(x)dx$$

Example

$$y = a.\sin(x)$$

$$\int y\,dx = a \int \sin(x) = -a.\cos(x)$$

Additivity Property

$$\int (\,f(x) + g(x))dx = \int f(x)dx + \int g(x)dx$$

Example

$$y = \cos(x) + \sin(x)$$

$$\int y\,dx = \int \cos(x) + \int \sin(x) = \sin(x) - \cos(x)$$

J.3 Integration by Parts

If the function to be integrated contains the key variable more than once, then integration by parts can be used. The basic formula for integration by parts is:

$$\int u.\left(\frac{dv}{dx}\right) dx = u.v - \int v\left(\frac{du}{dx}\right) dx$$

Example

$$\int x.\cos(x)dx$$

$$u = x \quad \frac{du}{dx} = 1$$

$$\frac{dv}{dx} = \cos(x) \quad v = \int \cos(x)dx = \sin(x)$$

$$u.v - \int v\left(\frac{du}{dx}\right) dx = x.\sin(x) - \int \sin(x).1\;dx = x.\sin(x) + \cos(x)$$

A prudent selection of which parts of the equation are designated u and $\frac{dv}{dx}$ can make the calculation process much easier.

J.4 Partial Derivatives

The functions differentiated so far have had one independent variable, such as $y = f(x)$, but on occasions the derivative is required of a function with more than one independent variable, for example, $z = f(x, y)$. To do this it is assumed that one of the variables changes, while the other variables are held constant. Therefore, for each variable, there is a corresponding equation representing the derivative for each independent variable. These derivatives are called partial derivatives, so for the function:

$$z = f(x, y)$$

The two partial derivatives are written as:

$$\frac{\partial z}{\partial x} \qquad \frac{\partial z}{\partial y}$$

Note the new symbol, ∂, which can be read as *partial dee*.

The process of partial differentiation can be illustrated by considering equation for computing the volume of a cylinder which depends on two variables (r, h):

$$v = \pi r^2 h$$

Then the partial derivative with respect to r is:

$$\frac{\partial v}{\partial r} = \pi\, 2r^1 h = 2\,\pi\, r\, h$$

and the partial derivative with respect to h is:

$$\frac{\partial v}{\partial h} = \pi\, r^2$$

Partial differentiation is subject to the same rules as ordinary differentiation. Partial derivatives are used in vector calculus, in differential geometry, and in error analysis.

J.5 Summary

To finish we can present functions and their integral and derivatives in Table J.5.

Table J.5 Integral and derivatives of some common function of the form $y = f(x)$

Integral	Function	Derivative
$\int f(x)dx$	$y = f(x)$	$\frac{dy}{dx}$ or $\frac{df(x)}{dx}$
$k\,x$	Any constant k	0
$\frac{1}{n+1}x^{n+1}$	x^n	nx^{n-1}
$k \int f(x)$	$k\,f(x)$	$k \cdot \frac{df}{dx}$
$-\cos(x)$	$\sin(x)$	$\cos(x)$
$\sin(x)$	$\cos(x)$	$-\sin(x)$
$-\ln|\cos(x)|$	$\tan(x)$	$\frac{1}{\cos^2(x)}$
$x\ln(x) - x$	$\ln(x)$	$\frac{1}{x}$
e^x	e^x	e^x

Note – the constant of integration has been omitted here

Appendix K Inertial Properties of Geometric Solids

Overview

Human body segments have been modeled as a series of geometric solids; therefore, this Appendix outlines how the inertial properties of a geometric solid can be determined and lists the equations which determine the inertial properties of some common geometric solids.

K.1 Determining Inertial Properties of a Geometric Solid

The inertial properties of a solid are determined by finding the triple integral of various functions. For example, for a rectangular cuboid of width in the x direction of w, length l in the y direction, and height h in the z direction (Fig. K.1):

Volume the volume (v) of the cuboid can be computed from:

$$v = \int_0^h \int_0^l \int_0^w dx\, dy\, dz = w\, l\, h$$

Mass the mass (m) of the cuboid, given its density (ρ), can be computed from:

$$m = \rho\, v = \rho\, w\, l\, h$$

Center of Mass the center of mass location (x_{CM}, y_{CM}, z_{CM}), assuming uniform density, can be computed from:

$$x_{cm} = \frac{1}{v} \int_0^h \int_0^l \int_0^w x\, dx\, dy\, dz = \frac{w}{2}$$

$$y_{cm} = \frac{1}{v} \int_0^h \int_0^l \int_0^w y\, dx\, dy\, dz = \frac{l}{2}$$

$$z_{cm} = \frac{1}{v} \int_0^h \int_0^l \int_0^w z\, dx\, dy\, dz = \frac{h}{2}$$

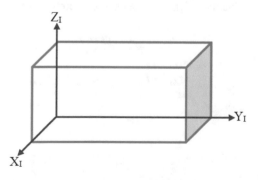

Fig. K.1 Rectangular cuboid, the axis system aligned with one edge, not through the principal axes of the cuboid

J. H. Challis, *Experimental Methods in Biomechanics*, https://doi.org/10.1007/978-3-030-52256-8

Inertia Tensor start with computing the moment of inertia about the X axis (I_{XX}):

$$I_{XX} = \int_0^h \int_0^l \int_0^w (y^2 + z^2)\rho \, dx \, dy \, dz$$

$$= \int_0^h \int_0^l (y^2 + z^2)w \, \rho \, dy \, dz$$

$$= \int_0^h \left(\frac{l^3}{3} + z^2 l\right) w \, \rho \, dz$$

$$= \left(\frac{hl^3}{3} + \frac{h^3 l}{3}\right) w \, \rho$$

As the mass of the solid is $m = \rho \, h \, w \, l$, this equation for the moment of inertia can be simplified to:

$$I_{XX} = \left(\frac{hwl^3}{3} + \frac{h^3 wl}{3}\right)\rho = \frac{\rho \, h \, w \, l}{3}(l^2 + h^2)$$

$$= \frac{m}{3}(l^2 + h^2)$$

Using the same principles, the moments of inertia about the other two axes can be determined:

$$I_{YY} = \frac{m}{3}(w^2 + h^2)$$

$$I_{ZZ} = \frac{m}{3}(w^2 + l^2)$$

The products of inertia (I_{XY}, I_{YZ}, I_{ZX}) can be computed using the same process:

$$I_{XY} = I_{YX} = \int_0^h \int_0^l \int_0^w (xy)\rho \, dx \, dy \, dz$$

$$= \int_0^h \int_0^l \left(\frac{w^2}{2}y\right)\rho \, dy \, dz$$

$$= \int_0^h \left(\frac{w^2 l^2}{4}\right)\rho \, dz$$

$$= \left(\frac{hw^2 l^2}{4}\right)\rho$$

$$= \frac{m}{4}(w \, l)$$

Using the same principles, the other products of inertia axes can be determined:

$$I_{YZ} = I_{ZY} = \frac{m}{4}(h \, l)$$

$$I_{ZX} = I_{XZ} = \frac{m}{4}(h \, w)$$

The inertia tensor (J) therefore is:

$$J = \begin{bmatrix} \dfrac{m}{3}\left(l^2 + h^2\right) & -\dfrac{m}{4}(w\,l) & -\dfrac{m}{4}(h\,w) \\[2ex] -\dfrac{m}{4}(w\,l) & \dfrac{m}{3}\left(w^2 + h^2\right) & -\dfrac{m}{4}(h\,l) \\[2ex] -\dfrac{m}{4}(h\,w) & -\dfrac{m}{4}(h\,l) & \dfrac{m}{3}\left(w^2 + l^2\right) \end{bmatrix}$$

Principal Moments of Inertia the inertia tensor has been determined about a set of axes, but it can be expressed relative to the principal axes. For this solid the principal axes require translation of the current set of axes to have their origin at the center of mass of the solid. The appropriate adjustments to give the inertia tensor about the principal axes (J_P) can be achieved using the parallel axis theorem. The relevant equations are:

$$I_{XX} = I_{XX,P} + m\left(y_{cm}^2 + z_{cm}^2\right)$$

$$I_{YY} = I_{YY,P} + m\left(x_{cm}^2 + z_{cm}^2\right)$$

$$I_{ZZ} = I_{ZZ,P} + m\left(x_{cm}^2 + y_{cm}^2\right)$$

$$I_{XY} - I_{YX} = I_{YX,P} - m\,x_{cm}y_{cm}$$

$$I_{YZ} = I_{ZY} = I_{ZY,P} - m\,z_{cm}y_{cm}$$

$$I_{ZX} = I_{XZ} = I_{XZ,P} - m\,x_{cm}z_{cm}$$

Therefore, inertia tensor relative to the principal axes is:

$$J_P = \begin{bmatrix} \dfrac{m}{12}\left(l^2 + h^2\right) & 0 & 0 \\[2ex] 0 & \dfrac{m}{12}\left(w^2 + h^2\right) & 0 \\[2ex] 0 & 0 & \dfrac{m}{12}\left(w^2 + l^2\right) \end{bmatrix}$$

The inertia tensor now only contains the principal moments of inertia.

Alternatively for a given inertia tensor relative to a set of axes, the inertia tensor about the principal axes can be determined because the eigenvalues of the original inertia tensor are the principal moments of inertia and the eigenvectors are the principal axes.

K.2 Computation of Segment Radii

For any shape with a circular cross section (e.g., cylinder), the radius is typically not measured but the distance around the outside the length of the perimeter. The perimeter is then assumed to be circular making it the circumference (C) so the radius (r) can be computed from:

$$r = \frac{C}{2\pi}$$

For an ellipse there are two radii, for example, in Fig. K.2 a is the semi-major axis associated with the X axis, and b is the semi-minor axis associated with the Y axis. Given the length of one axis and the length of the perimeter (P), the length of the other axis can be computed. For example, if a is calculated as half a diameter and the perimeter C is measured of the ellipsoid, then b can be calculated from:

$$b^2 = \frac{1}{2}\left(\frac{P}{\pi}\right)^2 - a^2$$

Fig. K.2 Basic anatomy of an ellipse

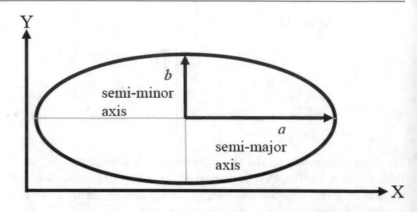

Table K.1 Inertial properties of a cylinder

Geometric solid	Cylinder
Figure	(figure)
Mass	$\rho . \pi . h . r^2$
Center of mass	$\frac{h}{2}$
Principal moments of inertia	$I_{xx} = I_{yy} = \frac{m.\left(3.r^2+h^2\right)}{12}$ $I_{zz} = \frac{m.r^2}{2}$

Table K.2 Inertial properties of an elliptical disk

Geometric solid	Elliptical disk
Figure	(figure)
Mass	$\rho . \pi . a . b . h$
Center of mass	$\frac{h}{2}$
Principal moments of inertia	$I_{xx} = \frac{m.\left(3.b^2+h^2\right)}{12}$ $I_{yy} = \frac{m.\left(3.a^2+h^2\right)}{12}$ $I_{zz} = \frac{m.\left(a^2+b^2\right)}{4}$

K.3 Equations for Geometric Solids

Tables K.1, K.2, K.3, K.4, K.5, and K.6 provide the equations for determining the inertial properties of some geometric solids: Fig. L.1 illustrates their dimensions. Many textbooks provide these equations for the moments of inertia but rarely those for mass and center of mass locations.

Table K.3 Inertial properties of a semi-ellipsoid of revolution

Geometric solid	Semi-ellipsoid of revolution
Figure	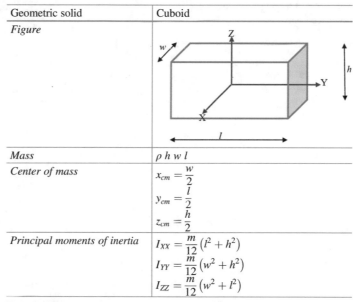
Mass	$\rho \cdot \dfrac{2.\pi.a.b.h}{3}$
Center of mass	$\dfrac{3.h}{8}$
Principal moments of inertia	$I_{xx} = \dfrac{m(b^2 + h^2)}{5} - m.\left(\dfrac{3.h}{8}\right)^2$ $I_{yy} = \dfrac{m(a^2 + h^2)}{5} - m.\left(\dfrac{3.h}{8}\right)^2$ $I_{zz} = \dfrac{2.m(a^2 + b^2)}{5}$

Table K.4 Inertial properties of a cuboid

Geometric solid	Cuboid
Figure	
Mass	$\rho\,h\,w\,l$
Center of mass	$x_{cm} = \dfrac{w}{2}$ $y_{cm} = \dfrac{l}{2}$ $z_{cm} = \dfrac{h}{2}$
Principal moments of inertia	$I_{XX} = \dfrac{m}{12}\left(l^2 + h^2\right)$ $I_{YY} = \dfrac{m}{12}\left(w^2 + h^2\right)$ $I_{ZZ} = \dfrac{m}{12}\left(w^2 + l^2\right)$

Note that these equations can be used to determine the inertial properties of a cube, by setting $w = l = h$

Note in the following:

1. ρ refers to the density of the solid.
2. Unless otherwise stated center of mass position is measured from the base of the geometric solid.
3. Moments of inertia are about the principal axes, with therefore the origin at the center of mass of the solid.
4. The axes are orientated so the Z axis corresponds with the longitudinal axis and the X and Y axes are about transverse axes.
5. For geometric solids which are circular in cross section, the moments of inertia about the X and Y axes are equal.

Table K.5 Inertial properties of an ellipsoid

Geometric solid	Ellipsoid
Figure	
Mass	$\rho \frac{4}{3}\pi\, a\, b\, c$
Center of mass	$x_{cm} = a$ $y_{cm} = b$ $z_{cm} = c$
Principal moments of inertia	$I_{XX} = \frac{m}{5}\left(b^2 + c^2\right)$ $I_{YY} = \frac{m}{5}\left(a^2 + c^2\right)$ $I_{ZZ} = \frac{m}{15}\left(a^2 + b^2\right)$

Note that these equations can be used to determine the inertial properties of a sphere, by setting $a = b = c$

Table K.6 Inertial properties of a truncated cone

Geometric solid	Truncated cone
Figure	
Mass	$\rho \frac{h}{3}\pi\left(r^2 + rR + R^2\right)$
Center of mass	$\frac{h}{2} \frac{\left(r^2 + 2rR + 3R^2\right)}{\left(r^2 + rR + R^2\right)}$
Principal moments of inertia	$A = \dfrac{r}{R}$ $B = \left(1 + A + A^2\right)^2$ $C = 1 + A + A^2 + A^3 + A^4$ $D = 1 + 4A + 10A^2 + 4A^3 + A^4$ $I_{XX} = I_{YY} = m\left(\frac{9m}{20\pi\rho h}\frac{C}{B} + \frac{3h^2}{80}\frac{D}{B}\right)$ $I_{ZZ} = \frac{3m}{10}\left(\frac{R^5 - r^5}{R^3 - r^3}\right)$

Note that these equations can be used to determine the inertial properties of a cylinder, by setting $r = R$. Here the term truncated cone has been used to indicate a cone with the top chopped off. On occasions the term frustum of a cone is used, this basically means that a portion of the solid which remains after its upper part has been cut off by a plane parallel to its base

K.4 Routh's Rule

A useful generalization for computing the moment of inertia about the longitudinal principal axis is Routh's rule. This rule states for a cuboid ($n = 3$), elliptical cylinder ($n = 4$), or ellipsoid ($n = 5$), the moment of inertia about the longitudinal principal axes is equal to:

$$I_L = \frac{m\left(a^2 + b^2\right)}{n}$$

where m is the mass of the body and a and b are the semi-axes perpendicular to the principal axis for which the moment of inertia is being computed.

K.5 Segmental Densities

To model the body segments as geometric solids requires segmental density values. Table K.7 gives segmental density values from three major cadaver studies.

Table K.7 Segment densities from three major cadaver studies for human body segments (units – kg.m^{-3} × 10^3)

Segment	Study		
	Dempster (1955)	Clauser et al. (1969)	Chandler et al. (1975)
Hand	1.16	1.109	1.080
Forearm	1.13	1.098	1.052
Upper arm	1.07	1.056	1.080
Foot	1.10	1.084	1.071
Shank	1.09	1.085	1.065
Thigh	1.05	1.044	1.0195
Trunk	1.03	1.019	0.853
Head	1.11	1.070	1.056

K.6 References

Chandler, R. F., Clauser, C. E., McConville, J. T., Reynolds, H. M., & Young, J. W. (1975). *Investigation of the inertial properties of the human body* (AMRL eTechnical Report 74–137). Ohio: Wright Patterson Air Force Base.

Clauser, C. E., McConville, J. T., & Young, J. W. (1969). Weight, volume and center of mass of segments of the human body (AMeRL Technical Report 69–70). Ohio: Wright Patterson Air Force Base.

Dempster, W. T. (1955). *Space requirements of the seated operator* (WADC Technical Report 55–159). Ohio: Aerospace Medical Research Laboratory, Wright-Patterson Air Force Base.

Appendix L Dot Product

Overview

The dot product is an operation on two vectors. It has applications in mechanics. For example, mechanical work can be computed from the dot product of force and displacement vectors; it can also be used to compute power which is the dot product of a force vector and a velocity vector.

L.1 Definition

The dot product is sometimes called the scalar product. The dot product of two vectors is a scalar; it is the sum of the products of the corresponding entries in two sequences of numbers (the vectors). It is written using a centered dot:

$$c = \mathbf{a} \cdot \mathbf{b}$$

Vectors **a** and **b** must be of the same length, and c is a scalar. The dot product represents the product of the Euclidean magnitudes of the two vectors and the cosine of the angle between them; therefore:

$$\mathbf{a} \cdot \mathbf{b} = \left\| a \right\| \left\| b \right\| \cos \theta_{ab}$$

Figure L.1 illustrates the features of the dot product.

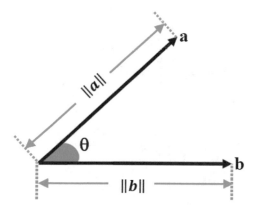

Fig. L.1 The dot product of vectors **a** and **b**, where $\|a\|$, $\|b\|$ are the magnitudes (lengths) of vectors **a** and **b**, respectively, and θ is the angle between the vectors

L.2 Computation

The dot product is computed as follows:

$$\mathbf{a} = (a_1, a_2, a_3) \qquad \mathbf{b} = (b_1, b_2, b_3)$$

$$\mathbf{a} \cdot \mathbf{b} = (a_1b_1 + a_2b_2 + a_3b_3)$$

Or for a two-element vector:

$$\mathbf{a} = (a_1, a_2) \qquad \mathbf{b} = (b_1, b_2)$$

$$\mathbf{a} \cdot \mathbf{b} = (a_1b_1 + a_2b_2)$$

L.3 Key Properties

The dot product has the following properties:

1. If $\mathbf{a}$ and $\mathbf{b}$ are perpendicular to one another, then $\mathbf{a} \cdot \mathbf{b} = 0$, since $\theta = 90°$ and $\cos\theta = 0$.
2. If $\mathbf{a} \cdot \mathbf{b} = 0$, then $\mathbf{a}$ and $\mathbf{b}$ are perpendicular or $\mathbf{a} = 0$ or $\mathbf{b} = 0$.
3. If $\mathbf{a}$ and $\mathbf{b}$ are parallel to one another, then $\mathbf{a} \cdot \mathbf{b} = \|a\|\,\|b\|$, since $\theta = 0°$ and $\cos\theta = 1$.
4. $\mathbf{a} \cdot \mathbf{a} = \|a\|\,\|a\| = \|a^2\|$ as $\mathbf{a}$ must be parallel to itself.
5. The dot product is commutative so $\mathbf{a} \cdot \mathbf{b} = \mathbf{b} \cdot \mathbf{a}$.

L.4 Examples

Here are some examples.

Example 1

$$\mathbf{a} = (5, 12) \qquad \mathbf{b} = (-6, 8) \qquad \mathbf{a} \cdot \mathbf{b} = 66$$

Since:

$$\mathbf{a} \cdot \mathbf{b} = \|\mathbf{a}\|\,\|\mathbf{b}\|\,\cos\theta_{ab}$$

Then:

$$\cos\theta_{ab} = \frac{\mathbf{a} \cdot \mathbf{b}}{\|a\|\,\|b\|} = \frac{66}{(13)(10)} \qquad \theta_{ab} = 59.5°$$

Example 2

$$\mathbf{a} = (2.3, 4) \qquad \mathbf{b} = (10, 15, 20) \qquad \mathbf{a} \cdot \mathbf{b} = 145$$

Then:

$$\cos\theta_{ab} = \frac{\mathbf{a} \cdot \mathbf{b}}{\|\mathbf{a}\|\,\|\mathbf{b}\|} = \frac{145}{(5.385)(26.926)} \qquad \theta_{ab} = 0°$$

Therefore, the two vectors are parallel.

L.5 Matrix Implementation

Matrix algebra can be used to compute the dot product. If the two vectors are column vectors, then:

$$\mathbf{a} \cdot \mathbf{b} = \mathbf{a}^T \mathbf{b}$$

where the superscript ^{T} refers to the transpose. If the two vectors are row vectors, then the transpose of the second term should be used.

Example

$$\mathbf{a} = (4, 3, 0) \qquad \mathbf{b} = (-3, 4, 0)$$

$$\mathbf{a} \cdot \mathbf{b} = \begin{bmatrix} -4 \\ 3 \\ 0 \end{bmatrix} \cdot \begin{bmatrix} -3 \\ 4 \\ 0 \end{bmatrix} = \begin{bmatrix} -4 & 3 & 0 \end{bmatrix} \begin{bmatrix} -3 \\ 4 \\ 0 \end{bmatrix} = 0$$

Appendix M Cardan and Euler Angle Matrices

Overview

Two popular ways of quantifying angular kinematics are Cardan and Euler angles. For each of these angle sets, there are six different options, which will be described in this Appendix.

M.1 Base Rotations

Both Cardan and Euler angles are based on rotations about the axes of a coordinate system. The matrices which perform rotations about the X, Y, and Z axes are:

$$R_x(\alpha) = \begin{bmatrix} 1 & 0 & 0 \\ 0 & \cos(\alpha) & -\sin(\alpha) \\ 0 & \sin(\alpha) & \cos(\alpha) \end{bmatrix}$$

$$R_y(\beta) = \begin{bmatrix} \cos(\beta) & 0 & \sin(\beta) \\ 0 & 1 & 0 \\ -\sin(\beta) & 0 & \cos(\beta) \end{bmatrix}$$

$$R_Z(\gamma) = \begin{bmatrix} \cos(\gamma) & -\sin(\gamma) & 0 \\ \sin(\gamma) & \cos(\gamma) & 0 \\ 0 & 0 & 1 \end{bmatrix}$$

where α, β, and γ are angles of rotation about the X, Y, and Z axes, respectively. In the following α, β, and γ will be used to designate the rotation about the first, second, and third axes, respectively, irrespective of which sequence of axes is used.

The influences of each of these rotation matrices cause non-linear transformations of data points, except for the coordinates associated with axis of rotation:

$$R_x(\alpha)\begin{bmatrix} x \\ y \\ x \end{bmatrix} = \begin{bmatrix} 1 & 0 & 0 \\ 0 & \cos(\alpha) & -\sin(\alpha) \\ 0 & \sin(\alpha) & \cos(\alpha) \end{bmatrix}\begin{bmatrix} x \\ y \\ z \end{bmatrix} = \begin{bmatrix} x \\ y\cos(\alpha) - z\sin(\alpha) \\ y\sin(\alpha) + z\cos(\alpha) \end{bmatrix}$$

$$R_y(\beta)\begin{bmatrix} x \\ y \\ z \end{bmatrix} = \begin{bmatrix} \cos(\beta) & 0 & \sin(\beta) \\ 0 & 1 & 0 \\ -\sin(\beta) & 0 & \cos(\beta) \end{bmatrix}\begin{bmatrix} x \\ y \\ z \end{bmatrix} = \begin{bmatrix} x\cos(\beta) + z\sin(\beta) \\ y \\ -x\sin(\beta) + z\cos(\beta) \end{bmatrix}$$

J. H. Challis, *Experimental Methods in Biomechanics*, https://doi.org/10.1007/978-3-030-52256-8

$$R_Z(\gamma)\begin{bmatrix} x \\ y \\ z \end{bmatrix} = \begin{bmatrix} \cos(\gamma) & -\sin(\gamma) & 0 \\ \sin(\gamma) & \cos(\gamma) & 0 \\ 0 & 0 & 1 \end{bmatrix}\begin{bmatrix} x \\ y \\ z \end{bmatrix} = \begin{bmatrix} x\,\sin(\beta) - y\,\sin(\beta) \\ x\,\sin(\beta) + y\,\cos(\beta) \\ z \end{bmatrix}$$

M.2 Properties of the Attitude Matrix

The rotation or attitude matrix has the following important properties:

1. The matrix is proper orthogonal, which means the rows are all orthogonal to one another, and similarly the columns.
2. As the matrix is proper orthogonal, the determinant of the attitude matrix is *1*.
3. The attitude matrix belongs to the special-orthogonal group of order three:

$$R \in SO(3)$$

4. As the attitude matrix belongs to the special-orthogonal group, the inverse of this matrix belongs to this group, as does the product of any matrices in this group.
5. The transpose of the attitude matrix is equal to its inverse; therefore:

$$R^T = R^{-1}$$

6. The attitude matrix has three eigenvalues.
   Case 1
   If $\lambda_1 = \lambda_2 = \lambda_3 = 1$, then $R = I$, so there is no rotation.
   Case 2
   If $\lambda_1 = 1$ and $\lambda_2 = \lambda_3 = -1$, then the angle of rotation about the helical axis described by the rotation matrix R is π.
   Case 3

$$\lambda_1 = 1 \; \lambda_2 = e^{i\phi} \; \lambda_3 = e^{-i\phi}$$

In which case, the general case, ϕ is the angle of rotation associated about the helical axis describing the rotation matrix R.
7. The sum of the eigenvalues of the attitude matrix is equal to the trace of the attitude matrix. Therefore:

$$tr(R) = 1 + e^{i\theta} + e^{-i\theta}$$
$$= 1 + 2\cos\theta$$
$$\cos\theta = \frac{1}{2}(r_{11} + r_{22} + r_{33} - 1)$$

M.3 Angle Sequences

Cardan and Euler angle sequences involve rotations about coordinate axes. If the rotations are completed in the following sequence, X and then Y and finally the Z, the matrix sequence is written as:

$$R_{XYZ} = R_Z(\gamma)R_Y(\beta)R_X(\alpha)$$

As matrix multiplication is not commutative, therefore:

$$R_{XYZ} \neq R_{ZYX}$$

So the sequence of rotations produces different sets of angles; therefore, when reporting angles, the angle sequence and the definition of the axes must be given.

Cardan angles can be described as an ordered sequence of rotations about each of the three coordinate axes, so rotations occur about each axis once. For Cardanic angles, the complete sets of potential sequences are X-Y-Z, X-Z-Y, Y-Z-X, Y-X-Z, Z-X-Y, and Z-Y-X. The matrix associated with each of these sequences is presented in the following section (M.4).

Euler angles can be described as an ordered sequence of rotations about three coordinate axes. In this case, the rotations use one axis twice (the initial and final rotation), that is, the Eulerian angles involve repetition of rotations about one particular axis. For example, the rotation sequence might be X-Y-X so the first and final rotation are about the same axis; of course, the final axis will not be in the same position for the final rotation as it will have moved due to previous the rotation about the middle axis. For Euler angles, the complete sets are X-Y-X, X-Z-X, Y-X-Y, Y-Z-Y, Z-X-Z, and Z-Y-Z. The matrix associated with each of these sequences is presented in section M.5.

M.4 Cardan Angle Sequences

$$R_{XYZ} = R_Z(\gamma)R_Y(\beta)R_X(\alpha)$$

$$\begin{bmatrix} c(\beta)c(\gamma) & c(\gamma)s(\beta)s(\alpha) - c(\alpha)s(\gamma) & c(\alpha)c(\gamma)s(\beta) + s(\alpha)s(\gamma) \\ c(\beta)s(\gamma) & s(\alpha)s(\beta)s(\gamma) + c(\alpha)c(\gamma) & c(\alpha)s(\beta)s(\gamma) - c(\gamma)s(\alpha) \\ -s(\beta) & c(\beta)s(\alpha) & c(\alpha)c(\beta) \end{bmatrix}$$

$$R_{XZY} = R_Y(\gamma)R_Z(\beta)R_X(\alpha)$$

$$\begin{bmatrix} c(\beta)c(\gamma) & s(\alpha)s(\gamma) - c(\alpha)c(\gamma)s(\beta) & c(\alpha)s(\beta) + c(\gamma)s(\alpha)s(\beta) \\ s(\beta) & c(\alpha)c(\beta) & -c(\beta)s(\alpha) \\ -c(\beta)s(\gamma) & c(\gamma)s(\alpha) + c(\alpha)s(\beta)s(\gamma) & c(\alpha)c(\gamma) - s(\alpha)s(\beta)s(\gamma) \end{bmatrix}$$

$$R_{YZX} = R_X(\gamma)R_Z(\beta)R_Y(\alpha)$$

$$\begin{bmatrix} c(\alpha)c(\beta) & -s(\beta) & c(\beta)s(\alpha) \\ s(\alpha)c(\gamma) + c(\alpha)c(\gamma)s(\beta) & c(\beta)c(\gamma) & c(\gamma)s(\alpha)s(\beta) - c(\alpha)s(\gamma) \\ c(\alpha)s(\beta)s(\gamma) - c(\gamma)s(\alpha) & c(\beta)s(\gamma) & c(\alpha)c(\gamma) + s(\alpha)s(\beta)s(\gamma) \end{bmatrix}$$

$$R_{YXZ} = R_Z(\gamma)R_X(\beta)R_Y(\alpha)$$

$$\begin{bmatrix} c(\alpha)c(\gamma) - s(\alpha)s(\beta)s(\beta) & -c(\beta)s(\gamma) & c(\gamma)s(\beta) + c(\alpha)s(\beta)s(\gamma) \\ c(\alpha)s(\gamma) + c(\gamma)s(\alpha)s(\beta) & c(\beta)c(\gamma) & s(\alpha)s(\gamma) - c(\alpha)c(\gamma)s(\beta) \\ -c(\beta)s(\alpha) & s(\beta) & c(\alpha)c(\beta) \end{bmatrix}$$

$$R_{ZXY} = R_Y(\alpha)R_X(\beta)R_Z(\alpha)$$

$$\begin{bmatrix} c(\alpha)c(\gamma) + s(\alpha)s(\beta)s(\gamma) & c(\alpha)s(\beta)s(\gamma) - c(\gamma)s(\alpha) & c(\beta)s(\gamma) \\ c(\beta)s(\alpha) & c(\alpha)c(\beta) & -s(\beta) \\ c(\gamma)s(\alpha)s(\beta) - c(\alpha)s(\gamma) & s(\alpha)s(\gamma) + c(\alpha)s(\gamma)s(\beta) & c(\beta)c(\gamma) \end{bmatrix}$$

$$R_{ZYX} = R_X(\gamma)R_Y(\beta)R_Z(\alpha)$$

$$\begin{bmatrix} c(\alpha)c(\beta) & -c(\beta)s(\alpha) & s(\beta) \\ c(\gamma)s(\alpha) + c(\alpha)s(\beta)s(\gamma) & c(\alpha)c(\gamma) - s(\alpha)s(\beta)s(\gamma) & -c(\beta)s(\gamma) \\ s(\alpha)s(\gamma) - c(\alpha)c(\gamma)s(\beta) & c(\alpha)s(\gamma) + c(\gamma)s(\alpha)s(\beta) & c(\beta)c(\gamma) \end{bmatrix}$$

N.B. – $\cos(\alpha)$ and $\sin(\alpha)$ are represented by $c(\alpha)$ and $s(\alpha)$, similarly for the other angles

M.5 Euler Angle Sequences

$R_{XYX} = R_X(\gamma)R_Y(\beta)R_X(\alpha)$

$$\begin{bmatrix} c(\beta) & s(\alpha)s(\beta) & c(\alpha)s(\beta) \\ s(\beta)s(\gamma) & c(\alpha)c(\gamma) - c(\beta)s(\alpha)s(\gamma) & -c(\gamma)s(\alpha) - c(\alpha)c(\beta)s(\gamma) \\ -c(\gamma)s(\beta) & c(\alpha)s(\gamma) + c(\beta)c(\gamma)s(\alpha) & c(\alpha)c(\beta)c(\gamma) - s(\alpha)s(\gamma) \end{bmatrix}$$

$R_{XZX} = R_X(\gamma)R_Z(\beta)R_X(\alpha)$

$$\begin{bmatrix} c(\beta) & -c(\alpha)s(\beta) & s(\alpha)s(\beta) \\ c(\gamma)s(\beta) & c(\alpha)c(\beta)c(\gamma) - s(\alpha)s(\gamma) & -c(\alpha)s(\gamma) - c(\beta)c(\gamma)s(\alpha) \\ s(\beta)s(\gamma) & c(\gamma)s(\alpha) + c(\alpha)c(\beta)s(\gamma) & c(\alpha)c(\gamma) - c(\beta)s(\alpha)s(\gamma) \end{bmatrix}$$

$R_{YXY} = R_Y(\gamma)R_X(\beta)R_Y(\alpha)$

$$\begin{bmatrix} c(\alpha)c(\gamma) - c(\beta)s(\alpha)s(\gamma) & s(\beta)s(\gamma) & c(\gamma)s(\alpha) + c(\alpha)c(\beta)s(\gamma) \\ s(\alpha)s(\beta) & c(\beta) & -c(\alpha)s(\beta) \\ -c(\alpha)s(\gamma) - c(\beta)c(\gamma)s(\alpha) & c(\gamma)s(\beta) & c(\alpha)c(\beta)c(\gamma) - s(\alpha)s(\gamma) \end{bmatrix}$$

$R_{YZY} = R_Y(\gamma)R_Z(\beta)R_Y(\alpha)$

$$\begin{bmatrix} c(\alpha)c(\beta)c(\gamma) - s(\alpha)s(\gamma) & -c(\gamma)s(\beta) & c(\alpha)s(\gamma) + c(\beta)c(\gamma)s(\alpha) \\ c(\alpha)s(\beta) & c(\beta) & s(\alpha)s(\beta) \\ -c(\gamma)s(\alpha) - c(\alpha)c(\beta)s(\gamma) & s(\beta)s(\gamma) & c(\alpha)c(\gamma) - c(\beta)s(\alpha)s(\gamma) \end{bmatrix}$$

$R_{ZXZ} = R_Z(\alpha)R_X(\beta)R_Z(\alpha)$

$$\begin{bmatrix} c(\alpha)c(\gamma) - c(\beta)s(\alpha)s(\gamma) & -c(\gamma)s(\alpha) - c(\alpha)c(\beta)s(\gamma) & s(\beta)s(\gamma) \\ c(\alpha)s(\gamma) + c(\beta)c(\gamma)s(\alpha) & c(\alpha)c(\beta)c(\gamma) - s(\alpha)s(\gamma) & -c(\gamma)s(\beta) \\ s(\alpha)s(\beta) & c(\alpha)s(\beta) & c(\beta) \end{bmatrix}$$

$R_{ZYZ} = R_Z(\gamma)R_Y(\beta)R_Z(\alpha)$

$$\begin{bmatrix} c(\alpha)c(\beta)c(\gamma) - s(\alpha)s(\gamma) & -c(\alpha)s(\gamma) - c(\beta)c(\gamma)s(\alpha) & c(\gamma)s(\beta) \\ c(\gamma)s(\alpha) + c(\alpha)c(\beta)s(\gamma) & c(\alpha)c(\gamma) - c(\beta)s(\alpha)s(\gamma) & s(\beta)s(\gamma) \\ -c(\alpha)s(\beta) & s(\alpha)s(\beta) & c(\beta) \end{bmatrix}$$

N.B. – $\cos(\alpha)$ and $\sin(\alpha)$ are represented by $c(\alpha)$ and $s(\alpha)$, similarly for the other angles

Appendix N Quaternions

Overview

Quaternions are generally attributed to William Rowan Hamilton (1805–1865) in 1843, because he had a sudden insight when walking with his wife on October 16, 1843. He was so excited by this insight, generalizing complex numbers into three-dimensional space, that he carved the key formula into the Broome Bridge in Dublin.

N.1 Format of Quaternion

The set of quaternions is associated with a four-dimensional vector space, so there are four elements associated with each quaternion. Somewhat confusingly the way in which quaternions are written varies, but they all mean the same thing and it depends on application and interpretation which approach is most appropriate. The three main ways are:

(i). A complex number with three imaginary parts; this is the three-dimensional version of two-dimensional complex numbers $(a + ib$, where $i = \sqrt{-1})$:

$$q = q_0 + iq_1 + jq_2 + kq_3$$

where q_0, q_1, q_2 and q_3 are all real and the imaginary components (i, j, k) have the following rules for multiplication:

$$i^2 = j^2 = k^2 = ijk = -1$$

which was the relationship carved on Broome Bridge by Hamilton. This representation is the Hamiltonian form of the quaternion.

(ii). A vector with four components:

$$q = (q_0, q_1, q_2, q_3)$$

This representation is the four-tuple form of the quaternion. It should be noted that on occasions the quaternion in this form moves the first element to the end (q_1, q_2, q_3, q_0).

(iii). A scalar (q) and a three-element vector $(\boldsymbol{q})$:

$$q = (q, \boldsymbol{q})$$

This representation is the scalar vector form of the quaternion.

The four-tuple form of the quaternion $(q = (q_0, q_1, q_2, q_3))$ will be used in the remainder of this Appendix.

N.2 Types of Quaternions

There are a number of types of quaternions; the primary ones are:

J. H. Challis, *Experimental Methods in Biomechanics*, https://doi.org/10.1007/978-3-030-52256-8

Pure quaternion	$q = (0, q_1, q_2, q_3)$
Identity quaternion	$q = (1, 0, 0, 0)$
Conjugate quaternion	$\bar{q} = (q_0, -q_1, -q_2, -q_3)$
Quaternion norm	$\|q\| = \sqrt{q_0^2 + q_1^2 + q_2^2 + q_3^2}$
Quaternion inverse	$q^{-1} = \frac{\bar{q}}{\|q\|^2}$
Unit quaternion	$q = (q_0, q_1, q_2, q_3)$ where $\|q\| = 1$

As for the unit quaternion its norm is equal to one, its inverse is therefore simply its conjugate. For describing rotations in three dimensions, unit quaternions have a number of advantages, due to their intrinsic properties.

N.3 Geometric Interpretation of Quaternions

There are two ways in which a quaternion can be visualized. The first is that each quaternion corresponds to a point on a hypersphere surface (Fig. N.1).

A quaternion, $q = (q_0, q_1, q_2, q_3)$, can be considered a rotation of angle Ω, about an axis $\underline{e}$, where:

$$q_0 = \pm \cos \frac{\Omega}{2} \text{ and } \begin{bmatrix} q_1 \\ q_2 \\ q_3 \end{bmatrix} = \pm \underline{e} \sin \frac{\Omega}{2}$$

where $0 \leq \Omega \leq \pi$.

Note these equations show that each rotation has two representations, (q_0, q_1, q_2, q_3) and $(-q_0, -q_1, -q_2, -q_3)$.

A quaternion can be directly visualized as a directed line in space about which there is a rotation. For example, see Fig. N.2; if a vector r_0 is transformed by a rotation matrix to vector r_1, then this transformation can be visualized as a rotation (Ω) about a line $(\underline{e})$.

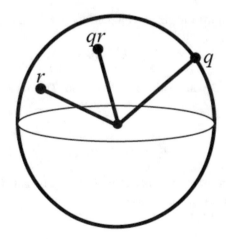

Fig. N.1 Quaternions represented on a hypersphere, where q and r are quaternions and qr is the quaternion resulting from their product

Fig. N.2 The transformation of a vector r_0 by a rotation of Ω about a line $\underline{e}$, to vector $r1$. The left image shows the general representation of the transformation, and the right image shows a view in a plane normal to the axis of rotation

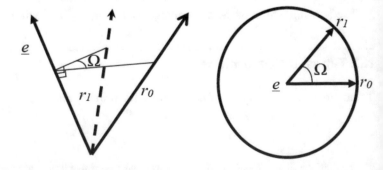

N.4 Direction Cosine Matrix from Quaternions

The direction cosine matrix (R) can be computed from a unit quaternion using:

$$R = (q_0^2 - \mathbf{q}^T\mathbf{q})I + 2\mathbf{q}\mathbf{q}^T - 2q_0S\{\mathbf{q}\}$$

which when expanded gives:

$$R(q) = \begin{bmatrix} q_0^2 + q_1^2 - q_2^2 - q_3^2 & 2(q_1q_2 - q_3q_0) & 2(q_1q_3 - q_2q_0) \\ 2(q_1q_2 + q_3q_0) & q_0^2 - q_1^2 + q_2^2 - q_3^2 & 2(q_2q_3 - q_1q_0) \\ 2(q_1q_3 - q_2q_0) & 2(q_2q_3 + q_1q_0) & q_0^2 - q_1^2 - q_2^2 + q_3^2 \end{bmatrix}$$

Note this matrix can be represented using quaternions in a number of different ways; they are all equivalent.

N.5 Quaternions from Direction Cosine Matrix

The quaternion can be computed from a direction cosine matrix ($[R]$) in a variety of ways. The following four vectors could each represent the quaternion of a direction cosine matrix:

$$x_0 = \begin{bmatrix} 1 + r_{1,1} + r_{2,2} + r_{3,3} \\ r_{2,3} - r_{3,2} \\ r_{3,1} - r_{1,3} \\ r_{1,2} - r_{2,1} \end{bmatrix} \quad x_1 = \begin{bmatrix} r_{2,3} - r_{3,2} \\ 1 + r_{1,1} - r_{2,2} - r_{3,3} \\ r_{1,2} + r_{2,1} \\ r_{1,3} - r_{3,1} \end{bmatrix}$$

$$x_2 = \begin{bmatrix} r_{3,1} - r_{1,3} \\ r_{2,1} + r_{1,2} \\ 1 + r_{2,2} - r_{3,3} - r_{1,1} \\ r_{2,3} + r_{3,2} \end{bmatrix} \quad x_3 = \begin{bmatrix} r_{1,2} - r_{2,1} \\ r_{3,1} + r_{1,3} \\ r_{3,2} + r_{2,3} \\ 1 + r_{3,3} - r_{1,1} - r_{2,2} \end{bmatrix}$$

The unit quaternion is found by normalizing any of these vectors:

$$q = \pm x_i / \left\| x_i \right\|$$

To minimize numerical errors, the vector selected for normalization should be the one with the largest norm. The norm for each vector does not need to be computed to make this selection; if the trace of the rotation matrix ($r_{1,1} + r_{2,2} + r_{3,3}$) is larger than $r_{i,i}$ (where $i = 1,3$), then the first listed vector (x_0) should be used; otherwise if $r_{1,1}$ is the largest, then use the second listed vector (x_1), similarly for the other vectors.

N.6 Quaternion Math

All normal laws of algebra apply to quaternions, except that multiplication is not commutative. Quaternions have three primary operations: addition, scalar multiplication, and multiplication. Table N.1 reveals the products of the imaginary numbers (i, j, and k).

Table N.1 Products of imaginary numbers comprising a quaternion

	i	j	k
i	-1	k	$-j$
j	$-k$	-1	i
k	j	$-i$	-1

The basic relationship can be visualized with a simple Fig. N.3.

Fig. N.3 Products of imaginary numbers comprising a quaternion. Start at any point and move in a counter-clockwise direction, with the arrows, and it gives you the results of the products of the imaginary numbers (e.g., $k.i = j$); going in a clockwise direction makes the product negative (e.g., $j.i = -k$)

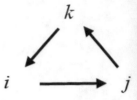

As examples we define quaternions q, r, and s.

$$q = (q_0, q_1, q_2, q_3) \quad r = (r_0, r_1, r_2, r_3) \quad s = (s_0, s_1, s_2, s_3)$$

Scalar Multiplication

$$\lambda q = (\lambda q_0) + i(\lambda q_1) + j(\lambda q_2) + k(\lambda q_3)$$

Addition:

$$q + r = (q_0 + r_0) + i(q_1 + r_1) + j(q_2 + r_2) + k(q_3 + r_3)$$

$q + (r + s) = (q + r) + s$
addition is associative
$q + r = r + q$
addition is commutative

Multiplication:

$$qr = (q_0 r_0 - q_1 r_1 - q_2 r_2 - q_3 r_3)$$
$$+ i(q_0 r_1 + q_1 r_0 + q_2 r_3 - q_3 r_2)$$
$$+ j(q_0 r_3 + q_2 r_0 + q_3 r_1 - q_1 r_3)$$
$$+ k(q_0 r_3 + q_3 r_0 + q_1 r_2 - q_2 r_1)$$

or in matrix form:

$$qr = Qr^T = \begin{bmatrix} q_0 & -q_1 & -q_2 & -q_3 \\ q_1 & q_0 & -q_3 & q_2 \\ q_2 & q_3 & q_0 & -q_1 \\ q_3 & -q_2 & q_1 & q_0 \end{bmatrix} \begin{bmatrix} r_0 \\ r_1 \\ r_2 \\ r_3 \end{bmatrix}$$

where the matrix Q can be formed from scalar vector form of the quaternion, $q = (q, \mathbf{q})$, assuming the vector is a column vector:

$$Q = \begin{bmatrix} q & -\boldsymbol{q}^T \\ \boldsymbol{q} & qI + P\{\boldsymbol{q}\} \end{bmatrix}$$

where I is the identity matrix and P is a skew-symmetric matrix which has the following format:

$$P\{q\} = \begin{bmatrix} 0 & -q_3 & q_2 \\ q_3 & 0 & -q_1 \\ -q_2 & q_1 & 0 \end{bmatrix}$$

$$q(r\,s) = (q\,r)\,s \text{ (multiplication is associative)}$$

$$qr \neq rq \text{ (multiplication is not commutative)}$$

Dot Product:

$$q \cdot r = (q_0 r_0 + q_1 r_1 + q_2 r_2 + q_3 r_3)$$

The angle between a pair of quaternions can be computed from its dot product:

$$\cos(\theta) = \frac{q \cdot r}{|q||r|}$$

If the quaternions are unit quaternions, then:

$$\cos(\theta) = q \cdot r$$

N.7 Vector Rotations

If we have a vector, for example, the location of a point on a rigid body, and we want to rotate it by an amount as indicated by a quaternion, there are two options. The first is to convert the quaternion into a direction cosine matrix and simply multiply the vector by the matrix:

$$p_1 = R(q)\, p_0$$

This transformation can be implemented in another way, which requires fewer mathematical operations:

$$p_1 = q\, p_0\, q^{-1}$$

The inverse of a unit quaternion is simply its conjugate so is simple to compute.

Two rotations can be combined by simply multiplying the quaternions, so if we have two quaternions (q, r), the result of the rotation caused by quaternion q followed by quaternion r is:

$$s = r\,q$$

Notice the order a rotation of q followed by a rotation of r is equivalent to a single rotation of s, where the first rotation is on the right-hand side of the multiplication. A number of rotations can easily be combined and represented by a single quaternion. As quaternion multiplication is not commutative, the order of the multiplication influences the resulting quaternion and the rotation it produces.

N.8 Cardan and Euler Angles to Quaternions

A Cardan angle set, or an Euler angle set, can be converted into a quaternion by converting each component matrix into a quaternion and then taking the product of these quaternions. Therefore, for a Cardan angle sequence, each of the component matrices can be expressed as quaternions:

$$
R_x(\alpha) = \begin{bmatrix} 1 & 0 & 0 \\ 0 & \cos(\alpha) & -\sin(\alpha) \\ 0 & \sin(\alpha) & \cos(\alpha) \end{bmatrix} \quad q_\alpha = \begin{bmatrix} \cos\left(\dfrac{\alpha}{2}\right) \\ \sin\left(\dfrac{\alpha}{2}\right) \\ 0 \\ 0 \end{bmatrix}
$$

$$
R_y(\beta) = \begin{bmatrix} \cos(\beta) & 0 & \sin(\beta) \\ 0 & 1 & 0 \\ -\sin(\beta) & 0 & \cos(\beta) \end{bmatrix} \quad q_\beta = \begin{bmatrix} \cos\left(\dfrac{\beta}{2}\right) \\ 0 \\ \sin\left(\dfrac{\beta}{2}\right) \\ 0 \end{bmatrix}
$$

$$
R_Z(\gamma) = \begin{bmatrix} \cos(\gamma) & -\sin(\gamma) & 0 \\ \sin(\gamma) & \cos(\gamma) & 0 \\ 0 & 0 & 1 \end{bmatrix} \quad q_\gamma = \begin{bmatrix} \cos\left(\dfrac{\gamma}{2}\right) \\ 0 \\ 0 \\ \sin\left(\dfrac{\gamma}{2}\right) \end{bmatrix}
$$

Therefore, for example, a Cardan angle sequence of:

$$
R_{ZYX} = R_X(\gamma)R_Y(\beta)R_Z(\alpha)
$$

can be expressed as a quaternion by taking the product of the component quaternions:

$$
q = q_\gamma q_\beta q_\alpha
$$

N.9 Interpolation of Quaternions

Quaternions can be interpolated to estimate values in between samples as can be performed with vectors. The difference is that vectors exist in linear space, but the quaternions exist on a curved space as each quaternion corresponds to a point on a unit hypersphere, so interpolation between pairs of quaternions must allow for the hypersphere surface (where its radius is equal to the quaternion norm). An appropriate approach will ensure a consistent angular velocity between a pair of quaternions. The procedure typically used for quaternion interpolation is called **Slerp**, a name which derived from **S**pherical **l**inear int**erp**olation.

The Slerp formula for interpolating between two quaternions q_1 and q_2 is:

$$
q = \frac{\sin\left((1-f)\theta\right)}{\sin\theta} q_1 + \frac{\sin(f\theta)}{\sin\theta} q_2
$$

where:

θ – angle between the two quaternions, which can be computed from their dot product
f – fraction of interval between the two quaternions ($0 < f < 1$) (Fig. N.4)

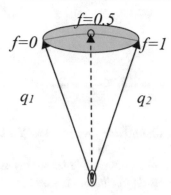

Fig. N.4 Spherical interpolation between the quaternions, q_1 and q_2. If simple linear interpolation was used, then the interpolated point would be based on a line passing through the hypersphere, but with spherical interpolation, the line passes along the surface

Table N.2 Some unit quaternions, with explanation of their action

q_0	q_1	q_2	q_3	Description
1	0	0	0	Identity quaternion, no rotation
0	1	0	0	180° rotation around X axis
0	0	1	0	180° rotation around Y axis
0	0	0	1	180° rotation around Z axis
$\sqrt{0.5}$	$\sqrt{0.5}$	0	0	90° rotation around X axis
$\sqrt{0.5}$	0	$\sqrt{0.5}$	0	90° rotation around Y axis
$\sqrt{0.5}$	0	0	$\sqrt{0.5}$	90° rotation around Z axis
$\sqrt{0.5}$	$-\sqrt{0.5}$	0	0	−90° rotation around X axis
$\sqrt{0.5}$	0	$-\sqrt{0.5}$	0	−90° rotation around Y axis
$\sqrt{0.5}$	0	0	$-\sqrt{0.5}$	−90° rotation around Z axis

N.B. – these are all unit quaternions so $\sqrt{q_0^2 + q_1^2 + q_2^2 + q_3^2} = 1$ and that $\cos^{-1}(\sqrt{0.5}) = 45°$, $\cos^{-1}(0) = 90°$

N.10 Angular Velocity

The angular velocity is the rate of change of the orientation of one reference frame with respect to another; therefore, the angular velocities cannot simply be computed from the differentiation of the orientation angles or the quaternions. Angular velocities can be computed from the quaternions using:

$$\begin{bmatrix} \omega_X \\ \omega_Y \\ \omega_Z \end{bmatrix} = \begin{bmatrix} -q_1 & -q_0 & -q_3 & q_2 \\ -q_2 & q_3 & q_0 & -q_1 \\ -q_3 & -q_2 & q_1 & q_0 \end{bmatrix} \begin{bmatrix} \dot{q}_0 \\ \dot{q}_1 \\ \dot{q}_2 \\ \dot{q}_3 \end{bmatrix}$$

N.11 Why Use Quaternions Over Other Angle Definitions?

There is an efficiency to using quaternions. Compared with other approaches (e.g., Euler angles, Cardan angles), the quaternion does not suffer from singularities and therefore avoids the gimbal lock. The quaternion represents the direction cosine matrix as a homogenous quadratic function of the components of the quaternion; unlike other approaches, it does not require trigonometric or other transcendental function evaluations. If storage is an issue, the direction cosine matrix can be represented by four parameters as opposed to all nine elements of the matrix.

N.12 Miscellaneous

Table N.2 presents some useful quaternions.

N.13 Useful References

Altmann, S. L. (1986). *Rotations, quaternions, and double groups*. New York: Clarendon.

Hanson, A. (2006). *Visualizing quaternions*. Boston: Elsevier.

Lagrange, J. L. (1773). Solutions analytiques de quelques problèmes sur les pyramides triangulaires. *Nouveaux mémoires de l'Académie royale des sciences et belles-lettres de Berlin, 3*, 661–692.

Appendix O Cross-Product

Overview

The cross-product was introduced by the Italian mathematician Joseph-Louis Lagrange (1736–1813) in 1773. It has applications in geometry and mechanics. For example, it can be used to compute angular momentum and moments.

O.1 Definition

The cross-product is sometimes called the vector product. The cross-product of two vectors in three-dimensional space is a vector which is perpendicular to both (and therefore by definition normal to the plane containing the original vectors). It can be written as:

$$\mathbf{c} = \mathbf{a} \times \mathbf{b} \qquad \text{or} \qquad \mathbf{c} = \mathbf{a} \wedge \mathbf{b}$$

Preferentially the symbol $\times$ is used to indicate the cross-product.

O.2 Computation

The cross-product is computed as follows:

$$\mathbf{a} = (a_1, a_2, a_3) \qquad \mathbf{b} = (b_1, b_2, b_3)$$

$$\mathbf{a} \times \mathbf{b} = [(a_2 b_3 - a_3 b_2), (a_3 b_1 - a_1 b_3), (a_1 b_2 - a_2 b_1)]$$

The cross-product is only defined for three-dimensional vectors.

O.3 Key Properties

The cross-product has the following properties:

1. $\mathbf{a} \times \mathbf{b}$ is perpendicular to both $\mathbf{a}$ and $\mathbf{b}$.
2. $\mathbf{a} \times \mathbf{b} = \|a\| . \|b\| . \sin\theta_{ab}$, where $\|a\|$ is the length of vector $\mathbf{a}$ and θ_{ab} is the angle between vectors $\mathbf{a}$ and $\mathbf{b}$.
3. If $\mathbf{a} \times \mathbf{b} = (0, 0, 0)$, then $\mathbf{a}$ and $\mathbf{b}$ are parallel (they lie in the same plane).
4. $\|a \times b\| = \|a\| . \|b\|$, then the two vectors, $\mathbf{a}$ and $\mathbf{b}$, are perpendicular.
5. The cross-product is not commutative so $a \times b \neq b \times a$, but $a \times b = -(b \times a)$.
6. The direction of $\mathbf{a} \times \mathbf{b}$ is specified by the right-hand rule. If your fingers point in the direction of the vectors, then your thumb is the direction of the vector resulting from their cross-product (Fig. O.1).

J. H. Challis, *Experimental Methods in Biomechanics*, https://doi.org/10.1007/978-3-030-52256-8

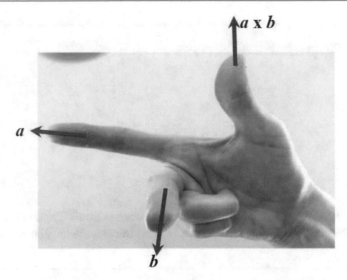

Fig. O.1 The right-hand rule. The direction of the cross-product of the two vectors **a** and **b** is the vector **c** which is perpendicular to both **a** and **b**

O.4 Examples

Here are some examples.

Example 1

$$\mathbf{a} = (2, 3, 4) \qquad \mathbf{b} = (10, 15, 20) \qquad \mathbf{a} \times \mathbf{b} = (0, 0, 0)$$

The two vectors are parallel.

Example 2

$$\mathbf{a} = (4, 3, 0) \qquad \mathbf{b} = (-3, 4, 0) \qquad \mathbf{a} \times \mathbf{b} = (0, 0, 25)$$

The two are perpendicular to one another in the XY plane; their cross-product vector is in the Z direction.

O.5 Matrix Implementation

Matrix algebra can be used to compute the cross-product:

$$\mathbf{a} \times \mathbf{b} = S\{a\}\, b$$

where $S\{\}$ generates a skew-symmetric matrix from a vector, for example, for vector $\mathbf{a} = (4, 3, 0)$:

$$S\{a\} = \begin{bmatrix} 0 & 0 & 3 \\ 0 & 0 & -4 \\ -3 & 4 & 0 \end{bmatrix}$$

Example

$$\mathbf{a} = (4, 3, 0) \qquad \mathbf{b} = (-3, 4, 0)$$

$$\boldsymbol{a} \times \boldsymbol{b} = S\{a\} \begin{bmatrix} -3 \\ 4 \\ 0 \end{bmatrix} = \begin{bmatrix} 0 & 0 & 3 \\ 0 & 0 & -4 \\ -3 & 4 & 0 \end{bmatrix} \begin{bmatrix} -3 \\ 4 \\ 0 \end{bmatrix} = \begin{bmatrix} 0 \\ 0 \\ 25 \end{bmatrix}$$

O.6 References

Lagrange, J. L. (1773). Solutions analytiques de quelques problèmes sur les pyramides triangulaires. *Nouveaux mémoires de l'Académie royale des sciences et belles-lettres de Berlin, 3*, 661–692.

Index

A

Acceleration, 11, 12
Accelerometers
    calibration, 96, 97
    definition, 94
    design, 94
    determining kinematics, 97, 98
    IMU, 99
    key capabilities, 95
    mounting, 99
Accuracy, 9, 15, 17–19, 197, 198, 209
    definition, 9
Aerial acrobatic activities, 139
Analog filter, 45
Analog-to-digital converter, 24, 43, 204, 232
Angular velocity, 175–177, 293
Animal studies, 147
Anisotropic errors, 209
Astronomy, 45
Attitude matrix, 153, 284
Autocorrelation-based procedure (ABP), 64–66
Automatic active marker motion analysis system, 99
Automatic motion analysis systems, 93
Avalanche noise, 201
Averaging angular kinematics
    mean pivot of rotation, 178, 179
    mean three-dimensional orientation angles, 177, 178
Axes of rotation, 172

B

Band reject filter, 49
Band-pass filter, 49, 51, 67, 72
Belmont Report, 5, 6
Biceps brachii, 72
Binary arithmetic, 231, 233
    concept, 231
Binary digits, 17
Biomechanics, 6, 15, 213, 227
    definitions, 1
    scope, 1
    tools, 5
Bipolar electrode configuration, 73
Bipolar electrodes, 71, 72
Body contours, 132
Body imaging, 133
Body landmarks, 146
Body segments, 125, 126
Bone-on-bone forces, 182
Borelli, Giovanni Alfonso, 103
Brown noise, 201

Buckingham's Π-theorem, 13
Burst noise, 201
Butterworth filter, 50, 51, 60, 64
    coefficients, 51
    cut-off frequency, 52, 54, 55
    data padding techniques, 53
    data processing, 52
    filter order, 52
    forward and reverse directions, 52
    fourth-order, 52
    frequency magnitude response, 52
    low-pass, 51
    phase lag, 52
    second-order, 51, 52
    signal derivatives, 55
    single pass and double pass, 52, 53
    transition band, 51

C

C++/C, 19
Cadavers, 131, 134
    anthropometry of women, 132
    height/mass, 131
    segment densities, 137, 138
    storage, 132
    whole body models, 138, 139
Calculus, 263
    derivative, 263–265
    function, 263, 267
    fundamental theorem, 266, 267
    integrals, 268
    integration, 268
Cardan and Euler angle sequences, 283, 284
Cardan angles, 285
    angle sequences, 163
    component matrices, 162
    direction cosine matrix, 163–165
    geometric interpretation, 164
    matrix multiplication, 163
    two rotation sequences, 163
Cardan angle sequences, 285
Cardan angle set, 292
Center of mass, 127, 129–132, 134–138, 140, 141
Center of pressure (COP)
    computation, 118, 119
    definition, 108
    gait, 116, 117
    location, 112, 113
    plots, 117
    quiet upright standing, 116, 120

respect for persons, 5
Euler angles, 151, 159, 163, 285
  component matrices, 165
  direction cosine matrix, 166, 167
  geometric interpretation, 166
  sequence of rotations, 165
Euler angle sequences, 286
Euler angle set, 292
Euler's formula, 238
Excel©, 19
Experimental errors, 3

F
Falsification, 1
Fast Fourier transform (FFT), 29, 30
Fermat's last theorem, 4
Fermi estimate, 16
FF fast fatigable, 70
Filter
  analog, 45
  application, 45
  band-pass filter, 49
  band reject filter, 49
  in biomechanics, 51
  Butterworth, 50 (*see also* Butterworth filter)
  cut-off frequency, 50, 64
  digital, 45
  FIR, 50
  first/second derivatives, 45
  frequency magnitude response, 50
  in hardware, 45
  high-frequency components, 49
  high-pass filter, 49
  low-pass filter, 49–51
  motion and force plate data, 64
  non-recursive filters, 50
  notch filter, 50
  passband, 49
  recursive filters, 50
  roll-off, 50
  signal components, 45
  stopband frequency, 50
  transforms, 45
  transition region, 50
  Type I Chebyshev filter, 50
Filtering (smoothing)
  ABP, 64–66
  assessments, 64
  derivatives, 64
  displacement data, 64
  Dowling data, 64, 65
  GCVQS, 64–66
Finite center of rotation (FCR), 156
Finite difference equations, 60–63, 246, 247
Finite helical axis (FHA), 160–162, 178, 208, 209
Finite impulse response (FIR) filters, 50
First law of thermodynamics, 190
First-order finite difference equations, 55
Force measurement
  applications, 107
  calibration, 106, 107
  piezoelectric crystals, 106
  springs, 103, 104

strain gauges, 105, 106
Force plate data
  frequency components, 113, 114
  impulse, 114, 115
  reaction force, 116
  signal peaks, 113, 114
Force plates, 103, 107–109, 122
  calibration, 110–112
  design, 108
  outputs, 112
  postural studies, 110
  problems, 110
  reaction forces, 107, 108
  strain gauge, 109
FORTRAN, 19
Forward difference equation, 61
Fourier analysis, 23
  coeffficients, determining, 29, 30
  Fourier transform, 27, 28
  methods, 24
  sine and cosine waves, 25–27
Fourier transform, 27
Fourth-order Butterworth filter, 52
FR fast fatigue-resistant, 70
Free moment, 112, 113
Frequency magnitude response, 46–53, 62, 63, 67
Frobenius norm, 225
Froude's number, 213, 214
Fundamental frequency, 32, 120

G
Gamma mass scanning (GM), 133, 141
Generalized cross-validation (GCV) spline, 57
Generalized cross-validation procedure (GCVQS), 64, 65
Generalized cross-validation quintic spline (GCVQS), 65, 66
Geometric model assessment, 140
Geometric scaling, 211, 212
Geometric solid models, 132, 135–137, 139, 140, 143
Geometric solids, 274
  inertia tensor, 273
  inertial properties, 271
  mass, 271
  moments, 272
  principal axes, 273
  relevant equations, 273
  volume, 271
Gibbs phenomenon, 32
Gimbal lock, 167, 170–172
Giuseppe Piazzi, 45
Global Positioning System (GPS), 99
Goldbach's theorem, 4
Goniometers
  advantages, 94
  approach, 94
  definition, 93
Gravity, 12

H
Helical axis
  format, 160
  geometric interpretation, 160, 161
  parameters, 160, 162
Henna tattoos, 72

Printed in the United States
by Baker & Taylor Publisher Services